Ecosystem Services and Agroforestry: Science and Practice

Ecosystem Services and Agroforestry: Science and Practice

Editor: Lester Bane

MURPHY & MOORE
www.murphy-moorepublishing.com

www.murphy-moorepublishing.com

MURPHY & MOORE

Cataloging-in-Publication Data

Ecosystem services and agroforestry : science and practice / edited by Lester Bane.
p. cm.
Includes bibliographical references and index.
ISBN 978-1-63987-751-5
1. Ecosystem services. 2. Agroforestry. 3. Forests and forestry--Environmental aspects.
4. Forest ecology. I. Bane, Lester.
QH541.15.E267 E26 2023
577--dc23

Murphy & Moore Publishing
1 Rockefeller Plaza,
New York City,
NY 10020, USA

ISBN 978-1-63987-751-5

Contents

Preface

Agroforestry refers to a land-use management system wherein trees or shrubs are grown among crops or pastureland. Agroforestry-based ecosystem services refer to human benefits that are achieved beyond the farm scale as a result of farm trees' interaction with soil and water, carbon storage, and biodiversity. This intentional combination of agriculture and forestry has multiple benefits, such as enhanced yields from staple food crops, enhanced farmer livelihoods from income generation, increased biodiversity, improved soil structure and health, reduced erosion, and carbon sequestration. The policies that promote agroforestry-based ecosystem services result in on-farm benefits, landscape and watershed benefits, and national and global benefits. This book is compiled in such a manner, that it will provide in-depth knowledge about agroforestry-based ecosystem services. It will also provide interesting topics for research, which interested readers can take up. The book is a vital tool for all researching or studying agroforestry.

The information contained in this book is the result of intensive hard work done by researchers in this field. All due efforts have been made to make this book serve as a complete guiding source for students and researchers. The topics in this book have been comprehensively explained to help readers understand the growing trends in the field.

I would like to thank the entire group of writers who made sincere efforts in this book and my family who supported me in my efforts of working on this book. I take this opportunity to thank all those who have been a guiding force throughout my life.

Editor

1

Unveiling Cacao Agroforestry Sustainability through the Socio-Ecological Systems Diagnostic Framework

Jilmar Castañeda-Ccori [1], Anne-Gaël Bilhaut [2], Armelle Mazé [3] and Juan Fernández-Manjarrés [1,*]

[1] Université Paris-Saclay, CNRS, AgroParisTech, Ecologie Systématique et Evolution, F-91405 Orsay, France; jilmarcas@gmail.com
[2] Institut Français d'Etudes Andines, Calle Alemania N° 32-188 y Guayanas, Quito 170519, Ecuador; agbilhaut@hotmail.com
[3] Université Paris-Saclay, INRAE, AgroParisTech, SAD-APT, 75005 Paris, France; armelle.maze@agroparistech.fr
* Correspondence: juan.fernandez@u-psud.fr

Abstract: Cacao cultivation is rapidly increasing in Latin America under the influence of public policies and external markets. In Ecuador, the cultivated surface of high quality cacao trees has doubled in the last 50 years, creating great expectations in neighboring countries. Here, we investigated the social-ecological sustainability of cacao-based agroforestry systems in four rural Amazonian highlands communities in eastern Ecuador, close to the region where cacao was once domesticated. Kichwa- and Shuar-speaking groups were interviewed by adapting Ostrom's institutional diagnostic framework for social-ecological systems. Through a set of specifically created indicator variables, we identified key interactions and outcomes to understand the fragility and the sustainability of those communities. The studied communities were fairly young, with land rights secured less than 30 years ago in most cases. Per-family surfaces were very restricted (typically one hectare) and plots were divided between cash producing crops and their own home food. The small production per household goes through a precarious commercialization by both intermediaries and cooperatives, making the cacao bean production merely sufficient for pocket money. Ties with specialist producers in one community close to the capital has promoted the use of native cacao lines. Elsewhere, improved varieties of high productivity are planted along native trees being commercialized indistinctly. The continuity of these communities currently depend on a reorganization of their demography with parts of the population working elsewhere, as cacao bean production alone will continue to be insufficient, and will compete with their food self-sufficiency.

Keywords: amazon highlands; agroforestry; self-organization; social-ecological systems

---

1. Introduction

Latin American countries have a long history of providing commodities to northern countries, a situation that has shaped and continues to shape the social, economic, and land use choices of the region. The climates, soils, and natural vegetation of Latin American countries favor tree-based production in many areas that are too humid or too warm for conventional staple foods [1–3]. The demand for commodities has occurred in successive waves involving quinine, rubber, coffee, and cacao, among others, during the past three centuries [4].

Almost invariably, the commodities involved in these waves are produced in some type of agroforestry systems. Agroforestry—or the integration of trees in farms and agricultural landscapes to diversify and sustain production [5]—is one of the most important land uses in Latin America, and predates European colonization [3]. Indeed, agroforestry may cover more than 300 million hectares in the region, mostly but not exclusively involving shaded tree-crop systems of coffee and cacao [6]. Agroforestry is currently an ecosystem management approach highly promoted in the region by public policies [see for example [7] and references therein], making it a popular ecosystem management objective. Currently, one of the most sought-after commodities from agroforestry systems is the cacao bean, because of the ever-increasing market for chocolate and chocolate-derived products.

Cacao tree cultivation, in agroforestry and intensive form, has been increasing in Latin America almost exponentially the since the 1960s (Peru 2456%, Colombia 367%, 312% Bolivia, 202% Ecuador, and Brazil 145%, [8]). In particular, Ecuador is regarded as an example for the region, because despite being relatively small in size, it provides the international markets with high quality cacao beans, thanks to the presence of well-established varieties and adequate transformation and commercialization. Large properties with direct access to international markets are common on the coast, coexisting with small producers [9]. On the other hand, on the Amazon basin on the east of the country at middle elevations (300–900 m above the sea level), cacao is grown on a totally different pattern of land tenure, including small households [10] and communities typically inhabited by autochthonous people.

In Ecuador, "communities" are a sociological and political concept that encompasses different structures. Communes (or "comunas" in Spanish) are a legal entity that represents, among other things, one type of land ownership and management. In our context, we assimilate these communities to social-ecological systems (SESs), as they are groups of people organized within clear geographic boundaries to make use of the Amazonian forest in particular ways that can be considered a complex adaptive system.

Though communities are typically associated with indigenous populations and collective land ownership, this is not always the case. In fact, the Ecuadorian Constitution (articles 56 to 60 in the 2008 text) is quite broad in defining collective rights for different types of communities, including people of African descent. Since the new Constitution of 2008, many indigenous peoples have been granted access to land if they organize themselves to claim their "ancestral territories" [11]. Thanks to their constitutional rights, indigenous peoples can, in principle, claim ancestral lands as their own if they are able to demonstrate a minimum level of organization through a long and complex process to obtain collective property land that is tax-exempt. The land is granted to the communities with conditions, which limits the extent of their use. Communities are sometimes granted large tracts of forests on the condition that the urban area is restricted. This restriction on the extent of land surface used for agriculture, agroforestry, or cattle ranching has profound consequences on the available income possibilities for these communities, as we will explain later. Hereafter, we use the word "communities" to describe the places visited, bearing in mind that their local organization and demographic composition may vary from place to place.

Of the 11 recognized Amazonian indigenous nations in Ecuador, two of them, the Kichwa and Shuar speaking communities, are of particular interest for understanding the current role of cacao production as a driver of community self-organization. The Kichwa communities are internationally recognized as growers of award-winning chocolate bars, thanks to the work of private companies that have commercialized their products overseas. By contrast, the Shuar include descendants of the people who originally domesticated cacao trees [12] about 5000 BP and inhabit archeological areas where ancestral trees of the *Nacional* type are suspected to occur [13]. The Shuar are not currently recognized as high-quality producers.

From a research point of view, Ecuador offers the opportunity to examine in a reduced geographical scale, different types of cultivation strategies and the different approaches to production by different sectors of the population. In fact, cacao bean production has been a staple export since the 19th century. In this regard, Ecuador is seen as a reference, and even as a myth by neighboring countries that seek to engage in successful export trade of fine quality cacao beans, mostly by Colombia where new cacao growers wish to emulate the 'fino de aroma' cacao from Ecuador (J.F.M., personal observation). In the latter country, cacao bean cultivation has been seen as key to providing alternative legal income for regions that have undergone decades of internal armed conflict and the cultivation of illegal crops. In consequence, hundreds, if not thousands, of families in Colombia are currently encouraged to plant cacao trees to produce beans for the national and international markets, an initiative that could benefit from the experience acquired in Ecuador.

In this study, we asked the central question: what are the inner workings of cacao-based agroforestry in the sustainability of upland Amazonian communities with different cultural backgrounds and different social contexts? To conduct our analysis, we used the Ostrom Institutional Analysis and Development (IAD) framework, which identifies the processes related to sustainability, the self-organization of people, and the role of different external drivers on the dynamics of SESs.

2. Materials and Methods

2.1. Study Sites

Napo Runa indigenous people. The Napo Runa, an Amazonian Kichwa-speaking group, has the largest population among the 11 Amazonian indigenous people recognized by the Ecuadorian government. They live in the eastern rainforest of the country's Amazon basin. This Kichwa population has essentially turned into agriculturalists, with sporadic hunting and fishing practices to meet their daily needs. The main farming activities are the planting and harvesting of *yuca* (manioc), banana, cacao, coffee, and *naranjilla* (*Solanum quitoense*). We selected two communities: "Pumayacu" and "Santa Rita", located in the northern Amazonian rainforest in the province of Napo (Figure 1, Table 1).

Table 1. Summary of political, physical and social characteristics of the study sites.

Language-Speaking Group	Amazonian Kichwa		Shuar Chicham	
Name of the Community	Pumayacu	Santa Rita	Yawi el Cisne	Tuntiak
Geographical region in Ecuador	Northern Amazon	Northern Amazon	Southern Amazon	Southern Amazon
Province	Napo	Napo	Morona Santiago	Zamora Chinchipe
Locality (canton)	Tena	Archidona	Gualaquiza	Centinela del Condor
Altitude above sea level	620 m	805 m	837 m	823 m
Population in 2017 (approximate)	150–200	748 (147 families)	43	600 (250 families)
Year of foundation	1991	1967	1997	1999
Surface of the collective community property	40 Ha	252 Ha	50 Ha	100 Ha
Surface of associated lands	1200 Ha	1400 Ha of primary forest	3120 Ha of Chumpias property	No additional lands
Main source of revenue	Cacao, fishing, mixed farming (crop, livestock), and recently charcoal as a complement	Cacao, heritage tourism (lodge, theme park)	Self-sufficient economy, marginal cacao, and land selling for mining	Fishing, livestock, artisanal mining, and charcoal production
Accessibility	5 h trip by public bus from Quito to Tena plus 45 min-walk	5 h trip by public bus from Quito to Tena, plus 30 min bus by unpaved road	15 h trip by public bus from Quito and 45 min in private 4 × 4 taxi	18 h away by public bus from Quito.
Telephone and internet	No	Yes	No	Yes

Shuar indigenous people. The Shuar people are settled in northern Peru and eastern Ecuador. They typically live in the headwaters and tributaries of the Marañon River, the tropical rainforest and savannahs of the Amazonian basin, and a few foothills of the Andes mountain range. They speak *shuar chicham*, and adopted Spanish as a second language [14] from the early 20th century, with the Salesian missions. They are well known for founding one of the first indigenous organizations in the world, the Shuar Federation, in 1962, and for developing a specific process of ethno-education through radio programs. The two Shuar communities visited were located in the southern Amazon provinces of Ecuador: "Yawi el Cisne" in Morona Santiago province and "Tuntiak" in Zamora Chinchipe province on the border with Peru (Table 1, Figure 1).

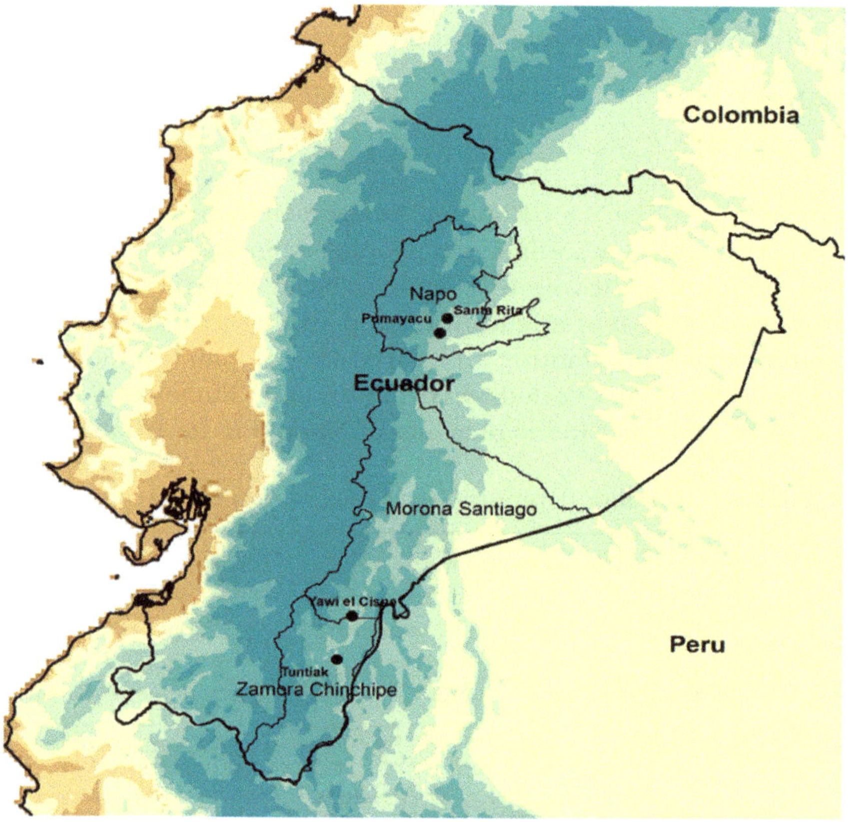

Figure 1. Study sites chosen for this study (black circles). The background colors of the map represent the altitude zones in this part of South America (brown < 50 m, light brown and sand 50 to 100 and 100 to 500 m, respectively; clear blue 500 to 1000 m; and darker blues 1000, 2000 to 3000, and upwards).

2.2. Data Collection and Survey Design

We used the Institutional Analysis and Development (IAD) framework developed by Elinor Ostrom for analyzing the sustainability of SESs [15,16], to design and organize data collection in the field. In this framework, eight types of umbrella categories (noted *S*, *RU*, *RS*, *GS*, *U*, *I*, *O* and *RE*), known as first-tier variables, are needed for the diagnostic (Table S1, first column). At one extreme Social, Economic and Political Settings variables (*S*) describe how current conditions impose boundaries to the economic activities in a given region, and limit the dynamics within the SES. In turn, the dynamics of the SESs are described through four interacting sub-systems: (i) Resource Units (*RU*), which, in our case, are the native and non-native cacao trees; (ii) Resource Systems (*RS*), describing how the harvested cacao seeds are sourced, transported, and commercialized; (iii) Governance System (*GS*), explicating how the access to the resource is regulated; and (iv) Users (*U*), describing the actors and their organization(s). As these four sub-systems interact, the Interactions (*I*), as well as their Outcomes

(*O*), need to be identified. Finally, as SESs belong to larger systems in their territories, a set of variables are defined to describe those interactions with Related Ecosystems (*RE*).

Each of these eight first tier variables has a pre-defined set of second-tier variables (Table S1, second column) that often need indicators specific to the study case. Hence, a set of 195 indicators (Table S1, third column) was developed, to construct the questions asked in the field. This set of indicators was used also as a checklist after the interviews, to ensure that enough information was gathered during the time spent with the actors. The full set of questions corresponding to the 195 indicators (in Spanish) can be requested directly from the first author.

The Ostrom approach relies on an institutional diagnosis looking initially at a large set of variables and afterwards, through an inductive process, the researcher focuses on variables that revealed conflicts between actors or specific challenges to the sustainability of the system. In general, but not always, these variables appear on the Interactions and Outcomes results. In our case, we decided to provide qualitative scores (very good, adequate, deficient, and completely lacking) of what was found in the field by consensus among the authors regarding the Interactions and Outcomes. These scores do not replace quantitative measures, which could have been done with other methods, but reflect an interpretative convergence [17] by the authors.

Interviews were conducted in an informal way avoiding any directive approaches and taking care to ask the questions in the most possible neutral way. When possible, the person conducting the interviews spent time with the local people, helping out in daily activities to build trust, while observing their activities. Voice and video recordings were accepted by most of the actors, allowing us to double check the field notes from the interviews if necessary.

Before visiting each community, arrangements were made with a local leader (the 'Presidente', in Spanish), which decided whom to visit and where to stay. The first community visited by the first two authors was Pumayacu. Previously, the second author had presented the goals of the study to the community and organized the one week stay. Without that initial contact, residents are reluctant to accept the presence of foreign people in their lands. In exchange for the interviews, a fee was arranged for housing and food, in a house designed by the president of the community. In each family household, both wife and husband work together in the production of food. Often, the interviews were done with a couple, but sometimes only the husband would remain, as he felt less shy with respect to the interview, and of having notes recorded on a digital recorder. In this first community, the age of the interviewed people was between 25 and 70 years old, as younger people were usually at school or working elsewhere. In this community, it was the children that showed the first author the cacao tree plantations, and are the ones in charge of gathering, selling, and being paid for the cacao in bulk.

About half of the interviews lasted between one and two hours when conducted at their homes, but the rest could last a complete working day, as the first author participated in their daily activities (fishing, charcoal production, husbandry, cacao cultivation training, etc.) These activities do not have a precise timing, but start very early in the morning and sometimes finish at dusk.

For the second visited community, Santa Rita, the first author contacted the president of the community directly by telephone. This community has more visitors because of the presence of external aid by the government, non-governmental organizations, and enterprises, so the members are more open for conversation. Likewise, the community leader indicated which families to contact and where to stay. Additionally, a modest fee for housing and food was requested.

In the case of the Tuntiak community, the organization of the visit proceeded through an agricultural technician working directly for the regional government, who has close ties with the people in this community. In contrast to the Kichwa speaking people, it is the women that are in charge of the food production, and of agricultural activities in general, hence, the interviews were carried out only with the women from each household. This was the only community where a hostel was used during the interviews. In contrast, for the Yawi El Cisne, a host family was designated by the community, and also a small fee for housing and food was requested. The decisions of who would be interviewed were taken collectively during the first contact with both Shuar communities. The interviews were conducted between March and April of 2017 on the four target communities as well with some scholars at Universities in the capital of Ecuador, Quito. Table 2 presents a summary of the actors interviewed in the four communities, as well as in two neighboring areas.

Table 2. Summary of actors interviewed during the field work. People or institutions not pertaining directly to the communities are marked as 'Other'.

Speaking Group	Province		Type of Organization	Type of Actor
Kichwa	Napo	Pumayacu	Community-based organization of cacao farmers and chocolate producers	Manager and legal representative (1)
			Amazonian Indigenous Community	Independent Small cocoa producers (3), commercial member of the local association (3), Small cocoa producers for the local association (2)
		Santa Rita	Amazonian Indigenous Community	Small cocoa producers (5)
		Other	Community-based Association	Manager and legal representative (1), Small cocoa producers (2)
			Independent producer	Small cocoa producer (1)
Shuar	Morona Santiago	Yawi el Cisne	Amazonian Indigenous Community	Small cocoa producers (6)
	Zamora-Chinchipe	Tuntiak	Amazonian Indigenous Community	Small cocoa producers (5)
		Other	Provincial government	Prefect of the province (1)
			Provincial municipality Government	Employee (1)
			Public enterprise for agricultural development	Agronomist (1)
			Federation of small organic agricultural	Manager, Tree nursery manager (1)
			Enterprise at the *Centinela del Condor* Municipality	Manager and legal representative (1)
			Private export enterprise at *El Pangui*	Manager and legal representative (1)

3. Results

3.1. Social, Economic and Political Settings (S)

One of the most striking results of our study is the fact that despite the visited communities are living in their own ancestral territories, their land ownership and self-decision-making capacities are quite recent. A review of the literature shows that, in the middle of last century, many indigenous communities in Ecuador settled in areas that were not necessarily their original settlement areas. These migrations, resulting from the colonization of their own territories after the agrarian reforms, obliged them to start afresh by incorporating new economic activities for which they had little or no experience into this process.

These rural movements were composed of indigenous organizations and marginalized communities from all regions in the country, but especially from the Amazon area. Collective organization and mobilization increased the pressure to secure land for indigenous peoples. Invasions, pressure from "colonos" landowners (colonos are people that claim de facto property of the land), and petitions were among the actions taken by community-based organizations to open forested land and enlarge their indigenous ancestral lands [18,19]. The protests scaled up until 1994, when the "Movilización por la Vida" forced the Ecuadorian government to negotiate the requests of the movement, represented by the "Confederacion de Nacionalidades Indigenas del Ecuador" (CONAIE, https://conaie.org/). A remarkable achievement of the movement was the constitutional recognition of the Amazonian indigenous communities' rights to their ancestral lands. Additionally, another triumph was a formal acceptance that communal farms are forms of productive organizations.

It is very likely that local ecological knowledge [20] regarding ecosystem management was lost during these migrations, thus, making these populations even more fragile, with respect to finding sustainable ways of living. Cacao trees were never kept in high densities neither during pre-Columbian times nor during post-colonization. These trees were grown as part of their family gardens, a sacred placed called ajas (pronounced ah-has) for the Shuar speaking people [21,22], and it was a plant kept mostly for medicinal properties, alongside several other species of shrubs and trees. A similar case occurs with the chakras of the Kichwa speaking people. High density cacao plantations to export beans began to flourish in many parts to the east of the Andes towards the Amazon basin from the 1990s, at the same time as when many rural movements arose in response to the abandonment of the central government, so any type of intensive cultivation is new for the four studied communities.

Interviews with the local actors in Pumayacu Napo Runa community revealed that land was claimed as ancestral territory quite recently, and the community was only created in 1991. Prior to this, the land was occupied and exploited by the religious Josephine mission. The Kichwa inhabitants of the area were exploited and underpaid by the mission (information extracted from interviews), until they decided to part ways. Rallies and protests led by one main family had spread out in the area by the end of the 1980s. After many clashes with the police, the religious mission gave up the land and abandoned the area. Since then, the descendants of these families have populated and worked in the community. Nowadays, the community leaders are taking steps to participate in the Socio Bosque (meaning "social forest" in Spanish) conservation program, a national Ecuadorian program that aims to pay for ecosystem services through forest conservation (http://sociobosque.ambiente.gob.ec/).

Nevertheless, the most extreme case of restricted communal land ownership that we observed was the Yawi El Cisne Shuar community. Interviews showed that the community was founded in 1997 by an extended single family that had previously lived in the nearest community a 30 min walk away. However, their agricultural lands were situated near the current location of the community. As a consequence, the patriarchs of this founding family and their eight children decided to move and create a community near their productive lands. The territory is collective, with Yawi being part of the 3120 hectares property known under the title of Chumpias. The community members are close relatives, and they have individual parcels and a division of land by inheritance. As a result, most of the members currently have between 1 and 4 hectares, which creates pressure over agricultural land. The community is surrounded by other communities, which are part of the global title mentioned earlier for the Shuar people. There were 43 permanent inhabitants at the time of the study, and all the households included one member of the patriarchal descendants. It is important to note that 27 other people from this community work elsewhere as school teachers in other Shuar communities, or are farm workers on the big plantations along the coast. The internal governance structure of the community is basic, being composed of the president and a few collaborators elected by family relationships.

Local leaders asserted that the government has recognized the communal ownership of the land by the members of the Yawi el Cisne, and it is guaranteed by its internal governance system. They cannot sell the land, but the inhabitants have formal individual rights for land tenure for agricultural and productive purposes. Additionally, their individual land possession is subject to intergenerational division by inheritance systems between the new members of the household, which means that less land is available for later generations. Currently, most of the households are still descendants of the founder family, and they have organized their internal governance structure by newer family ties. We observed that the members of the community referred to the founders and the oldest people of the community in a respectful manner. Each member has well-defined roles in the community and its internal governance structure.

In the case of the Santa Rita community, a complete lack of access to land ownership did not trigger self-organization but rather a growing population of children without education facilities in proximity. According to the people interviewed, the community was officially founded in 1967 (the oldest case studied here), although the area has always been populated by Kichwa Napo Runa people but without any communal structure. The number of children during the 1960s increased in the area, and they had to walk about 7 km through the forest to reach the nearest school. Therefore, with the support of a Catholic priest, some inhabitants of the area decided to create a school, an event that triggered the land rights claim process. The community has 252.2 hectares, with the land considered to be an ancestral territory surrounded by primary forest and other communities. Nowadays, it has the same property rights governance as the other Kichwa community, because the government has recognized the communal ownership of the land, which is guaranteed by its internal governance system with defined rules. They cannot sell the land, but the inhabitants have formal individual rights in land tenure for agricultural and productive purposes. Additionally, their individual land possession is subject to intergenerational division by heritance, which decreases the land available for future generations. Most of the interviewees have between 3 to 7 hectares of agricultural land.

The Tuntiak Shuar community was created 18 years ago in a territory that was previously held by the Ecuadorian army; it is the largest community visited in this study with more than 800 people. In the early 2000s, some Shuar families were looking for a place to settle down, so the army and the Shuar population agreed to share the territory. The land is also considered to be "ancestral territory". They therefore have communal property rights over the land, as well as individual property rights with intergenerational heritage, as in the three former communities. However, according to its internal governance structure, no single person is directly responsible for land sharing or cacao production activities (see the next section).

3.2. Resource Units (RU), Resource Systems (RS) and Users (S)

The most outstanding finding of our survey (our own observation) is that trees belonging to native *Nacional* or *Criollo* are only found in areas with strong links to niche markets, like the Santa Rita community, and are not necessarily abundant in more remote areas of the Amazon provinces of Ecuador (Table 3).

> *" ... here, we didn't have that many trees of cacao before, but when the coffee market prices reached 1 cent the pound, we said that's it, and we cut all the trees ... then we had people coming here telling us to plant cacao trees ... ".* cacao producer in Santa Rita (our own translation)

In stark contrast, all the other visited communities exhibit a mixture of trees including the highly productive clone CCN-51, which was distributed by the government (information provided by locals), probably a result of the large public policy of the reactivation of *fino de aroma* varieties that was started in 2012 (https://www.agricultura.gob.ec/magap-impulsa-proyecto-de-reactivacion-del-cacao-fino-y-de-aroma/) and had a target of offering help for 100 thousand families. The second most striking finding, unrelated to cacao production but related to the overall land use, is the presence of associated lands (Table 1), which can represent between three and 60 times the urban and exploitation area granted to each community. Only one of the visited communities lacks additional lands (Tuntiak in Zamora Chinchipe), but it theoretically has the option of being granted forested areas to manage outside the community through the *Socio Bosque* program.

In all the visited communities, cacao bean production is not the main activity, but a source of supplementary income, as the main activity is agricultural subsistence, a common response by all the interviewed people. Surplus banana plantain, manioc, and miscellaneous fruits are sold in the local markets. The most salient result in terms of the marketing of cacao by these communities is that organic produce is not necessarily the most profitable commodity (Table 4). In fact, all types confounded of dried beans could be sold in 2017 to intermediaries at prices up to USD 0.60 (note that Ecuador uses the US dollar as currency), but as low as USD 0.30 if sold to unscrupulous intermediaries. Organic dried beans are sold at intermediate prices (USD 0.45). The opportunity cost of accessing the more stable prices of organic cacao (more work on the trees, strict selection of pods, careful fermentation, and drying) means that the "on the pulp" or on-the-spot selling of cacao beans to an intermediary is an attractive option, as standards are much lower, and the production is less time-consuming.

Table 3. Cacao resource units (*RU*) characteristics in the four communities.

Language-Speaking Group	Amazonian Kichwa		Shuar Chicham	
Name of the Community	**Pumayacu**	**Santa Rita**	**Yawi el Cisne**	**Tuntiak**
Governmental program known as "fino de aroma" reactivation project (2012)	Yes	Yes, with additional funding from the FAO and local government	No	Yes for the plant provisions, but alternative programs exist with the provincial government
Date of cacao development	15 years	More than 15 years	Not known by the interviewed people	Not known by the interviewed people
Cacao producer's collective association	Yes (community-based cooperative)	Yes, community-based cooperative	None, but one person started a *Criollo* nursery at the time of the interview	incipient cooperation of few families
Cacao varieties	Mainly *Nacional* intermixed with CCN51, *Criollos*, mountain, and other native varieties	Mainly *Nacional*, with some *Criollos* and native varieties	Mainly old "ancestral" *Criollos*, a few *Nacional* and CCN51	Mainly *Nacional*, with a high presence of native varieties. A few CCN51 in community plantations
Plantation type	1–3 ha, 300–600 trees maximum	3–7 ha, 600 trees	1–4 ha with heterogeneous plantations	High-density *Nacional* with 800–1000 trees; use of natural fertilizers, disease control, and technical assistance
Technical assistance	Local cooperative	Local cooperative	None	Agronomist paid by the provincial government
Production issues	Disease on cacao pods observed, but no management	Well-managed cacao plantations	Disease on cacao pods observed, but no management	Basic management of disease on cacao pods

Table 4. Cacao resource system (*RS*) characteristics in the four communities.

Language-Speaking Group	Amazonian Kichwa		Shuar Chicham	
Name of the Community	**Pumayacu**	**Santa Rita**	**Yawi el Cisne**	**Tuntiak**
Plantation location in the communities	15–60 min walk	Close to the community	Close to the community	Close to the community
Technical guidelines for harvesting	Defined by local cooperative: beans collected and dried for one night before being collected by truck	Specific requirements of cooperative 1: 3–5 days in advance without opening under the trees, then pod opening and bean harvesting. Cooperative 2 is less rigid	No specific guidelines, local know-how practices	No specific guidelines, local know-how practices
Purchase planning	Every 2 weeks for 4–5 months by local cooperative	Close coordination with cooperative 1 and local cooperative 2	None	None
Post-harvesting facilities	Association facilities	In association with cooperative 1	None	None
Cacao marketing	Fresh beans with pulp (USD 0.45): local cooperative (80%) and intermediaries (20%). "Dried bad cacao" all types confounded: sold to intermediaries (USD 0.20–0.60)	Fresh beans with pulp sold to cooperatives 1 (USD 0.35) and 2 (USD 0.45) for organic beans only. Market for dried beans all types confounded (30 min bus with prices of USD 0.35–0.60)	Dried beans sold to intermediaries in marketplaces (40 min walk, 20 min by bus). Trading prices unknown but probably not different from other places	Dried beans sold to intermediaries; no price difference for organic beans (USD 0.55)

3.3. Governance System (GS) of Amazonian Shaded Cacao Agroforestry

Our study found that the governance system of native cacao trees is quite weak or non-existent and that it depends more on external markets that actively search a stable production of *Criollo* or *Nacional*, like in Santa Rita. For the other communities, we could not detect any pattern of local protection or promotion of their spontaneous trees. Growers are aware that different species exist, like *Theobroma bicolor* and *T. grandiflorum* among others, and that even the closely related *Herrania* species, from which chocolate can also be made, may have some interest for commercialization in the future, but they prefer to focus on their current rudimentary markets. Locals seem to be willing to accept improved varieties like CCN51, distributed by the central government with the hopes that some marketing scheme will follow. However, lots of skepticism was perceived during the interviews:

> "*... in my opinion, the cacao policies are just cheap policies from the central government ... but how else are we going to earn 60 or 70 dollars to pay for basic needs?*" Independent cacao producer in Pumayacu

As for the prices, all interviewed people acknowledged they did not have any control on the price of cacao beans, whatever their quality. In fact, one of the sources of conflict found was not directly related to land ownership, as one would expect, but related to what represents an acceptable price to be paid for by the cooperatives or the intermediaries (Table 5).

> "*... each time they lower the price of the cacao beans by the pound they say it is because the international market prices have dropped ...* ". Independent cacao producer in Santa Rita

The second source of conflict that could affect the native cacao as a resource, and ecologically sustainable agroforestry, is the emergence of alternate activities, like charcoal production (Pumayaco community), the opening of forests for cattle ranching, or selling tracts of land for little controlled mining operations (Shuar speaking people, Table 5). The geographical position of the Shuar speaking territories farther away from the capital city or from regional capitals allow land clearing and mining activities to be performed with little supervision. These activities affect not only cacao production, but possibly too the availability of lands for growing crops for their own basic needs.

> "*... we cannot live from cacao alone, it is seasonal ... , we rely on charcoal production in our own parcels ...* ". Woman and husband from the same household in Pumayacu

Table 5. Conflicts issues affecting the governance system (*GS*) of the studied communities.

Language-Speaking Group	**Amazonian Kichwa**		**Shuar Chicham**	
Name of the Community	**Pumayacu**	**Santa Rita**	**Yawi el Cisne**	**Tuntiak**
Main sources of disputes or debates	Cacao prices	Cacao prices. Disagreement about how to manage cacao trees and harvest	Cacao prices. Disputes about land selling for mining, clearing for cattle ranching, or cutting trees for charcoal production	Cacao prices. Disputes about land selling for mining, clearing for cattle ranching, or cutting trees for charcoal production
Alternative sources of revenue to cacao	Increasing charcoal production, because a 50 kg bag is worth $7–8 at the nearest market	A few people employed in cacao-related ecotourism	Land selling for mining activities	Agriculture and livestock supported by the local prefecture; artisanal mining

3.4. Interactions (I) and Outcomes (O)

We observed that the visited communities can be considered to have clear physical boundaries, external and internal, where the interactions of actors between them, and between the actors and the ecosystem, have been clearly defined with the creation of each community. Urban/housing areas, per family plots and adjacent forest reserves have all clear boundaries that seem to be respected. The current generation of inhabitants are aware that they can only leave as inheritance a fraction of what they could secure during the creation of the communities and that the next generation will need to rely on jobs elsewhere to establish their own family, for which they need an education:

> *" ... our heritage is more on giving the means for studying than in giving lands to our children ... I have six children and seven hectares, ... after they finish high school, I will give each one, one hectare to build their own house, that is all what I can offer ... "*. Associated producer to the cooperative in Santa Rita

Outcomes seem more positive for communities close to the capital city and with the large involvement of the public and private sector (Figure 2).

A clear outcome has also emerged, as the less organized communities use more of the clonal varieties distributed by the Ecuadorian government, because of their higher yield and robustness against diseases compared to the communities that work closely with companies using organic cacao beans for national and international markets. Paradoxically, the areas of the Shuar communities that are in the direction of the archeological sites showing the early domestication and use of cacao [23] seem to be more fragile regarding their disorganized mixing of clonal and local varieties. Isolated communities struggle to make a living, so they have little choice but to accept any kind of external aid from the national or regional government.

3.5. Related Ecosystems

From the information gathered in this study, only three of the four studied communities have access to external tracts of land. As the access is relatively restricted, the communities can only benefit indirectly if there is a program like '*Socio Bosque*', which can give funds to the community if they engage in a sustainable use of the resource. It is still too early to know, in our opinion, if communities with adjacent lands would be more sustainable than those without, as income and crops for their own food are not grown in the adjacent areas. In addition, there is a demographic growth and any external aid for forest management with the community would need to be shared between more people. A clear opportunity exists if the adjacent lands (old and new) will be considered as repositories of native cacao trees, as their management and conservation could be funded externally by the commerce of high specialty cacao beans.

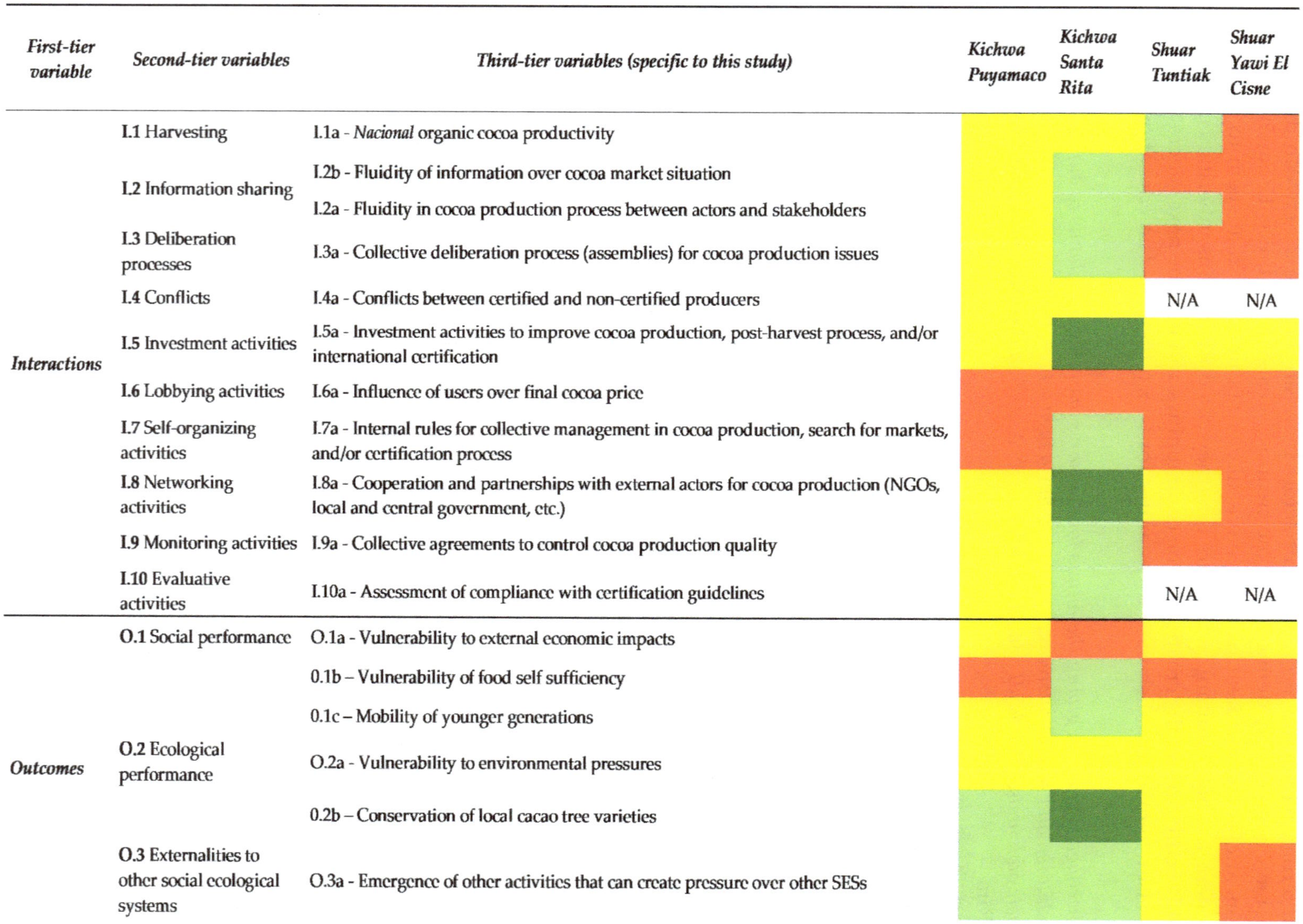

Figure 2. Summary of outcomes (*O*) resulting from local interactions (*I*). The scale is qualitative, and colors represent conditions as assessed by the authors from the interviews: very good (dark green), adequate (green), deficient (yellow), and completely lacking (red). The white cells with N/A represent instances where the variable is not applicable.

4. Discussion

Our results show that agroforestry centered on cacao is an activity that is constrained by the community's food security needs, even in communities where apparent signs of success stories can be shown to external observers. In simple terms, it does not produce enough and constant revenue for the farmers to rely on them as their main source of income. As evidenced with the Kichwa and Shuar communities, alternative activities, like producing charcoal and mining are attracting more inhabitants because of their urgent need for revenue sources. Moreover, the level of self-organization observed was not dependent on the size of the community (large communities were loosely organized), but rather on the external connections that allowed for more organized interactions in the communities.

Communities have a weak tolerance to accept the opportunity costs imposed by organic markets. A frequent response in our study was that if the buyer of organic cacao did not show up for whatever reason, the alternative was to sell in bulk their dried beans on the local market, or if cash was needed urgently, to sell them not dried but still with the pulp on them that is even less profitable. The presence of local associations or small niche companies that seek to provide fair prices to family producers only slightly changes the amount of income earned from selling beans. In this regard, despite all the good intentions of associations and buyers of organic beans willing to pay a premium, there is simply not enough production per family to encourage an activity that goes beyond extra spending money. In general, the lack of accessibility to pre-processing facilities to export cacao paste (also called cacao liquor) instead of raw beans, impedes producers from escaping a poverty trap along the cacao value chain [24,25]. The inability to directly access export markets, because cacao beans pass through at least two levels of re-sale within the country [26], keeps small cacao bean producers at the base of an asymmetrical value chain for which more work does not necessarily represent higher revenue [27]. Not even temporary positive price fluctuations favor the small producers, as they rush to sell their produce at the expense of quality standards [28]. Unfortunately, all these processes described by the literature were observed in one way or another in the four communities.

In Santa Rita, almost all our qualitative indicators were considered positive, as there is clear evidence of discussion and governance regarding the amount and extent of organic cacao using non-clonal varieties (*Nacional* and *Criollo*), with clear objective-driven production for the local companies that purchase their produce. However, we would not conclude that the somewhat sophisticated level of self-organization around the production of cacao beans and the promotion of chocolate culture means that this community has reached economic sustainability. A recent evaluation of their territorial management goals [29] clearly suggests that the cacao bean industry is not yet the main source of revenue, and that activities like eco-tourism and chocolate promotion provide only a few local people with wages, so it has therefore not replaced basic agricultural activities. More importantly, this community like others has undergone a rapid population growth, which has led to the emigration of young people to look for jobs in other areas of the country and not necessarily in the agriculture sector. Still, being close to the capital city of Quito offers the younger generation with more study opportunities and possibilities of future jobs outside of the agricultural sector.

A second factor possibly influencing the degree of local interactions (self-organization) is the age of the community (Santa Rita is at least 50 years old, while the others have existed for barely 20 years), but Santa Rita's proximity to a main national road linking the site to the capital easily confounds what is the result of self-organization and what is the result of external inputs. Five decades have allowed the communities to interact with external actors, and test the different ways of securing income, including through the chocolate industry, for the past 15 years. More isolated communities are associated with simpler social structures and more fragile production means, an observation that is especially evident in the southern Shuar communities.

With the information gathered, we hypothesize that if the above pattern holds for the many indigenous communities that have recently become established, the emerging land use created by granting ancestral territory will create a series of difficulties because of: (a) small but densely populated urban centers of up to 800 people; (b) patches of land for agroforestry used in ways that depend on the amount of external help (i.e., organic with *Nacional* trees in shade zones or mixtures of non-shade and shade zones for different varieties); and (c) a forested area not used for any kind of agroforestry. On account of this land use, the zones that can be used for agroforestry or subsistence agriculture are limited by the contract with the government, which may induce migration out of these communities after a period of time, as has already been observed, even in well-developed communities like Santa Rita [29]. "Fair trade" initiatives may also not offer a stable market environment, as fierce competition between labels may induce a relaxation of standards that directly impacts the revenue and working conditions of farmers [30]. In general, it has been observed that the creation of these niche markets is possible, requiring the presence of strong cooperatives [31], as well as a long, difficult, and continuous learning process [32]. In our study cases, cooperatives were present mostly in the Kichwa communities, but they act in a similar mode as a private company, raising a mitigated interest in the members of the communities.

It is unknown whether the Shuar and Kichwa peoples would turn to monocultures out of exasperation with the difficulties associated with organic cacao and its low profitability in the fair trade value chains. In analyzing 50 households in coastal Ecuador where both native cacao agroforestry and intensive improved varieties co-exist [33], it was observed that, if sources of revenue are scant, the additional effort in time required by organic farming is unaffordable for households, a situation that is perfectly understood by intermediaries and works to their advantage. A second study on coastal Ecuador with 188 households showed that without access to specialty markets with a differentiated value chain, it is very difficult for families struggling for increased income to maintain organic production [9]. In fact, a decade-long analysis in Ecuador has shown that, despite the presence of new markets, many farmers switch to monocultures of improved cacao varieties on account of their greater productivity, because their prices are similar to the sought-after *Nacional*, *Arriba*, or *Criollo* varieties [34].

The possibility of maintaining food and medicinal plants in the same plot may prevent households from implementing monoculture practices. This may well be the case of the Shuar and Kichwa people, who value forested areas as sources of food, fibers, and medicinal plants, through their traditional way of gardening—the chakra system for the Kichwa and *aja* for the Shuar—being the women's responsibilities. Thus, certification requirements of "organic farming" and shade farming are common grounds for continuing to use the forested areas, as diverse and multifunctional vegetation not needing external inputs for nutrients or plague control [35]. However, innovative structures need to be created for allowing economic growth not based on monocultures, as happens in coastal Ecuador.

Can the effects of the limits on the communal lands be solved with current policy tools? A bolder approach to helping these communities would incorporate not only payments for the forest conservation of adjacent lands but also for maintaining the different wild cacao trees present in their own community lands (*Theobroma* spp., *Herrania* spp.). The produce of these very valuable trees could then be commercialized at a premium price, allowing the local people to skip unscrupulous intermediaries. From the different programs set up to pay for ecosystem services in the country, the literature reports that the Ecuadorian *Socio Bosque* program has proved to be successful in other parts of the country [36]. The interesting aspect of this program is that it is not centered on the actual lands used by a community for agricultural purposes, but on adjacent lands. Currently, only one community benefits of economic help for the management of the forests, but all the communities could benefit from an extended program covering an adjacent forest, in addition to their high diversity biodiversity agroforestry systems.

5. Conclusions

It can be argued that, no matter what "miracle" tree species is proposed in these tropical areas to provide income to rural communities, as many public policies do in the region [37], two basic problems will always haunt these strategies. First of all, land tenure in Latin America, even if granted in a very unique way in Ecuador, does not give enough space for growing communities to rely on agriculture alone. Second, as raw producers at the start of the value chain, the margins of commodities will always be low for the communities if they remain fragile and preyed upon by intermediaries that profit from their lack of organization.

Land tenure restrictions, like in most Latin American countries, continues to be the main factor that favors or prevents the self-organization of communities [38]. In the particular case of eastern Amazonian Ecuador, we observe the dynamics of a very recent process, which tests the abilities of these rapidly growing communities to earn enough income within the granted lands [29]. Unfortunately, for native cacao beans to become a game changer, a very sophisticated commercialization scheme needs to reach these communities, which is not yet the case, as several layers separate them from the international markets.

Countries like neighboring Colombia and Peru can learn from the outcomes of these communities and their interaction with the cacao market. There is a clear risk of losing very valuable and unique genetic resources if only a handful of improved varieties are used to start widespread programs of cacao production. Wild and semi-wild cacao trees grow in many Latin American countries, and peasants and indigenous peoples may, in the long run, lose access to niche markets, if everyone produces undifferentiated cacao trees from breeding programs. Additionally, it is clear that some degree of association is needed, so that households can negotiate the price of the cacao beans in a given region, avoiding the tyranny of intermediaries or companies holding a local monopoly. Under current conditions, the economic sustainability of households seems very difficult to achieve if centered only on a single resource if available surfaces remain restricted, both for Amazonian Ecuador and neighboring countries.

Future programs in the Amazon promoting the development of niche markets should bear in mind that household agriculture is mostly managed by women in Shuar and Kichwa speaking communities. While Kichwa women are active members of cooperative and boards, Shuar women have weak participation in the producer's organizations, and the commercial aspects are still largely handled by men. Hence, conflicts may arise if this added complexity of the agroforestry production and cacao bean market in Shuar communities is not taken into account by organizations seeking to promote these kinds of niche products.

Author Contributions: Conceptualization, A.-G.B., A.M. and J.F.-M.; funding acquisition, A.-G.B., A.M. and J.F.-M.; investigation, J.C.-C. and A.-G.B.; methodology, J.C.-C., A.M. and J.F.-M.; project administration, J.F.-M.; supervision, J.F.-M.; writing—original draft, J.C.-C.; writing—review & editing, J.C.-C., A.-G.B., A.M. and J.F.-M. All authors have read and agreed to the published version of the manuscript.

Acknowledgments: We would first like to thank all the people from the communities in Ecuador who provided their valuable time to answer our questions with complete honesty and that showed their day-to-day reality. We would also like to thank the engineer T. Ontaneda-Hidalgo that introduced the first author to the Shuar communities and helped with the transport within the area. Additionally, two students already conducting research in the Kichwa and Shuar communities, A. Pech and A. Wilmort, helped the first author to be integrated with the families.

References

1. Maezumi, S.Y.; Alves, D.; Robinson, M.; de Souza, J.G.; Levis, C.; Barnett, R.L.; Almeida de Oliveira, E.; Urrego, D.; Schaan, D.; Iriarte, J. The legacy of 4500 years of polyculture agroforestry in the eastern Amazon. *Nat. Plants* **2018**, *4*, 540–547. [CrossRef] [PubMed]
2. McNeil, C.L. Deforestation, agroforestry, and sustainable land management practices among the Classic period Maya. *Quat. Int.* **2012**, *249*, 19–30. [CrossRef]

3. Stahl, P.W.; Pearsall, D.M. Late pre-Columbian agroforestry in the tropical lowlands of western Ecuador. *Quat. Int.* **2012**, *249*, 43–52. [CrossRef]
4. Voeks, R.; Greene, C. God's healing leaves: The colonial quest for medicinal plants in the Torrid Zone. *Geogr. Rev.* **2018**, *108*, 545–565. [CrossRef]
5. FAO. Agroforestry Definition. Available online: http://www.fao.org/forestry/agroforestry/80338/en/ (accessed on 1 March 2020).
6. Somarriba, E.; Beer, J.; Alegre-Orihuela, J.; Andrade, H.J.; Cerda, R.; DeClerck, F.; Detlefsen, G.; Escalante, M.; Giraldo, L.A.; Ibrahim, M.; et al. Mainstreaming Agroforestry in Latin America. In *Agroforestry—The Future of Global Land Use*; Nair, P.K.R., Garrity, D., Eds.; Springer: Dordrecht, The Netherlands, 2012; pp. 429–453. [CrossRef]
7. WorldBank-Group. *Future of Food: Shaping a Climate-Smart Global Food System*; World Bank: Washington, DC, USA, 2015; p. 32.
8. FAOSTAT. Available online: http://www.fao.org/faostat/en/#data/Qc (accessed on 1 March 2020).
9. Diaz-Montenegro, J.; Varela, E.; Gil, J.M. Livelihood strategies of cacao producers in Ecuador: Effects of national policies to support cacao farmers and specialty cacao landraces. *J. Rural Stud.* **2018**, *63*, 141–156. [CrossRef]
10. Viteri Salazar, O.; Ramos-Martín, J.; Lomas, P.L. Livelihood sustainability assessment of coffee and cocoa producers in the Amazon region of Ecuador using household types. *J. Rural Stud.* **2018**, *62*, 1–9. [CrossRef]
11. Acevedo, V.L.; Ulloa, J.; Osejo, J.A. *Cartografía Histórica de Áreas Naturales Protegidas y Territorios Indígenas de la Amazonía Ecuatoriana*; FLACSO: Calle, Guatemala, 2016.
12. Zarrillo, S.; Gaikwad, N.; Lanaud, C.; Powis, T.; Viot, C.; Lesur, I.; Fouet, O.; Argout, X.; Guichoux, E.; Salin, F.; et al. The use and domestication of Theobroma cacao during the mid-Holocene in the upper Amazon. *Nat. Ecol. Evolut.* **2018**, *2*, 1879–1888. [CrossRef]
13. Loor, R.G.; Lachenaud, P.; Fouet, O.; Argout, X.; Peña, G.; Macias, J.C.; Puyutaxi, A.; Marcelo, F.; Valdez, F.; Hurtado, J. Rescue of cacao genetic resources related to the nacional variety: Surveys in the Ecuadorian Amazon (2010–2013). *Rev. ESPAMCiencia* **2015**, *6*, 7–15.
14. Salazar, E. *An Indian Federation in Lowland Ecuador*; IWGIA: Copenhagen, Danmark, 1977; Volume 28.
15. Ostrom, E. A diagnostic approach for going beyond panaceas. *Proc. Natl. Acad. Sci. USA* **2007**, *104*, 15181–15187. [CrossRef]
16. Ostrom, E. A General Framework for Analyzing Sustainability of Social-Ecological Systems. *Science* **2009**, *325*, 419–422. [CrossRef]
17. Ratajczyk, E.; Brady, U.; Baggio, J.; Barnett, A.; Perez-Ibarra, I.; Rollins, N.; Rubiños, C.; Shin, H.; Yu, D.; Aggarwal, R.; et al. Challenges and opportunities in coding the commons: Problems, procedures, and potential solutions in large-N comparative case studies. *Int. J. Commons* **2016**, *10*. [CrossRef]
18. De Zaldivar, V.B.S. From agrarian reform to ethnodevelopment in the highlands of Ecuador. *J. Agrar. Chang.* **2008**, *8*, 583–617. [CrossRef]
19. Schaefer, T. Engaging modernity: The political making of indigenous movements in Bolivia and Ecuador, 1900–2008. *Third World Q.* **2009**, *30*, 397–413. [CrossRef]
20. Berkes, F.; Colding, J.; Folke, C. Rediscovery of traditional ecological knowledge as adaptive management. *Ecol. Appl.* **2000**, *10*, 1251–1262. [CrossRef]
21. Caballero-Serrano, V.; Onaindia, M.; Alday, J.G.; Caballero, D.; Carrasco, J.C.; McLaren, B.; Amigo, J. Plant diversity and ecosystem services in Amazonian homegardens of Ecuador. *Agric. Ecosyst. Environ.* **2016**, *225*, 116–125. [CrossRef]
22. Descola, P. *In the Society of Nature: A native Ecology in Amazonia*; Cambridge University Press: Cambridge, UK, 1994; Volume 93, p. 372.
23. Valdez, F. Ressources anciennes et enjeux contemporains. In *Une Archéologie pour le Développement*; Galipaud, J.-C., Guillaud, D., Eds.; Les Editions La Discussion: Marseille, France, 2014; pp. 119–130.
24. Barrezueta-Unda, S.; Paz-Gonzalez, A. Indicators of sustainability social and economic: Case cocoa farmers of El Oro, Ecuador. *Ciencia Unemi* **2018**, *11*, 20–29. [CrossRef]
25. Perdomo, C.; Barzola, W.J.F.; Acosta, O.N.R. Open innovation, an alternative for the strengthening of Ecuadorian agriculture. *Rev. Univ. Soc.* **2018**, *10*, 263–269.
26. Nelson, V.; Tallontire, A.; Collinson, C. Assessing the benefits of ethical trade schemes for forest dependent people: Comparative experience from Peru and Ecuador. *Int. For. Rev.* **2002**, *4*, 99–109. [CrossRef]

27. Purcell, T.F.; Fernández, N.; Martínez, E. The uneven geographical development of cacao commodity chains in Ecuador: From differential to monopoly ground-rent. *CENEDET Working Paper #6* **2016**, 1–20.
28. Purcell, T.F. "Hot chocolate": Financialized global value chains and cocoa production in Ecuador. *J. Peasant Stud.* **2018**, *45*, 904–926. [CrossRef]
29. Cummins, I.; Alavarado, E.; Andi, B. *Actualización del Plan de Manejo de Comunidad Santa Rita 2016–2026*; Runa Foundation: New York, NY, USA, 2016; p. 33.
30. Larrauri, O.D.; Neira, D.P.; Montiel, M.S. Indicators for the Analysis of Peasant Women's Equity and Empowerment Situations in a Sustainability Framework: A Case Study of Cacao Production in Ecuador. *Sustainability* **2016**, *8*, 1231. [CrossRef]
31. Rueda, X.; Paz, A.; Gibbs-Plessl, T.; Leon, R.; Moyano, B.; Lambin, E.F. Smallholders at a Crossroad: Intensify or Fall behind? Exploring Alternative Livelihood Strategies in a Globalized World. *Bus. Strategy Environ.* **2018**, *27*, 215–229. [CrossRef]
32. Dulcire, M. From passivity to collaboration: Changes in the links between cocoa farmers and a manufacturer in Ecuador. *Cah. Agric.* **2010**, *19*, 249–254. [CrossRef]
33. Useche, P.; Blare, T. Traditional vs. modern production systems: Price and nonmarket considerations of cacao producers in Northern Ecuador. *Ecol. Econ.* **2013**, *93*, 1–10. [CrossRef]
34. Melo, C.J.; Hollander, G.M. Unsustainable development: Alternative food networks and the Ecuadorian Federation of Cocoa Producers, 1995–2010. *J. Rural Stud.* **2013**, *32*, 251–263. [CrossRef]
35. Coq-Huelva, D.; Higuchi, A.; Alfalla-Luque, R.; Burgos-Moran, R.; Arias-Gutierrez, R. Co-Evolution and Bio-Social Construction: The Kichwa Agroforestry Systems (Chakras) in the Ecuadorian Amazonia. *Sustainability* **2017**, *9*, 1920. [CrossRef]
36. Mohebalian, P.M.; Aguilar, F.X. Additionality and design of forest conservation programs: Insights from Ecuador's Socio Bosque Program. *For. Policy Econ.* **2016**, *71*, 103–114. [CrossRef]
37. FAO. *Advancing Agroforestry on the Policy Agenda: A Guide for Decision-Makers*; Food and Agriculture Organization of the United Nations: Rome, Italy, 2013; p. 37.
38. Ceddia, M.G.; Gunter, U.; Corriveau-Bourque, A. Land tenure and agricultural expansion in Latin America: The role of Indigenous Peoples' and local communities' forest rights. *Global Environ. Chang.* **2015**, *35*, 316–322. [CrossRef]

2

Sustainability Assessment of Smallholder Agroforestry Indigenous Farming in the Amazon

Marco Heredia-R [1,2,*], Bolier Torres [2,3], Jhenny Cayambe [4], Nadia Ramos [5], Marcelo Luna [6] and Carlos G. H. Diaz-Ambrona [1]

[1] AgSystems, Ceigram, itdUPM, Centro de Innovación en Tecnología para el Desarrollo, Universidad Politécnica de Madrid (UPM), 28040 Madrid, Spain; carlosgregorio.hernandez@upm.es
[2] Universidad Regional Amazónica IKIAM, Km 7 Vía Muyuna, Tena 150150, Ecuador; bolier.torres@ikiam.edu.ec
[3] Departamento de Ciencias de la Vida, Universidad Estatal Amazónica (UEA), Pastaza 160101, Ecuador
[4] School of Agricultural and Environmental Sciences, Pontificia Universidad Católica del Ecuador Sede Ibarra (PUCESI), Imbabura 100112, Ecuador; jmcayambe@pucesi.edu.ec
[5] Andean University Simon Bolivar, Toledo N22-80, Quito 170143, Ecuador; nadiaramosr@gmail.com
[6] Departamento de Ciencias de la Tierra, Universidad Estatal Amazónica (UEA), Pastaza 160101, Ecuador; mluna@uea.edu.ec
* Correspondence: mageher@gmail.com

Abstract: In the Amazon, the Yasuní Biosphere Reserve (YBR) is considered a natural and cultural diversity hotspot. It is populated by several indigenous groups, including the Kichwa, who are characterized by their traditional systems of production, which are a means of subsistence and socio-ecological integration. The objective of this research was to evaluate the sustainability of small farmers who use a traditional agroforestry system (*chakra*) within the buffer, transition, and core zones of the YBR. We conducted 133 interviews with Kichwa heads of households. The socio-demographic structure and distribution were identified, and the response-inducing sustainability evaluation (RISE) methodology was used to evaluate chakra sustainability according to social, economic, and ecological dimensions, expressed using 10 indicators from 50 parameters, valued from 0 (worst case) to 100 (best case). The results are expressed in a polygon, defined by the areas: (1) good performance, (2) medium performance, and (3) poor performance. We employed the multivariate classification hierarchical cluster technique and analysis of variance (ANOVA) to identify dissimilarities between groups of chakras and the existence of statistical differences, respectively. Among the studied indigenous Kichwas, a pyramidal structure progressive type was identified, which is characteristic of young populations and the nonexistence of significant differences between the RISE indicators and chakras. The lowest-scoring indicators using the RISE guidelines were: use of materials and environmental protection, animal production, economic viability and chakra administration. We provide suggestions for decision makers who support Kichwa populations in socio-productive management with sustainability goals. We to taking actions on the indicators identified with high priority to improve the sustainability in the chakras and sociodemographic dynamics.

Keywords: hotspot; natural resources; sustainable agriculture; Yasuní

---

1. Introduction

The biological and cultural richness of the tropics is recognized globally. Worldwide, it is estimated that there are approximately 300 million indigenous people who belong to 5000 different cultures, representing between 80% and 90% of the world's cultural diversity [1]. They are also called tribal, aboriginal, or autochthonous peoples; they are located in various ecosystems, particularly in the tropics [2]. In many cases, national minorities may meet one or more of the following criteria: (a) they are descendants of the original inhabitants of an ancestral territory; (b) they are peoples of the ecosystem, such as cultivators, shepherds, hunter-gatherers, fishers, permanent or itinerant artisans, etc.; (c) they practice a form of small-scale rural production; (d) they organize their lives at a community level; (d) they share common traditional clothing, moral values, and identifying characteristics; (e) they have a custodial and non-materialistic attitude based on a symbolic exchange with nature; (f) some are subjugated by a dominant culture, society, and economy, and others remain isolated from market economies, preserving their culture and social dynamics; and (g) they consist of individuals who subjectively consider themselves to be indigenous [3].

The Western Amazon includes parts of Bolivia, Colombia, Ecuador, Peru, and western Brazil. It is one of the most biodiverse areas on the planet for many taxa, which include plants, insects, amphibians, birds, and mammals [4–9]. The region maintains large areas of intact tropical humid forest and has a high probability of stable climatic conditions in the face of global warming [10]. The Western Amazon is also home to many indigenous ethnic groups, including some of the world's last peoples living in voluntary isolation [11–13]. Ecuador has a cultural heritage with an estimated one million people self-identifying as indigenous [14]. In the Ecuadorian Amazon Region (EAR), there are currently 11 officially identified nationalities, one of which is the Amazonian Kichwa, the most populous ethnic group in this region. The EAR has been inhabited since before the Hispanic conquest by the Kichwa and other indigenous peoples [15,16]. Since the second half of the 19th century, these populations have increased due to rubber plantations [17]. However, their large-scale environmental transformation began in the 1970s with the intensification of oil exploitation, highway expansion, and colonization [18], which led to the division of ancestral lands and the displacement and disintegration of indigenous nationalities [19]. Later, the construction of roads, initially to facilitate oil activities, brought the small-scale agricultural colonization of migrant settlers, who were also pushed by government land tenure policies [20]. These processes transformed the cultural, social, and, to some extent, the agricultural context of the Kichwa populations given their territorial displacement as well as their contact with societies based on the market economy [21]. With this intervention by external agents in the territory, new strategies for living were created such as: participation in salaried jobs, the purchase of manufactured products, the use of government services, and participation in political activism. In the agricultural context, access to markets provoked new agricultural practices [22–24], such as the expansion of monocultures like cacao (*Theobroma bicolor* L.) and especially for the sale of small-scale agricultural and livestock products [25]. This caused the rapid expansion of the agricultural frontier and consequent deforestation [26], diminishing the quantity and quality of ecosystem services [27], which have been used ancestrally as a means of livelihood.

The Amazonian Kichwa populations have also been characterized by their agricultural systems, having practiced the traditional system called *chakra* for thousands of years. This system was initially oriented toward subsistence, integrated with the cultivation of basic foodstuffs, such as manioc (*Manihot esculenta* Crantz), plantain banana (*Musa paradisiaca* L.), peach palm (*Bactris gasipaes* Kunth) etc., as well as medicinal plants [28,29]. It is characterized by its high level of diversity [30] and its ability to provide security and sovereignty in terms of food and health [31–33]. The chakra plots in the Northern Ecuadorian Amazon range from 0.05 to 3.0 ha [30,34,35], have high levels of ecological and social integration [36–38], and can mitigate both the impact of population growth in the Amazon [29] and the effects of climate change [39].

The need is urgent for greater agricultural sustainability [40]; comprehensive responses are required to understand the complex dynamics between social, economic, and ecological sustainability [41,42]. In 2015, the United Nations plenary approved the 2030 agenda, converging the development, environment, and climate agendas. The agenda includes 17 sustainable development goals (SDGs) that must be met by 2030, unlike the previous sectorial and regional goals, such as the Millennium

Development Goals for the period 2000–2015, which only directly affected developing countries. This new agenda is for everyone and for a global world. This means that in the coming years, all political actions and business strategies, and especially private ones, must be aligned with the achievement of those objectives. In the field of agriculture, agricultural and environmental policies will be aligned with the agenda, especially in the field of sustainability. The specific related goals include SDG 15: Life of terrestrial ecosystems, in which the Amazon biome is a main exponent; the scope of SDG 1: End poverty in all its forms; and SDG 2: Zero hunger should be allowed, in coexistence with local populations [43]. Assessing agricultural sustainability is one of these responses. Although capturing the systemic complexity of sustainability through assessment is difficult, it is warranted by the attention of policy makers beyond crop productivity to include dimensions of human well-being and ecological soundness [44,45]. Sustainability is a multidimensional concept [46] of a dignified life for the present without compromising a dignified life for future generations or threatening the natural environment and endangering the global ecosystem [47]; its evaluation is an important process in promoting the concept of sustainable agricultural systems [48,49].

To understand the transition toward more sustainable production, a variety of tools have been developed to obtain information on the sustainability performance of production [50,51]. Indicator-based sustainability assessment tools require the management of a wide variety of information types, parameters, and uncertainties [52], and vary in scope (geographic and sectorial), target group (farmers or policy makers), selection of indicators, aggregation, weighting methods, and time required for their execution [50,51,53]. Although many emphasize the importance of integrating social, economic, and environmental issues into sustainability assessment tools, environmental issues and tools generally receive more attention [50,51,54–56]. The results of sustainability assessment tools should be seen as a starting point for discussion, reflection, and learning [57].

Several tools are available for the evaluation of sustainability based on indicators: farm sustainability indicators (IDEA) [58], sustainability assessment of food and agriculture systems (SAFA) [59], response-inducing sustainability evaluation (RISE) [60], and others [61,62]. The IDEA tool was developed in France in 1998. Its analysis is at the farm level and it represents an approach based on principles; methodological indicators can be added [58]. SAFA is an open process methodology. According to the guidelines of the Food and Agriculture Organization of the United Nations (FAO), it is used as a self-assessment tool for food producers and manufacturers [59]. The justifications for the selection of indicators have been documented, and the integration problems between scales have been considered. It is not possible to add indicators, and the validation of the indicators is based on comparison and expert evaluation [63]. RISE is an easy-to-use tool that covers all dimensions of sustainability [57] and uses principles-based approaches; RISE presents both numerical and graphical results [51]. RISE is the best compared to other tools for sustainability assessment in terms of scientific robustness, feasibility, utility, influence, spatial applicability, and adaptability [52]. RISE has been used in various countries such as Poland, Spain, Kenya, and Canada, and in different evaluation scenarios (Table 1).

Table 1. Evaluation of sustainability using the response-inducing sustainability evaluation (RISE) methodology.

Evaluation Scenario	Country	Indigenous Peoples and/or Small Producers	References
Family farming	Sweden	No	[64]
Organic and conventional farms	Poland	No	[65]
Indigenous agriculture	Ecuador	Yes	[66]
Organic farms	Denmark	No	[67]
Agricultural and livestock production	Ethiopia	Yes	[68]
Tea plantation	India	Yes	[69]
Agricultural production	Spain	No	[70]
Coffee cultivation	Brazil	Yes	[71]

Sustainability assessment tools can provide support for decision making in production systems and may, therefore, significantly impact their sustainable development [53,72]. The availability and quality of data, time and budget requirements, as well as factors related to unknown terminology and ease of use of use and accessibility of the tool, influence the perception of small producers about the relevance of the tool and, consequently, whether they decide to use it [53,73,74].

In this study, our aim was to assess the sustainability of small farms using the traditional agroforestry system (chakra) in the Ecuadorian Amazon, with the objective of understanding traditional knowledge about the farming system and to explore and discuss possible improvements.

2. Materials and Methods

2.1. Geographical Location

We focused on the Kichwa populations settled along the Napo River, in the north of the Yasuní Biosphere Reserve (YBR). This indigenous group is the result of inter-ethnic relations with ancestral populations in the area: Quijos, Záparos, Omaguas, Tucanos, Shuar, Achuar, Siona, Secoya, and Highland Kichwas and Peruvian Quechuas [36]. The Kichwa of this region are the most numerous indigenous populations in the EAR (60,000 inhabitants) [37,38]. Those residing in the Yasuní National Park (YNP) present a type of grouping based on associations that cohabit communal territories, associations, and federations [39]. They speak *Runa Shimi*, a dialect based on Kichwa and Spanish. Their identities are complex and overlap with other indigenous and white settler groups.

The study was conducted in two sectors, A and B, located in the YBR, which is considered one of the areas with the greatest biological and cultural biodiversity on the planet [75–78], including the YNP, Waorani Ancestral Territory (WAT), Tagaeri Taromenane Intangible Zone (TTIZ), and the Fringe of Diversity and Life (FDL), located in the EAR. Ecuador is 1 of 17 megadiverse countries [79,80] (Figure 1). The YBR was announced by UNESCO in 1989; it is located in the provinces of Orellana (51.96%), Pastaza (39.40%), and Napo (8.64%), between the Napo and Curaray Rivers [81]. The predominant ecosystem is lowland evergreen forest in the Napo–Curaray (BsTa02) [82], the temperature ranges from 24 to 27 °C, rainfall is 3200 mm/year, relative humidity ranges from 80% to 94%, and the soils are relatively young and are generated by fluvial sediments from the erosion of the Andes [83].

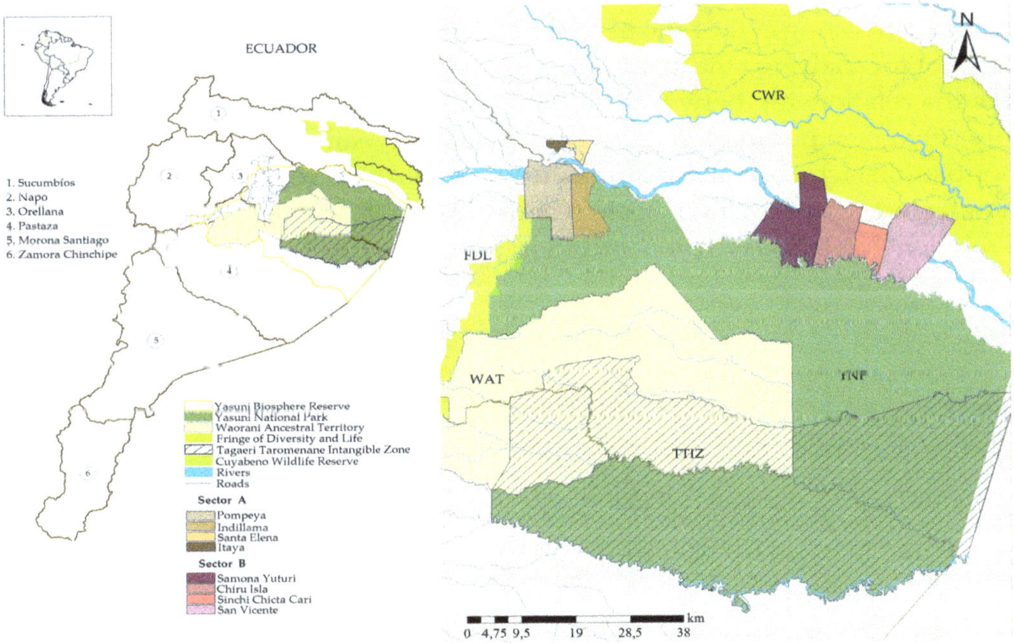

Figure 1. Sectors A and B located in the north of the Yasuní Biosphere Reserve (YBR) in the Ecuadorian Amazon region.

2.2. Sampling and Data Collection

In the sectors identified, sampling was performed using the reference chain or snowball methodology [84], given the difficulty of constructing a sampling frame due to the scarcity of demographic information in the intervention zone and the complex logistics for traveling to and mobilization through the communities. This method begins with one interviewee, who gives the researcher the name of at least one other potential interviewee. This interviewee, in turn, provides the name of at least one other potential interviewee, etc., with the sample growing like a rolling snowball if more than one reference is provided per interviewee [85,86]. In the initial phase of the survey, participants expressed their concern that local and international institutions would cut social and agricultural programs due to participation in the study. In this situation, a signed informed consent form was not used, but with the support of the German Agency for Development Cooperation (GIZ-Ecuador), all approaches were made to producers and the principles of ethical research were explicitly applied [87], discussing the objectives, risks, methodology, and schedule of the study with members of each community. One of the pillars of our positive relationship with the participants was the presentation of the results of the questionnaire. The restitution of results is an ethical duty that is often overlooked when conducting research on human beings, especially when indigenous communities participate [88,89].

2.3. Characterization and Sociodemographic Indices

Surveys were conducted with 133 Kichwa households with chakras. They were distributed by sector. In sector A (61), the surveys occurred in the communities of Pompeya (35), Indillama (10), Santa Elena (4), and Itaya (12). In sector B (72), they were conducted in the communities of Samona Yuturi (14), Chiru Isla (31), Sinchi Chicta Cari (15), and San Vicente (12). The average area of the chakras was 0.5 ha in 60 ha of titled land per household.

We studied the Kichwa population structure and its distribution by sex and age from a population pyramid (statistical representation) [90] to examine its implications with traditional production systems [91]. We calculated the following indices: (1) proportion of young population (<14 years) ($P_{young\ people}$); (2) proportion of adult population (between 15 and 64 years, P_{adults}); (3) ratio of children to women, defined as the number of children under 5 years of age for each woman of reproductive age *(R)*; (4) the ratio of men, consisting of the ratio of men for every 100 women in a given population, considered as the first indicator for analyzing the distribution by sex in the population (*R.M.*); (5) youth dependency ratio, which is the relationship between the potentially dependent age population (<15 years) and the potentially active age population (between 15 and 64 years, (I^{t}_{dj}); (6) the structure index of the active age population, which is the relationship between the population from 40 to 64 years and the population from 15 to 39 years (I_r); and (7) the rate of change of the active age population (I^{t}_{r}), which is the relationship between the population from 60 to 64 years and the population from 15 to 19 years [90–94], as follows:

$$P_{young\ people} = \frac{P_{0-14}}{P} \times 100 \quad (1)$$

$$P_{adults} = \frac{P_{15-64}}{P} \times 100 \quad (2)$$

$$R = \frac{P_{0-4}}{P_{M.\ 15-49}}\ children\ per\ woman \quad (3)$$

$$R.M. = \frac{P_{men}}{P} \times 100 \quad (4)$$

$$I^{t}_{dj} = \frac{P^{t}_{<15}}{P^{t}_{15-64}} \times 100 \quad (5)$$

$$I_r = \frac{P^t_{15-64}}{P^t_{15-39}} \times 100 \quad (6)$$

$$I^t_r = \frac{P^t_{60-64}}{P^t_{15-19}} \times 100 \quad (7)$$

2.4. Assessment of Agricultural Sustainability

The RISE methodology was applied to holistically evaluate the sustainability of the traditional agroforestry system (chakra). The dimensions considered were economic, social, and ecological [95–99], which allowed us to analyze and compare the degree of sustainability between the chakras. This methodological tool is characterized by the balance between the simplicity of the analysis, the complexity of reality and the transparency of the results [60,97]. RISE seeks to create a tangible evaluation based on science that allows for the beginning of the creation of measures to improve sustainability [98,99] and to initiate a constructive dialogue between producers and processors to spread the philosophy of sustainable production [96–98]. The methodological process began with an interview of the owner of a chakra. The RISE questionnaire was designed with three types of questions: open, drop-down list, and Boolean. The structural index of the questionnaire was divided into three stages: (A) preparation of the field visit to the chakras, containing general questions; (B) questions for the field visit, containing questions with qualitative and quantitative data that were collected during the RISE interview; and (C) comments, a space where notes can be transcribed about the questions or observations during the visit to the chakra. The duration of the questionnaire was 95 min [98].

For the systematization and analysis of the data of the conglomerate of the communities and the holistic evaluation, we used RISE 3.0 Software [99], developed by the Swiss College of Agriculture (SHL), based on the 10 standard indicators according to 50 parameters (Table 2), valued from 0 (worst case) to 100 (best case). As a result, a sustainability polygon was issued, defined by the following areas: (1) good yield, green coloration (66.66–100); (2) average yield, yellow coloration (33.34–66.65); and (3) bad yield, red coloration (0–33.33). The qualification values in the RISE method are fixed and cannot be changed. The red line superimposed on the polygon indicates the degree of sustainability per indicator, which is based on the arithmetic average of four to seven parameters that have the same weight [98].

Table 2. Indicators and parameters used in this study based on response-inducing sustainability evaluation (RISE) 3.0 methodology.

Indicators	Subtopics	
1. Land use	1.1. Land management 1.2 Crop productivity 1.3 Soil organic matter 1.4 Soil reaction	1.5 Soil contamination 1.6 Soil erosion 1.7 Soil compaction
2. Livestock production	2.1 Cattle management 2.2 Cattle productivity 2.3 Behavioral opportunity according to species	2.4 Quality of animal housing 2.5 Animal health
3. Use of materials and environmental protection	3.1 Nitrogen balance 3.2 Phosphorous balance 3.3 Self-sufficiency of nitrogen and phosphorus	3.4 Ammonia emissions (risk) 3.5 Waste management
4. Water use	4.1 Water management 4.2 Water supply	4.3 Intensity of water use 4.4 Risks to water quality
5. Energy and climate	5.1 Energy management 5.2 Intensity of energy used in agricultural production	5.3 Balance of greenhouse gases
6. Biodiversity	6.1 Management of crop protection 6.2 Areas of ecological priority 6.3 Intensity of agricultural production	6.4 Quality of landscape 6.5 Diversity of agricultural production

Table 2. *Cont.*

Indicators	Subtopics	
7. Working conditions	7.1 Staff management 7.2 Working hours	7.3 Health and safety 7.4 Salary and income levels
8. Quality of life	8.1 Occupation and education 8.2 Financial situation 8.3 Social relations	8.4 Personal freedom and values 8.5 Health
9. Economic viability	9.1 Liquidity reserve 9.2 Level of debt 9.3 Economic vulnerability 9.4 Secureness of household livelihoods	9.5 Cash flow, sales volume ratio 9.7 Exhaustion of the capacity to serve the capital of others (payment of interest and amortization)
10. Administration	10.1 Strategies and planning 10.2 Guarantee of supply and performance	10.3 Tools for planning and documenting 10.4 Administration of quality 10.5 Cooperation with others

2.5. The State and Driving Force for Calculating the Degree of Sustainability

Each indicator contains parameters that describe the state of the system (S) and the driving force (D) within the system, leading it in a certain direction of development. S indicates the current condition of each specific indicator, and D is a measure of estimated pressure exerted by the agricultural system on the specific indicator [60]. The state parameters have a value between 0 (worst case) and 100 (best case). The driving force parameters are also calculated using a range of 0 to 100, but since they are valued as a negative pressure in the system, 0 represents the best case and 100 the worst case (the highest pressure).

The degree of sustainability (DS) is calculated as the difference between the state (S) and the driving force (D); DS = S–D (Figure 2). Considering the S and D parameters not only allows for the creation of a static picture of the current situation; it reveals the development trends of the system. Therefore, the RISE evaluation provided an analytical snapshot of the chakra situation while describing some aspects of the dynamics that change the system over time. The assessment of the driving force allowed us to understand and highlight the trends and threats that can be decisive for the concept of operational sustainability.

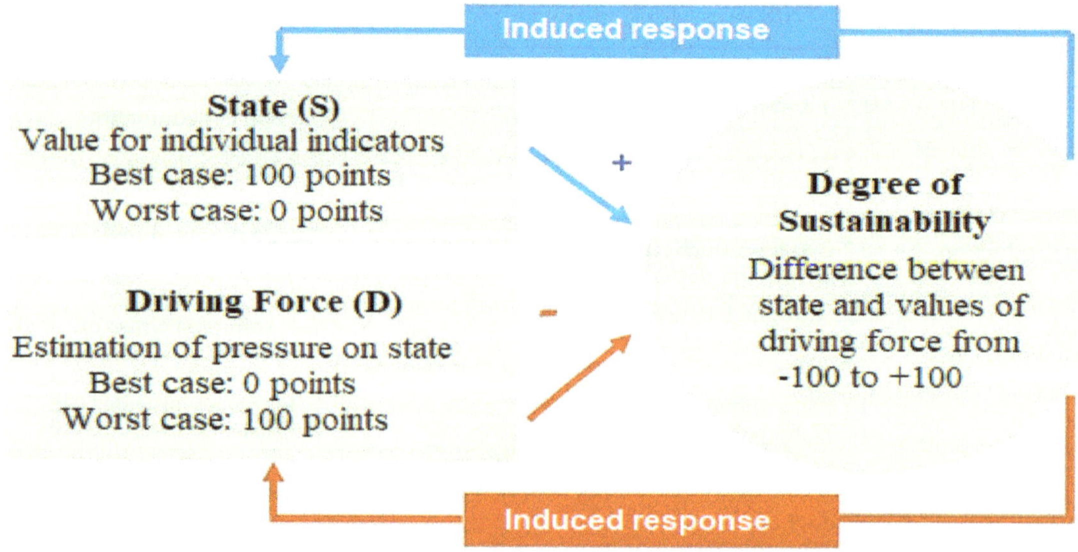

Figure 2. Methodology for calculating the degree of sustainability according to the RISE methodology.

2.6. Hierarchical Cluster Analysis

The statistical analysis was conducted using the multivariate classification hierarchical cluster technique. This is a mathematical method that is included in multivariate statistics. Multivariate analysis is the branch of statistical analysis that studies, analyzes, represents, and interprets the data resulting from the observation of $p > 1$ statistical variables on a sample of n individuals. It focuses on the simultaneous investigation of two or more characteristics (variables), measured on a set of individuals [100]. Hierarchical cluster analysis is mainly used for the formation of groups starting with the basic unit of the objective (BUO), according to the similar characteristics from the similarities or dissimilarities that are presented between pairs of these BUOs in the evaluated characteristics [101]. The hierarchical methodology deals with the grouping of clusters to form a new group, starting with groups as existing individuals in the study and grouping them to form all the cases in the same group [102].

All the sustainability variables defined by the RISE evaluation framework were used, applying the Euclidean squared distance as a similarity criterion, which is a measurement that is affected by the differences in metrics between the variables [100]. It is a measurement of dissimilarity: its minimum value is 0, but it has no maximum value [101]. We also employed the Ward grouping method for hierarchically forming the groups and agglomerates by minimizing the intra-group variation. It tends to generate small but very balanced conglomerates [103] and, at each step of the agglomeration process, uses the distance between classes that meets the objective of joining [104,105].

The data were organized in an Excel spreadsheet (version 19.0, Microsoft, Redmond, United States), using the SPSS version 22 statistical software (IBM, Armonk, United States) [106] to evidence the history of clusters, where the groups to be formed were selected according to the dissimilarity index. The resulting dendrograms are presented for analysis of the sectors in an individual and united way according to the owner of the chakra and analysis of variance (ANOVA) was used to identify the existence of significant differences (0.05) between groups of means.

3. Results

3.1. Sociodemographic and Agricultural Characterization

The Kichwa population in the study area was 48.5% men and 51.5% women (726 total). The average age for men was 27 years and the average for women was 29. The proportion of young men was 17.7% and women, 22.8%; adult men comprised 30.7% and women, 28.6% (Figure 3). The ratio of children to women was 0.45%, while the ratio of men was 48%. The youth dependency index was 68%, the structural index of the working age population was 40%, and the rate of change of the working age population was 4%.

The following pattern was found in the evaluated sites: in sector A, the household composition varied between four and six people. In this sector, the Pompeya and Itaya communities reported at least one household with 10 individuals. In sector B, the household composition ranged from four to seven (Table 3) and the Sinchi Chicta Cari community reported a household with 11 individuals. The average household size in the communities studied was five individuals.

The average age of the heads of household was 42 years. The youngest head of the household (25 years) lived in the Sinchi Chicta Cari community. The youngest age range was between 25 and 34 years. In the communities of Pompeya, Chiru Isla, and San Vicente resides, the oldest head of the household was 60 years. The oldest age ranged from 44 to 60 years.

In the communities of Pompeya and Chiru Isla EL 8.57% and 3.23%, respectively, completed the third level of education. In Itaya and San Vicente, 8.33% of the heads of household entered the third level of education but had not finished it. In sectors A and B, 19 (31.31%) and 28 (38.24%) heads of household had completed secondary education, respectively; in sectors A and B, 43 and 32 heads of household had completed basic education, respectively.

With regards to land tenure, of the 133 heads of household surveyed, 88% (117 Kichwa indigenous) had deeds (legal land tenure) to their chakras. In the community of Santa Elena (sector A), all had deeds, whereas in the Chiru Isla community (sector B), 23% did not have legal documentation for their land. On average in sectors A and B, 4% and 8% of the chakra owners, respectively, did not have deeds to their lands.

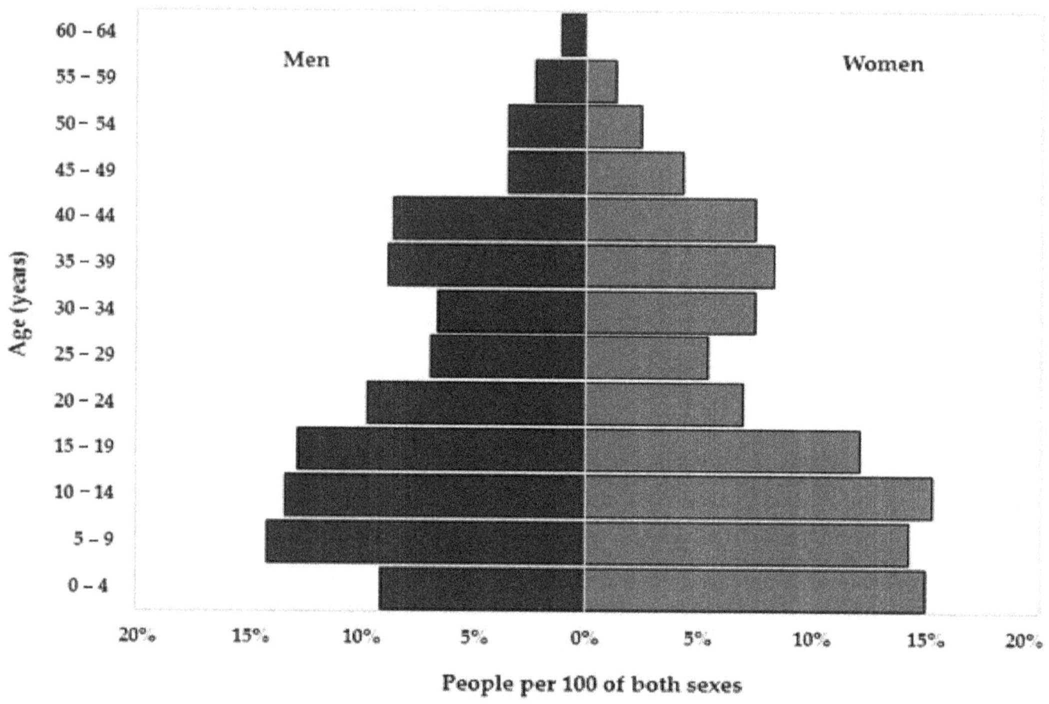

Figure 3. Population pyramid (progressive type) of Kichwas superimposed in the YBR in January 2018.

Table 3. Socio-demographic and agricultural characteristics of the Kichwa population by sector and communities in the north of the Yasuní Biosphere Reserve in the Ecuadorian Amazon region.

Variables		Sector A				Sector B			
		Pompeya	Indillama	Santa Elena	Itaya	Samona Yuturi	Chiru Isla	Sinchi Chicta Cari	San Vicente
Household Size	Average (SD)	5 (2.18)	4 (2.37)	4 (2.16)	6 (2.20)	6 (2.16)	5 (2.17)	4 (2.91)	7 (2.47)
	Max	10	9	7	10	9	9	11	9
	Min	2	1	2	3	3	2	1	2
Age of head of the household (years)	Average (SD)	43 (10.80)	45 (11.67)	47 (10.24)	35 (4.98)	41 (11.69)	44 (9.34)	39 (8.04)	44 (11.83)
	Max	60	57	56	44	64	60	56	60
	Min	29	26	34	28	29	27	25	27
Education of head of the household (%)	Higher education	8.57	0.00	0.00	8.33	0.00	3.23	0.00	8.33
	High school	48.57	60.00	0.00	16.67	35.71	32.26	60.00	25.00
	To middle school	17.14	0.00	100.00	50.00	42.86	29.03	40.00	50.00
	None	25.71	40.00	0.00	25.00	21.43	35.48	0.00	16.67
Literate?	% (Yes)	83	90	100	92	93	77	87	83
Existing species in the chakras	Cacao (*Theobroma bicolor L.*), coffee (*Caffea* sp.), manioc (*Manihot esculenta*), plantain banana (*Musa paradisiaca*), corn (*Zea mays*), sugar cane (*Saccharum officinarum*), annatto (*Bixa orellana*) and pacay (*Inga* spp.)								
Animals	Poultry (laying hens, broilers, ducks)								

Sector, Community set; SD, Standard deviation; Max, Maximum; Min, Minimum.

3.2. Cluster Analysis and Sustainability Indicators

3.2.1. Values in Sector A

Using the indicators resulting from the evaluation of the sustainability of traditional agroforestry systems (chakra) in sector A using hierarchical cluster analysis, three groups were obtained in a dendrogram (Figure 4) with a Euclidean distance (measurement of association): in groups 1 and 2, we gathered 24 chakras; in group 3, there were 13 chakras.

In the three groups (Table 4), the indicators identified by level of importance (highest to lowest average scores) were: energy and climate (86.82), working conditions (77.89), water use (74.97), and land use (72.62). The quality of life indicator in groups 2 and 3 had a difference of 17.77. The indicators that scored medium performance were biodiversity and use of materials and environmental protection, with average values of 64.52 and 41.94, respectively. In groups 1 and 3, the indicator animal production had a difference of 3.56 and the quality of life in group 1 has a score of 64.25. The lowest performing indicators were: economic viability and administration, presenting averages of 29.19 and 23.66, respectively; and in group 2, the indicator animal production was 30.38.

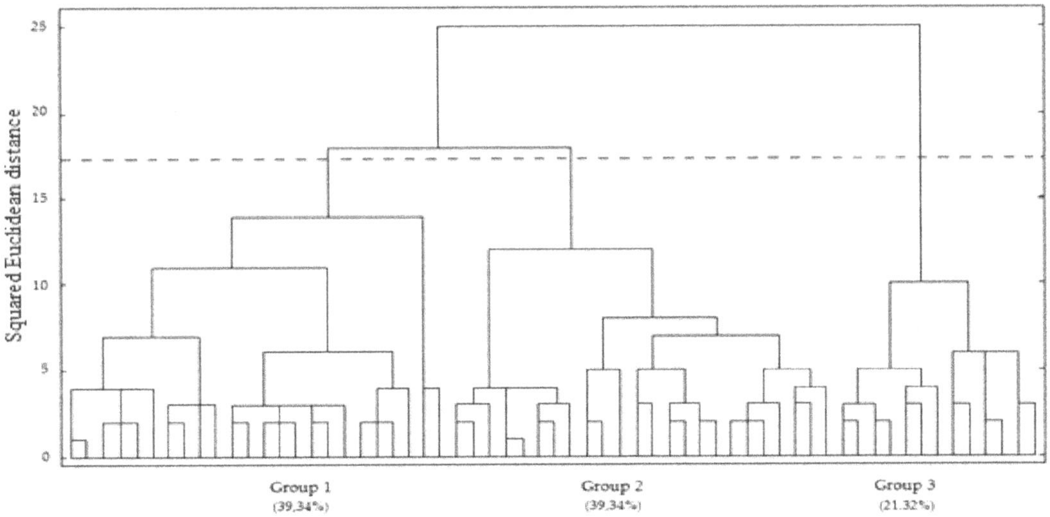

Figure 4. Dendrogram using a Ward linkage to establish three groups within the Kichwa communities in sector A (Pompeii, Indillama, Santa Elena, and Itaya) in the northern Yasuní Biosphere Reserve (YBR) of the Ecuadorian Amazon (percentage of *chakras* in each group).

Table 4. Means and standard deviation of sustainability indicators among clusters (groups) of the chakra systems in sector A in the north of the Yasuní Biosphere Reserve in the Ecuadorian Amazon region.

Indicators	Groups **						Overall Mean of Sector A
	Group 1 (24)		Group 2 (24)		Group 3 (13)		
	Mean	SD	Mean	SD	Mean	SD	
Land use	70.83	4.07	73.50	5.48	73.54	3.04	72.62
Animal production	38.25	4.41	30.38	10.38	34.69	7.27	34.44
Use of materials and environmental protection	36.54	7.60	46.58	7.15	42.69	9.68	41.94
Water use	75.08	3.83	74.67	4.24	75.15	4.22	74.97
Energy and climate	87.00	0.00	86.46	2.65	87.00	0.00	86.82
Biodiversity	65.54	5.45	64.71	4.27	63.31	4.44	64.52
Working conditions	76.83	4.95	80.46	2.06	76.38	5.41	77.89
Quality of life	64.25	6.35	69.08	5.45	86.85	4.88	73.39
Economic viability	29.79	6.95	28.71	3.70	29.08	6.74	29.19
Administration of the chakra	24.33	2.76	22.79	2.48	23.85	2.88	23.66
Global indicator of sustainability	56.84		57.73		59.25		57.94

** There were no significant statistical differences between the evaluated groups.

The overall sustainability indicator in the three groups was medium. The best score was obtained in group 3, whose differences between group 2 and group 1 were 1.52 and 2.41, respectively. Sustainability dynamics by group and color were blue for group 1, orange for group 2, and red for group 3 (Figure 5).

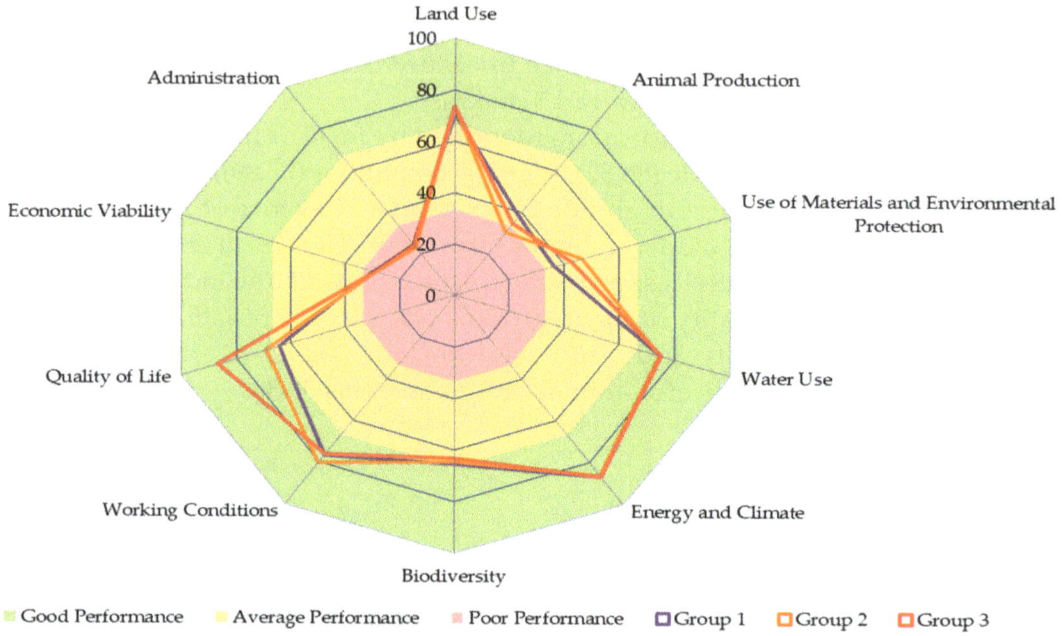

Figure 5. Degree of sustainability of the Kichwa indigenous communities for groups 1, 2, and 3 of sector A in the north of the Yasuní Biosphere Reserve in the Ecuadorian Amazon region.

3.2.2. Values in Sector B

Similarly, the results of the chakras in sector B were obtained and grouped into a dendrogram (Figure 6) with a Euclidean distance (measurement of association) of 14. Three groups were obtained: in group 1, there were 21 chakras; group 2 had 24; and group 3 had 27.

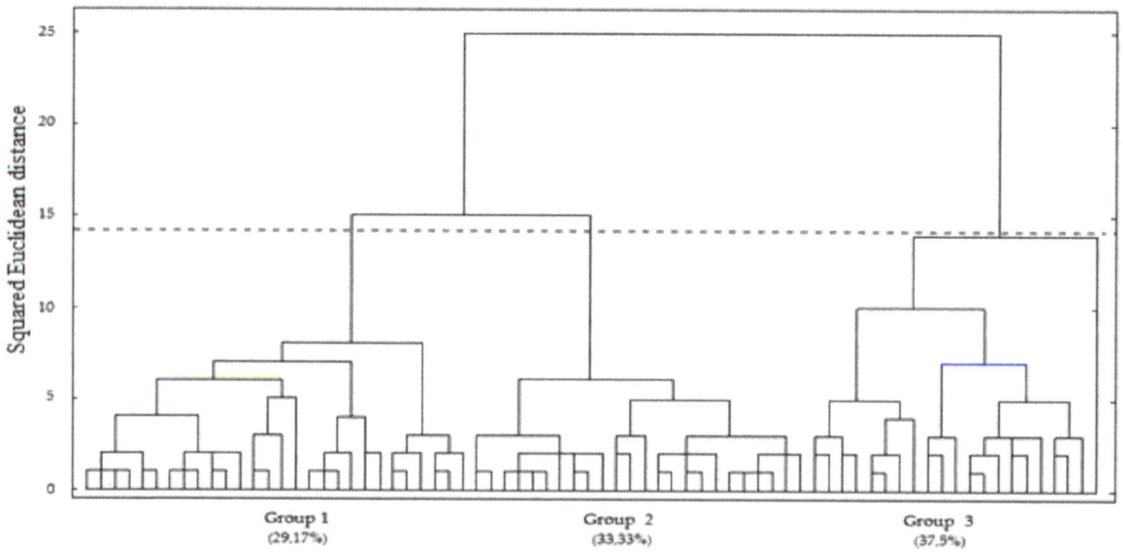

Figure 6. Dendrogram using Ward's linkage: Communities in sector B (Samona Yuturi, Chiru Isla, Sinchi Chicta Cari, and San Vicente) in the north of the Yasuní Biosphere Reserve in the Ecuadorian Amazon region (percentage of *chakras* in each group).

The indicators with good performance (Table 5) from highest to lowest average scores in the three groups were: energy and climate (92.39), water use (80.18), working conditions (75.57), and land use

(67.23). The Bbodiversity indicator in group 1 had 2.68 points more than in group 2; the quality of life indicator in group 3 had 12.52 and 7.9 points more than groups 1 and 2, respectively. The indicators with medium performance were: Animal production, which had an average of 37.90 in the three groups; use of materials and environmental protection in group 1 was 15.78 and 14.17 points higher than groups 2 and 3, respectively. In group 3, biodiversity had 4.53 and 1.85 fewer points than groups 1 and 2, respectively. The quality of life indicator in group 2 was 4.62 points higher than group 1, and in management, group 3 had the best score with a difference of 8.48 and 6.12 from groups 1 and 2, respectively. The lowest performing indicators were: economic viability in all three groups, having an average of 29.81. Group 1 had the best score with 5.05 and 2.82 more points than groups 2 and 3, respectively. Materials use and environmental protection was 1.61 points higher in group 3 than in group 2, and management was higher in group 2 than group 1 by 2.36 points.

Table 5. Means and SDs of sustainability indicators among clusters (groups) of the chakra systems in sector B in the north of the Yasuní Biosphere Reserve in the Ecuadorian Amazon region.

Indicators	Groups **						Overall Mean of Sector B
	Group 1 (21)		Group 2 (24)		Group 3 (27)		
	Mean	SD	Mean	SD	Mean	SD	
Land use	67.38	3.22	67.13	2.11	67.18	4.99	67.23
Animal production	41.00	9.23	38.58	3.11	34.11	5.93	37.90
Use of materials and environmental protection	43.24	10.47	27.46	3.15	29.07	5.74	33.26
Water use	80.29	3.89	80.83	2.55	79.43	2.6	80.18
Energy and climate	92.14	2.15	92.04	3.00	93.00	0.00	92.39
Biodiversity	70.10	5.58	67.42	3.48	65.57	3.49	67.70
Working conditions	74.95	3.07	72.58	5.04	79.18	2.68	75.57
Quality of life	61.05	14.68	65.67	3.42	73.57	6.47	66.76
Economic viability	32.43	8.58	27.38	3.55	29.61	3.65	29.81
Administration of the chakra	27.52	7.39	29.88	5.07	36.00	5.93	31.13
Global indicator of sustainability	59.01		56.90		58.67		58.19

** There were no significant statistical differences between the evaluated groups.

The overall sustainability indicator in all three groups was medium, with the best score being in group 1, which was 2.11 and 0.34 higher than in groups 2 and 3, respectively. The dynamics of the degree of sustainability by groups and color were blue for group 1, orange for group 2, and red for group 3 (Figure 7).

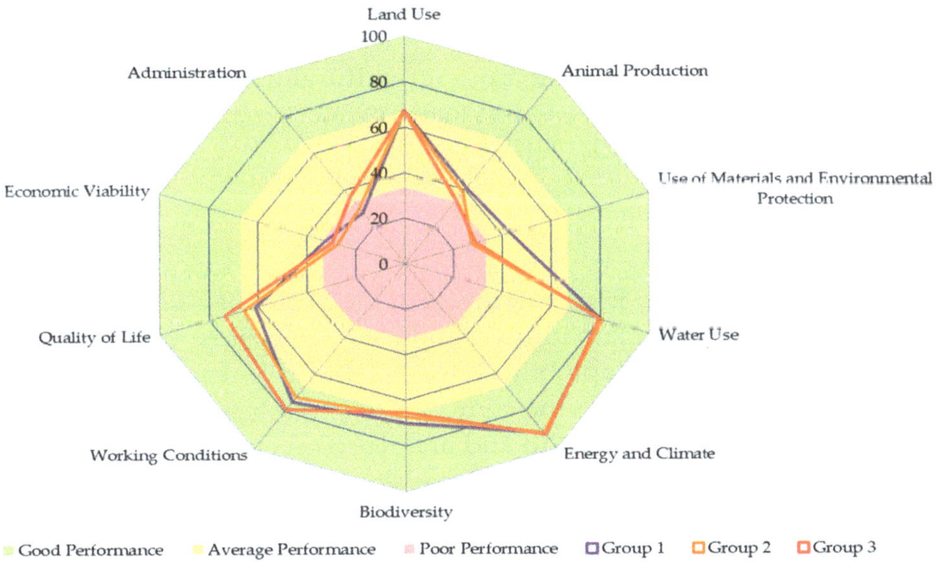

Figure 7. Degree of sustainability of the Kichwa indigenous communities for groups 1, 2, and 3 of sector B in the north of the Yasuní Biosphere Reserve in the Ecuadorian Amazon region.

4. Discussion

The two aspects that were studied in this research: (1) the population and agricultural dynamics of the Kichwa population, and (2) the indicators of the evaluation of sustainability in the chakras. The maximum size per Kichwa household was 11 people, and of the heads of households, approximately 21% had no educational. In terms of sustainability, the chakras did not present significant differences, although only 5 of the 10 analyzed indicators showed good performance, and the economic viability and administration indicators were performing poorly.

4.1. *Socio-Demographic and Agricultural Characterization of the Indigenous Kichwa Population*

Currently, defining who counts as a resident of the communities is difficult because many young people often come into and out of the study area, mainly for education and marriage, which is a typical dynamic of indigenous people in the north of the Ecuadorian Amazon [107]. Therefore, we describe the socio-demographic characteristics of the population interviewed in 2018.

Figure 3 shows that there was little difference (3.03%) in the population proportion of Kichwa men and women (48.48 and 51.51%, respectively). The proportions of youth and adults were 40.63% and 59.37%, respectively. The ratio of Kichwa children to women (0.45%) was much lower than that of Waorani (0.91%) and Shuar (0.98%) indigenous women [108]. The youth dependency population was 28% higher than the working age population, which corroborates the high birth rates in indigenous populations [109]. In the 40–64-year age range (birth year from 1978 to 1958), only 17.1% of the indigenous Kichwa population is of working age; this value may be influenced by migratory processes [107].

The maximum number of people per household (Table 3) ranged between 7 and 11 members including two parents, which is probably related to higher fertility rates [80] driven by the lack of access to family planning and reproductive health services among indigenous peoples in the Amazon [110–112]. Although it has been debated whether the high fertility among indigenous populations is linked to pronatalism and the desire to repopulate indigenous lands [113–115], the rapid changes in the northern Amazon region have had profound impacts on the demographic dynamics of local indigenous people, including changing patterns of settlement and livelihood strategies, and increasing interactions with markets [116–122].

The heads of household did not complete higher education, despite starting their higher studies; 35.34% (sector A) and 56.39% (sector B) of the heads of household had completed secondary and basic education, respectively. In the Northern Amazon region of Ecuador, the most educated households are likely to have higher non-agricultural incomes than the least educated households in mostly migrant settler populations [120,121]. Participation in non-agricultural employment may be driven by *push factors* such as declining soil fertility, low yields, small production areas, and isolation from markets, which negatively affect farm income [122,123].

4.2. *Evaluation of Sustainability Based on the RISE Methodology*

According to the ANOVA using the results of the RISE, we found no significant differences between groups and sectors evaluated in the northeast of the YBR. The RISE tool, compared with other frameworks for the evaluation of sustainability, produced the best results in some categories such as scientific soundness, usefulness, adaptability, etc. [52], but it lacks a cultural approach based on the location, which can be detrimental to the communities under evaluation [123,124]. Biocultural approaches are those that start explicitly with and are based on cultural perspectives on the basis of the place and the embodied values, knowledge and needs, and recognize the feedback between the ecological state and human well-being [125,126]. These approaches, in combination with the guidelines of the different sustainability assessment tools, can be used to develop indicators to more accurately depict reality [127]. The exchange between in situ and ex situ actors facilitates the identification of crucial problems and solutions that are currently lacking in many regional and international sustainability tools [126,128,129].

The indigenous populations in this zone are associated with sustainable agricultural practices [130] that have little environmental impact and are compatible with resource conservation [47,131]. This is corroborated by the excellent scores for land use and water use within the Kichwa chakra dynamics, similar to the Shumahuani people (Peruvian Amazon), who keep most of their land in the forest and use sustainable management practices in contrast to the migrant settlers who cut down the forest to establish pastures [132]. In the Ecuadorian Amazon Region, most soils are not suitable for agriculture [133,134] and the settlers tended to compensate for the low fertility by clearing more land for cultivation. As a consequence, deforestation rates in the Ecuadorian Amazon Region are considered one of the highest in the world [135].

Chakras are complex systems that adapt to and resist the effects of climate change and development [104,105], have high structural complexity [136,137], and present high levels of carbon sequestration [28] and tree diversity [30] compared with other forms of land use. Chakras also allow for the sustainable use of forests by combining the cultivation of fine and aromatic Ecuadorian cocoa, controlled wood extraction, production of basic food items, and the conservation of medicinal plants. In Amazonian communities, chakras contribute to food security as well as to the well-being and conservation of the region's high biodiversity [28]. The adequate management of the chakras increases their potential to recover part of the carbon released into the atmosphere due to deforestation [138]. This is contrary to the circumstances in the evaluated chakras, since the indicator for use of materials and environmental protection had critical and average scores, influenced by the poor management of the chakra: (1) nitrogen and phosphorus balance, (2) ammonia emissions, and (3) waste management [63,100,101], as the sustainability objective of the indicator is the use of natural cycles [97]. Inadequate waste management can endanger human, animal, and environmental health. Inherited pollution from oil activity can also harm future generations and, therefore, violates the principle of sustainability [139]., which is influenced by the non-agricultural income received from providing unskilled labor services to oil companies, as well as external assistance from the government and international non-governmental organizations (NGOs). All this results in the chakras not being adequately cared for [140].

The Kichwa people obtain their livelihood mainly from subsistence agriculture, collection of forestry products, hunting, and fishing [141,142], all of which are related to the diversity of species in the chakras, but it had a critical score within the biodiversity indicator [56,57]. Scientific evidence shows that some Kichwa have adopted migrant-settler-style agricultural systems and engage in commercial agriculture, such as cattle ranching, logging, and off-farm wage employment in areas near roads and urban centers [23–25,143,144]. The integration of Kichwas into the market economy has received attention in the literature because of its implications for both human well-being and ecological viability. Animal production has often been driven by oil companies and embodied in livestock breeding projects [145], which justifies the low values in the animal production indicator, since it is not part of the socio-productive dynamics of the Kichwa people.

In terms of quality of life, the studied Kichwa population received outstanding and medium scores; the best scores were expressed in the parameters of social relations and personal freedom and values; the worst score was in the parameter of health [95–97]. Notably, the studied Kichwa population lives in an area overlapped by oil blocks, and in the period from 2001 to 2011, 464 accidental oil spills were registered [146], resulting in a high level of environmental contamination [147]. In 2018, the presence of 1-hydroxypyrene (1-OHP) and creatinine (chemical products from oil activities) was confirmed in the urine of Kichwa indigenous people (inhabitants of the banks of the Napo River) due to the consumption of contaminated food. In addition, the toxicity of drinking water has been revealed, which has caused health problems ranging from dizziness to cancer [148–150]. Within the excellent scores of the working conditions indicator in the chakras, an average score was received for the staff management parameter. An important variable of this parameter is child and adolescent work in the

RISE methodology, as established by the Universal Declaration of Human Rights [151], understanding that the RISE guidelines do not consider underage work. This stands in opposition to the Kichwan cultural dynamics in the management of the chakras since, from a social perspective, it is the result of a set of values, deeply ingrained in the Kichwa cosmovision [51], where ancestral knowledge is transmitted from generation to generation through harmonious relationships, based on solidarity and reciprocity [152]. The contributions of indigenous and local knowledge (ILK) to research is increasingly considered in the science of sustainability [153,154]. ILK is defined as a cumulative body of knowledge, practices, and beliefs that have evolved through processes of adaptation and are transmitted from generation to generation by cultural transmission, through the relationship of living beings (including humans) with each other and with their environment [155–157]. The precarious economic viability and management of the chakras, which have both been obtained, make it more likely that timber will be sold illegally and that income will be earned outside the farm [158]. Given the dynamic evidence for each group in each sector, there is a need to integrate and promote new economic models and reconfigurations of the territory, which may involve changes in governance systems that generate an integration of essential indigenous rules that positively influence chakra management [159].

The sustainability assessments in the Kichwa *chakras* showed certain problems related to the dimensions of the RISE methodology, which must be resolved to improve the level of sustainability. There are priorities (high, medium, and low) for the management of the territory that could be adopted by local political agents to support the Kichwa populations and to strengthen the studied indicators (Table 6).

Table 6. Recommendations to strengthen the indicators resulting from the evaluation of sustainability.

Indicators	Recommendation	Examples	Priority
Land use	1. Increase species biodiversity in the chakras. 2. Incorporate species to increase nitrogen fixation and prevent soil erosion.	[160,161]	Medium
Animal production	1. Seek advice on the proper handling of poultry and animals.	[162]	High
Use of materials and environmental protection	1. Improve weed management. 2. Use more organic fertilizers to improve the balance of nutrients. 3. Use cover crops. 4. Use certified seeds.	[163–165]	Medium
Water use	1. Incorporate technologies to save water and make an inventory of good practices.	[166]	Low
Energy and climate	1. Strengthen the governance system of natural resources. 2. Incorporate solar panels for electrical use and a solar water heating. 3. Evaluate the perception of the changing climate at the rural, academic, and local public policy levels.	[167–170]	Low
Biodiversity	1. Disseminate knowledge about ecological infrastructure. 2. Tools for capacity building. 3. Promote the use of local varieties.	[171–174]	Medium
Working conditions	1. Training to maintain or improve working conditions.	[175]	Low
Quality of life	1. Improve education and training levels. 2. Promote talks and courses to improve health conditions in the family. 3. Promote sustainable behavior workshop based on scientific evidence.	[176–178]	Medium

Table 6. *Cont.*

Indicators	Recommendation	Examples	Priority
Economic viability	1. Develop revenues from projects for reducing emissions from deforestation and degradation (REDD+). 2. Generate opportunities to create marketing chains 3. Promote the diversification of economic income.	[143,177,179]	High
Administration of the traditional agroforestry system (*Chakra*)	1. Encourage the strengthening of field schools for potential of the traditional agroforestry system (chakra) administration. 2. Identify model chakras for training processes 3. Help Kichwa people access information 4. Strengthen capacities to increase strategic alliances among Kichwa people 5. Identify opportunities for proposing biodiversity conservation areas to generate another type of economic income, e.g., sustainable agro-ecotourism 6. Identify the change in land use, according to existing classes or categories	[180–185]	High

5. Conclusions

The average degree of sustainability of the studied Kichwa chakras had a medium score for both sectors. In general terms, we argue that the Kichwa chakras located along the banks of the Napo River have certain weaknesses for the following indicators based on the RISE guidelines: use of materials and environmental protection, animal production, economic viability, and administration of the traditional chakra system. Given the very low values for the economic indicators and for administration in the traditional agroforestry system (chakras), losses are generated compared to the neocolonial models belonging to the migrant settlers and potential structural changes in the landscape agroforestry. However, we also identified strengths in the following indicators: energy and climate, water use, working conditions, biodiversity, land use, and quality of life.

The holistic evaluation of the chakras' sustainability using the RISE tool provides a system of early warning and can be used to help identify potential measures to improve the situation of the indigenous Kichwa people in the Yasuní Biosphere Reserve. In addition, the results provide a scientific basis for the people in charge of formulating public policies to improve strategies for the rescue and management of the chakras. Consequently, this would improve quality of life, strengthen food security and sovereignty, bolster self-employment, and increase the production of sustainable products among Kichwa populations in the Ecuadorian Amazon region.

Although the RISE methodology was not designed to be used as an instrument of certification, it could offer additional value for some process of certification, for example, organic certification of the chakras, conservation of the biodiversity of species in productive spaces, and so on.

Author Contributions: M.H.-R., J.C., and N.R. conducted the field work in this study. M.H.-R., B.T., J.C., N.R., M.L. and C.G.H.D.-A., analyzed the data, compiled the literature, prepared the text, and approved the final manuscript. All authors have read and agreed to the published version of the manuscript.

Acknowledgments: We would like to thank the German Development Cooperation Agency (GIZ-Ecuador) for their collaboration in the field logistics. We are also grateful to the Yasuní National Park rangers for their help during the data collection, as well as the 133 Kichwa households who shared valuable information with us during the visits about their socio-demographic characteristics and chakra production systems.

References

1. Burger, J. *Report from the Frontier: The State of the World's Indigenous Peoples*; Cultural Survival Inc.: London, UK, 1987.
2. Toledo, V.M. Indigenous Peoples and Biodiversity. *Encycl. Biodivers.* **2013**, *3*, 269–278. [CrossRef]
3. Persoon, G.A.; Minter, T.; Slee, B.; van der Hammen, C. *The Position of Indigenous Peoples in the Management of Tropical Forests*; Tropenbos: Wageningen, The Netherlands, 2004.
4. Mittermeier, R.A. Biodiversity. In *Primate Diversity and the Tropical Forest. Case Studies from Brazil and Madagascar and the Importance of the Megadiversity Countries*; Wilson, E.O., Peter, F.M., Eds.; National Academy Press: Washington, DC, USA, 1998; pp. 145–154.
5. Myers, N. Threatened biotas:"Hot spots" in tropical forests. *Environmentalist* **1988**, *8*, 187–208. [CrossRef] [PubMed]
6. Dangles, O.; Nowicki, F. *Biota Maxima: Ecuador Biodiverso*; Pontificia Universidad Catolica del Ecuador (PUCE): Quito, Ecuador; Institut de Recherche pour le Developpment: Quito, Ecuador, 2010.
7. Steege, H.T.; Pitman, N.C.A.; Sabatier, D.; Castellanos, H.; van der Hout, P.; Daly, D.C.; Silveira, M.; Phillips, O.; Vasquez, R.; Van Andel, T.; et al. A spatial model of tree α-diversity and β-density for the Amazon Region. *Biodivers. Conserv.* **2003**, *12*, 2255–2276. [CrossRef]
8. Erwin, T.L.; Pimienta, M.C.; Murillo, O.E.; Aschero, V. Mapping patterns of β-diversity for beetles across the western Amazon Basin: A preliminary case for improving conservation strategies. *Proc. Calif. Acad. Sci.* **2004**, *56*, 72–85.
9. Ceballos, G.; Ehrlich, P.R. Global mammal distributions, biodiversity hotspots, and conservation. *Proc. Natl. Acad. Sci. USA* **2006**, *103*, 19374–19379. [CrossRef]
10. Killeen, T.J.; Douglas, M.; Consiglio, T.; Jørgensen, P.M.; Meika, J. Dry spots and wet spots in the Andean hotspot. *J. Biogeogr.* **2007**, *34*, 1357–1373. [CrossRef]
11. Napolitano, D.A.; Ryan, A.S. The dilemma of contact: Voluntary isolation and the impacts of gas exploitation on health and rights in the Kugapakori Nahua Reserve, Peruvian Amazon. *Environ. Res. Lett.* **2007**, *2*, 1–12. [CrossRef]
12. Gamboa Balbín, C.; Santillán Bartra, A. *Régimen Especial Transectorial de Protección a Favor de Pueblos Indígenas en Aislamiento y en Contacto Inicial*; Bellido Ediciones EIRL: Lima, Peru, 2006.
13. Heredia Rengifo, M.G.; Hernández Díaz-Ambrona, C.G. Comportamiento demográfico: Dinámico–Probabilístico de los pueblos indígenas en aislamiento de la amazonía ecuatoriana. *Revista Científica Axioma* **2019**, *20*, 25–34.
14. Inec. Ecuadorencifras.Gob.Ec. (En Línea). 2010. Available online: http://www.ecuadorencifras.gob.ec/estadisticas/ (accessed on 20 March 2020).
15. Torres, B.; Günter, S.; Acevedo-Cabra, R.; Knoke, T. Livelihood strategies, ethnicity and rural income: The case of migrant settlers and indigenous populations in the Ecuadorian Amazon. *For. Policy Econ.* **2018**, *86*, 22–34. [CrossRef]
16. Uzendoski, M.; Whitten, N.E., Jr. From "Acculturated Indians" to "Dynamic Amazonian Quichua-Speaking Peoples. *Tipití J. Soc. Anthropol. Lowl. South Am.* **2014**, *12*, 1–13.
17. Muratorio, B. *Rucuyaya Alonso y la Historia Social y Económica del Alto Napo*; Docutech: Quito, Ecuador, 1998; pp. 1850–1950.
18. Finer, M.; Jenkins, C.N.; Pimm, S.L.; Keane, B.; Ross, C. Oil and Gas Projects in the Western Amazon: Threats to Wilderness, Biodiversity, and Indigenous Peoples. *PLoS ONE* **2008**, *3*, e2932. [CrossRef] [PubMed]
19. Juteau-Martineau, G.; Becerra, S.; Maurice, L. Ambiente, petróleo y vulnerabilidad política en el oriente ecuatoriano:¿ hacia nuevas formas de gobernanza energética? *Am. Lat. Hoy* **2014**, *67*, 119–137. [CrossRef]
20. Bilsborrow, R.E.; Barbieri, A.F.; Pan, W. Changes in population and land use over time in the Ecuadorian Amazon. *Acta Amazonica* **2004**, *34*, 635–647. [CrossRef]
21. Lu, F.; Bilsborrow, R.E. A Cross-Cultural Analysis of Human Impacts on the Rainforest Environment in Ecuador. In *Human Population*; Springer: Berlin, Germany, 2011; pp. 127–151.
22. Vasco, C.; Bilsborrow, R.; Torres, B.; Griess, V. Agricultural land use among mestizo colonist and indigenous populations: Contrasting patterns in the Amazon. *PLoS ONE* **2018**, *137*, e0199518. [CrossRef]

23. Torres, B.; Vasco, C.; Günter, S.; Knoke, T. Determinants of agricultural diversification in a hotspot area: Evidence from colonist and indigenous communities in the Sumaco Biosphere Reserve, Ecuadorian Amazon. *Sustainability* **2018**, *10*, 1432. [CrossRef]
24. Gray, C.L.; Bilsborrow, R.E.; Bremner, J.L.; Flora, L. Indigenous Land Use in the Ecuadorian Amazon: A Cross-cultural and Multilevel Analysis. *Hum. Ecol.* **2008**, *36*, 97–109. [CrossRef]
25. Lu, H.F. Integration into the market among indigenous peoples: A cross-cultural perspective from the Ecuadorian Amazon. *Curr. Anthropol.* **2007**, *48*, 593–602. [CrossRef]
26. Ferrer Velasco, R.; Köthke, M.; Lippe, M.; Günter, S. Scale and context dependency of deforestation drivers: Insights from spatial econometrics in the tropics. *PLoS ONE* **2020**, *151*, e0226830. [CrossRef]
27. Barlow, J.; Lennox, G.D.; Ferreira, J.; Berenguer, E.; Lees, A.C.; Mac Nally, R.; Thomson, J.R.; de Barros Ferraz, S.F.; Louzada, J.; Oliveira, V.H.F.R.; et al. Anthropogenic disturbance in tropical forests can double biodiversity loss from deforestation. *Nature* **2016**, *535*, 144. [CrossRef]
28. Torres, B.; Maza, O.J.; Aguirre, P.; Hinojosa, L.; Günter, S. The Contribution of Traditional Agroforestry to Climate Change Adaptation in the Ecuadorian Amazon: The Chakra System. In *Handbook of Climate Change Adaptation*; Springer: Berlin, Germany, 2015; pp. 1973–1994.
29. Coq-Huelva, D.; Higuchi, A.; Alfalla-Luque, R.; Burgos-Morán, R.; Arias-Gutiérrez, R. Co-evolution and bio-social construction: The Kichwa agroforestry systems (chakras) in the Ecuadorian Amazonia. *Sustainability* **2017**, *9*, 1920. [CrossRef]
30. Vera, V.R.R.; Cota-Sánchez, J.H.; Grijalva Olmedo, J.E. Biodiversity, dynamics, and impact of chakras on the Ecuadorian Amazon. *J. Plant Ecol.* **2019**, *12*, 34–44. [CrossRef]
31. Irvine, D. Indigenous federations and the market: The Runa of Napo, Ecuador. In *Indigenous People and Conservation Organizations Experiences in Collaborations*; Weber, R., Butler, J., Larson, P., Eds.; World Wildlife Fund: Washington, DC, USA, 2000; pp. 21–46.
32. Lu, F.; Bilsborrow, R.E.; Oña, A. *Demography, Household Economics, and Land and Resource Use of Five Indigenous Populations in the Northern Ecuadorian Amazon: A Summary of Ethnographic Research*; Occasional Paper; Carolina Population Centre, University of North Carolina: Chapel Hill, NC, USA, 2004.
33. Whitten, N.E.; Whitten, D.S. *Puyo Runa: Imagery and Power in Modern Amazonia*; University of Illinois Press: Champaign, IL, USA, 2008.
34. Coq-Huelva, D.; Torres-Navarrete, B.; Bueno-Suárez, C. Indigenous worldviews and Western conventions: Sumak Kawsay and cocoa production in Ecuadorian Amazonia. *Agric. Hum. Values* **2018**, *35*, 163–179. [CrossRef]
35. GESOREN–GIZ. *Fomento de la Cadena de Valor de Cacao en Organizaciones de Pequeños Productores de Esmeraldas y Napo*; Análisis de Impactos del Programa, Fascículo 5; GIZ: Quito, Ecuador, 2011; p. 93.
36. Lehmann, S.; Rodríguez, J. *La Chakra Kichwa: Criterios para la Conservación y Fomento de un Sistema de Producción Sostenible en la Asociación Kallari y sus Organizaciones Socias*; Serie de Sistematización 7, 201; GIZ: Quito, Ecuador, 2013; pp. 20–31.
37. Reeve, M.E. *Los Quichua del Curaray: El Proceso de Formación de la Identidad*; Editorial Abya Yala: Quito, Ecuador, 2002; ISBN 9978-22-020-8.
38. Santafe Troncoso, V.; Loring, P.A. Indigenous food sovereignty and tourism: The Chakra Route in the Amazon region of Ecuador. *J. Sustain. Tour.* **2020**, *29*, 1–20. [CrossRef]
39. Andy, P.; Calapucha, A.; Calapucha, L.; López, H.; Shiguango, K.; Tanguila, A.; Tanguila, A.; Yasacama, C. *Sabiduría de la Cultura Kichwa de la Amazonía Ecuatoriana*, 1st ed.; Tomo, I., Ed.; Universidad De Cuenca: Cuenca, Ecuador, 2012; ISBN 978-9978-14-000-0.
40. Hingh, S.J.; Haas, W.; Fraňková, E. Conclusions: Promises and Challenges for Sustainable Agri-Food Systems. In *Socio-Metabolic Perspectives on the Sustainability of Local Food Systems*; Springer: Cham, Germany, 2017; pp. 345–356.
41. Godfray, H.C.J.; Beddington, J.R.; Crute, I.R.; Haddad, L.; Lawrence, D.; Muir, J.F.; Pretty, J.; Robinson, S.; Thomas, S.M.; Toulmin, C. Food Security: The Challenge of Feeding 9 Billion People. *Science* **2010**, *327*, 812–818. [CrossRef]
42. Foley, J.A.; Ramankutty, N.; Brauman, K.A.; Cassidy, E.S.; Gerber, J.S.; Johnston, M.; Mueller, N.D.; O'Connell, C.; Ray, D.K.; West, P.C.; et al. Solutions for a cultivated planet. *Nature* **2011**, *478*, 337–342. [CrossRef]

43. Gil, J.D.B.; Reidsma, P.; Giller, K.E.; Todman, L.C.; Whitmore, A.P.; Van Ittersum, M.K. Sustainable development goal 2: Improved targets and indicators for agriculture and food security. *Ambio* **2019**, *48*, 685–698. [CrossRef]
44. Pope, J.; Annandale, D.; Morrison-Saunders, A. Conceptualising sustainability assessment. *Environ. Impact Assess. Rev.* **2004**, *24*, 595–616. [CrossRef]
45. Purvis, B.; Mao, Y.; Robinson, D. Three pillars of sustainability: In search of conceptual origins. *Sustain. Sci.* **2019**, *14*, 681–695. [CrossRef]
46. Valizadeh, N.; Hayati, D. Development and validation of an index to measure agricultural sustainability. *J. Clean. Prod.* **2021**, *280*, 123797. [CrossRef]
47. Dufour, D.L. Use of Tropical Rainforests by Native Amazonians. *Bioscience* **1990**, *40*, 652–659. [CrossRef]
48. Astier, M.; García-Barrios, L.; Galvan-Miyoshi, Y.; González-Esquivel, C.E.; Masera, O.R. Assessing the Sustainability of Small Farmer Natural Resource Management Systems. A Critical Analysis of the MESMIS Program (1995–2010). *Ecol. Soc.* **2012**, *17*. [CrossRef]
49. Ebel, R. Are Small Farms Sustainable by Nature?—Review of an Ongoing Misunderstanding in Agroecology. *Challenges Sustain.* **2020**, *8*, 17–29. [CrossRef]
50. Binder, C.R.; Feola, G.; Steinberger, J.K. Considering the normative, systemic and procedural dimensions in indicator-based sustainability assessments in agriculture. *Environ. Impact Assess. Rev.* **2010**, *30*, 71–81. [CrossRef]
51. Schader, C.; Grenz, J.; Meier, M.S.; Stolze, M. Scope and precision of sustainability assessment approaches to food systems. *Ecol. Soc.* **2014**, *19*. [CrossRef]
52. Talukder, B.; Blay-Palmer, A. Comparison of Methods to Assess Agricultural Sustainability. *Sustain. Agricult. Rev.* **2017**, *47*, 149–168. [CrossRef]
53. Marchand, F.; Debruyne, L.; Triste, L.; Gerrard, C.L.; Padel, S.; Lauwers, L. Key characteristics for tool choice in indicator-based sustainability assessment at farm level. *Ecol. Soc.* **2014**, *19*. [CrossRef]
54. Halog, A.; Manik, Y. Advancing Integrated Systems Modelling Framework for Life Cycle Sustainability Assessment. *Sustainability* **2011**, *3*, 469–499. [CrossRef]
55. Rasmussen, L.V.; Bierbaum, R.; Oldekop, J.A.; Agrawal, A. Bridging the practitioner-researcher divide: Indicators to track environmental, economic, and sociocultural sustainability of agricultural commodity production. *Glob. Environ. Chang.* **2017**, *42*, 33–46. [CrossRef]
56. De Olde, E. Sustainable Development of Agriculture: Contribution of Farm-Level Assessment Tools. Ph.D. Thesis, Wageningen University and Research, Wageningen, The Netherlands, 2017.
57. De Olde, E.M.; Oudshoorn, F.W.; Sørensen, C.G.; Bokkers, E.A.M.; De Boer, I.J. Assessing sustainability at farm-level: Lessons learned from a comparison of tools in practice. *Ecol. Indic.* **2016**, *66*, 391–404. [CrossRef]
58. Zahm, F.; Viaux, P.; Vilain, L.; Girardin, P.; Mouchet, C. Assessing farm sustainability with the IDEA method–From the concept of agriculture sustainability to case studies on farms. *Sustain. Dev.* **2008**, *16*, 271–281. [CrossRef]
59. FAO. *Sustainability Assessment of Food and Agriculture Systems SAFA. Guidelines*; Version 3.0; Food and Agriculture Organization of the United Nations: Rome, Italy, 2014.
60. Häni, F.; Braga, A.; Stämpfli Keller, T.; Fischer, M.; Porsche, H. RISE, a tool for holistic sustainability assessment at the farm level. *Int. Food Agribus. Manage. Rev.* **2003**, *6*, 78–90.
61. Meul, M.; Passel, S.; Nevens, F.; Dessein, J.; Rogge, E.; Mulier, A.; Hauwermeiren, A. MOTIFS: A monitoring tool for integrated farm sustainability. *Agron. Sustain. Dev.* **2008**, *28*, 321–332. [CrossRef]
62. Paracchini, M.L.; Bulgheroni, C.; Borreani, G.; Tabacco, E.; Banterle, A.; Bertoni, D.; Rossi, G.; Parolo, G.; Origgi, R.; De Paola, C. A diagnostic system to assess sustainability at a farm level: The SOSTARE model. *Agric. Syst.* **2015**, *133*, 35–53. [CrossRef]
63. Jawtusch, J.; Schader, C.; Stolze, M.; Baumgart, L.; Niggli, U. Sustainability Monitoring and Assessment Routine: Results from pilot applications of the FAO SAFA Guidelines. In Proceedings of the Symposium International sur L'Agriculture Biologique Méditerranénne et Les Signes Distinctifs de Qualité liée à l'Origine, Agadir, Morocco, 2–4 December 2013.
64. Röös, E.; Fischer, K.; Tidåker, P.; Källström, H.N. How well is farmers' social situation captured by sustainability assessment tools? A Swedish case study. *Int. J. Sustain. Dev. World Ecol.* **2019**, *26*, 268–281. [CrossRef]

65. Berbeć, A.K.; Feledyn-Szewczyk, B.; Thalmann, C.; Wyss, R.; Grenz, J.; Kopiński, J.; Stalenga, J.; Radzikowski, P. Assessing the Sustainability Performance of Organic and Low-Input Conventional Farms from Eastern Poland with the RISE Indicator System. *Sustainability* **2018**, *10*, 1792. [CrossRef]
66. Heredia-R, M.; Bravo, C.; Vargas-B, J. *Sostenibilidad y Manejo de Recursos Naturales a Nivel de Finca: Marco de Evaluación RISE*; Torres, B., Radice, M., Ochoa-Moreno, S., Cueva, K., Eds.; Universidad Estatal Amazónica: Puyo, Ecuador, 2017; pp. 120–133.
67. De Olde, E.M.; Oudshoorn, F.W.; Bokkers, E.A.M.; Stubsgaard, A.; Sørensen, C.G.; De Boer, I.J.M. Assessing the Sustainability Performance of Organic Farms in Denmark. *Sustainability* **2016**, *8*, 957. [CrossRef]
68. Pineau, M.; Ayalneh, W.; Studer, C.; Peden, D. Response-inducing sustainability evaluation (RISE) linking agricultural practices and water productivity. *CGIAR Chall. Progr. Water Food* **2009**, *2*, 110–114.
69. Häni, F.; Gerber, T.F.; Stämpfli, A.; Porsche, H.; Thalmann, C.; Studer, C. RISE: A tool for improving sustainability in agriculture: A case study with tea farmers in southern India. In *Sustainable Agriculture: From Common Principles to Common Practice*; Häni, F.J., Pintér, L., Herren, H.R., Eds.; International Institute of Sustainable Development and Swiss College of Agriculture: Bern, Switzerland, 2007; pp. 121–148.
70. Thalmann, C.; Urutyan, V.; Porsche, H.; Grigoryan, A.; Studer, C. Sustainability of Agricultural Production in Armenia. Interim Report. España. 2007. Available online: http://citeseerx.ist.psu.edu/viewdoc/download?doi=10.1.1.618.809&rep=rep1&type=pdf (accessed on 4 April 2007).
71. Häni, F.; Stämpfli, A.; Tello, J.R.; Braga, F. *Sustainability Assessment of Six Brazilian Coffee Farms in Bahia and Minas Gerais*; Swiss College of Agriculture: Zollikofen, Switzerland; University of Applied Sciences Bern: Bern, Switzerland; University of Guelph: Guelph, ON, Canada, 2005.
72. Le Gal, P.-Y.; Dugué, P.; Faure, G.; Novak, S. How does research address the design of innovative agricultural production systems at the farm level? A review. *Agric. Syst.* **2011**, *104*, 714–728. [CrossRef]
73. De Olde, E.M.; Sautier, M.; Whitehead, J. Comprehensiveness or implementation: Challenges in translating farm-level sustainability assessments into action for sustainable development. *Ecol. Indic.* **2018**, *85*, 1107–1112. [CrossRef]
74. Van Meensel, J.; Lauwers, L.; Kempen, I.; Dessein, J.; Van Huylenbroeck, G. Effect of a participatory approach on the successful development of agricultural decision support systems: The case of Pigs2win. *Decis. Support Syst.* **2012**, *54*, 164–172. [CrossRef]
75. Villaverde, X.; Ormaza, F.; Marcial, V.; Jorgenson, J. *Parque Nacional Yasuní: Historia, Problema y Perspectivas*; WCS–Programa Ecuador: Quito, Ecuador, 2005.
76. Uzendoski, M. *The Napo Runa of Amazonian Ecuador*; University of Illinois Press: Champaign, IL, USA, 2005; Volume 57.
77. Bass, M.S.; Finer, M.; Jenkins, C.N.; Kreft, H.; Cisneros-Heredia, D.F.; McCracken, S.F.; Pitman, N.C.A.; English, P.H.; Swing, K.; Villa, G.; et al. Global Conservation Significance of Ecuador's Yasuní National Park. *PLoS ONE* **2010**, *5*, e8767. [CrossRef] [PubMed]
78. Finer, M.; Babbitt, B.; Novoa, S.; Ferrarese, F.; Pappalardo, S.E.; De Marchi, M.; Saucedo, M.; Kumar, A. Future of oil and gas development in the western Amazon. *Environ. Res. Lett.* **2015**, *10*, 024003. [CrossRef]
79. Mittermeier, R.A.; Mittermeier, C.; Gil, P.R. *Megadiversity: Naciones Biológicamente más ricas de la Tierra*; Cemex: Ciudad de México, Mexico, 1997.
80. Myers, N.; Mittermeier, R.A.; Mittermeier, C.G.; Da Fonseca, G.A.B.; Kent, J. Biodiversity hotspots for conservation priorities. *Nature* **2000**, *403*, 853–858. [CrossRef] [PubMed]
81. UNESCO. MAB Directorio de Reservas de Biosfera. 1989. Available online: http://www.unesco.org/mabdb/br/brdir/directory/biores.asp?mode=gen&code=ecu+02 (accessed on 29 November 2019).
82. Guevara, J.; Pitman, N.; Mogollón, H.; Cerón, C.; Palacios, W. *Ministerio del Ambiente del Ecuador 2012. Sistema de Clasificación de los Ecosistemas del Ecuador Continental*; Ministerio del Ambiente del Ecuador: Cuenca, Ecuador, 2013; pp. 178–180.
83. Donoso, A. Parque Nacional Yasuní ha Despertado Gran Interés Científico. Ecuador. 2008. Available online: http://www.ambiente.gob.ec/wp-content/uploads/downloads/2012/07/ParqueNacional-Yasuní.pd (accessed on 4 December 2019).
84. Atkinson, R.; Flint, J. Accessing hidden and hard-to-reach populations: Snowball research strategies. *Soc. Res. Update* **2001**, *33*, 1–4.
85. Cohen, N.; Arieli, T. Field research in conflict environments: Methodological challenges and snowball sampling. *J. Peace Res.* **2011**, *48*, 423–435. [CrossRef]

86. Kirchherr, J.; Charles, K. Enhancing the sample diversity of snowball samples: Recommendations from a research project on anti-dam movements in Southeast Asia. *PLoS ONE* **2018**, *13*, e0201710. [CrossRef]
87. Vanclay, F.; Baines, J.T.; Taylor, C.N. Principles for ethical research involving humans: Ethical professional practice in impact assessment Part I. *Impact Assess. Proj. Apprais.* **2013**, *31*, 243–253. [CrossRef]
88. Smith, L.T. *Decolonizing Methodologie: Research and Indigenous Peoples*; Zed Books: London, UK, 1999.
89. Vasquez-Fernandez, A.M.; Hajjar, R.; Shuñaqui Sangama, M.I.; Lizardo, R.S.; Pérez Pinedo, M.; Innes, J.L.; Kozak, R.A. Co-creating and Decolonizing a Methodology Using Indigenist Approaches: Alliance with the Asheninka and Yine-Yami Peoples of the Peruvian Amazon. *ACME Int. J. Crit. Geogr.* **2017**, *17*, 720–749.
90. Wilson, T. Visualising the demographic factors which shape population age structure. *Demogr. Res.* **2016**, *35*, 867–890. [CrossRef]
91. Milovanovic, V.; Smutka, L. Populating Aging in Rural India: Implication for Agriculture and Smallholder Farmers. *J. Popul. Ageing* **2019**, *13*, 305–323. [CrossRef]
92. Book Review: Donald Rowland, Demographic Methods and Concepts, Oxford: Oxford University Press, 2003, xiv + 546 pp. *Eur. J. Popul.* **2004**, *20*, 291–292. [CrossRef]
93. Hinde, P.R.A. Demographic Methods. *Population* **2001**, *56*, 478. [CrossRef]
94. Holdsworth, C.; Finney, N.; Marshall, A.; Norman, P. *Population and Society*; Sage: London, UK, 2013.
95. Häni, F.; Stämpfli, A.; Tello, J.R.; Braga, F. *Sustainability Assessment of Six Brazilian Coffee Farms in Bahia and Minas Gerais*; Working Paper; Swiss College of Agriculture: Zollikofen, Switzerland; University of Applied Sciences Bern: Bern, Switzerland; University of Guelph: Guelph, ON, Canada, 2004.
96. Porsche, H.; Fischer, M.; Braga, F.; Häni, F. Introduction of the sustainability assessment tool RISE into Canadian agriculture. *Work. Pap. J. Univ. Guelph* **2004**, *11*, 11–19.
97. Grenz, J.; Thalmann, C.; Stämpfli, A.; Studer, C.; Häni, F. RISE—A method for assessing the sustainability of agricultural production at farm level. *Rural Dev. News* **2009**, *1*, 5–9.
98. Grenz, J.; Mainiero, R.; Schoch, M.; Sereke, F.; Stalder, S.; Thalmann, C.; Wyss, R. *RISE 3.0—Manual. Sustainability Themes and Indicators*; HAFL: Zollikofen, Switzerland, 2016; p. 96.
99. RISE, Response-Inducing Sustainability Evaluation. Version 3.0. 2000. Available online: http://rise.shl.bfh.ch (accessed on 10 October 2018).
100. Pardo, C.E.; Del Campo, P.C. Combinación de métodos factoriales y de análisis de conglomerados en R: El paquete FactoClass. *Revista Colombiana Estadística* **2007**, *30*, 231–245.
101. Afifi, A.; Clark, V.; May, S. *Practical Multivariate Analysis*; Chapman & Hall/CRC Texts in Statistical Science: London, UK, 2012.
102. González de Miguel, C.; Hernández Díaz-Ambrona, C.G.; Postigo Sierra, J.L. *Evaluación de la sostenibilidad agraria. El caso de La Concordia (Nicaragua)*; Ingeniería Sin Fronteras, Asociación para el Desarrollo: Barcelona, Spain, 2009.
103. Ward, J.H., Jr. Hierarchical grouping to optimize an objective function. *J. Am. Stat. Assoc.* **1963**, *58*, 236–244. [CrossRef]
104. Perez, C. *Técnicas Estadísticas con SPSS*; Prentice Hall: New Jersey, NJ, USA, 2001.
105. Cuadras, C. *Nuevos Mètodos de Analisis Multivariante*; CMC Editions: Vancouver, BC, Canada, 2007.
106. Leech, N.L.; Barrett, K.C.; Morgan, G.A. *IBM SPSS for Intermediate Statistics: Use and Interpretation*; Routledge: New York, NY, USA, 2014.
107. Davis, J.; Sellers, S.; Gray, C.; Bilsborrow, R. Indigenous Migration Dynamics in the Ecuadorian Amazon: A Longitudinal and Hierarchical Analysis. *J. Dev. Stud.* **2017**, *53*, 1849–1864. [CrossRef]
108. INEC. *Censo de Población y Vivienda*; INEC: Quito, Ecuador, 2010.
109. Jarrín-V, P.S.; Carrillo, L.T.; Zamora, G. Demografía y transformación territorial: Medio siglo de cambio en la región amazónica de Ecuador. EUTOPÍA. *Revista Desarrollo Econ. Territorial* **2017**, *12*, 81–100. [CrossRef]
110. McSweeney, K.; Arps, S. A "demographic turnaround": The rapid growth of indigenous populations in lowland Latin America. *Lat. Am. Res. Rev. JSTOR* **2005**, *40*, 3–29. [CrossRef]
111. Bremner, J.; Bilsborrow, R.; Feldacker, C.; Holt, F.L. Fertility beyond the frontier: Indigenous women, fertility, and reproductive practices in the Ecuadorian Amazon. *Popul. Environ.* **2009**, *30*, 93–113. [CrossRef]
112. Wurtz, H. Indigenous women of Latin America: Unintended pregnancy, unsafe abortion, and reproductive health outcomes. *Pimatisiwin* **2012**, *10*, 271. [PubMed]

113. Engelman, R.; Bremner, J.; De Souza, R.M.; Mogelgaard, K. Indigenous population, fertility, and reproductive intention in the lowland neotropics: Response to McSweeney. *Conserv. Biol.* **2006**, *20*, 1315–1317. [CrossRef] [PubMed]
114. McSweeney, K. Indigenous population growth in the lowland Neotropics: Social science insights for biodiversity conservation. *Conserv. Biol.* **2005**, *19*, 1375–1384. [CrossRef]
115. Borman, R. Cofan: Story of the forest people and the outsiders. *Cult. Surv. Q.* **1999**, *23*, 48–50.
116. Macdonald, T., Jr. *Ethnicity and Culture amidst New "Neighbors": The Runa of Ecuador's Amazon Basin*; Allyn and Bacon: Boston, MA, USA, 1999.
117. Rival, L.M. *Trekking through History: The Huaorani of Amazonian Ecuador*; Columbia University Press: New York, NY, USA, 2002.
118. Rudel, T.K.; Bates, D.; Machinguiashi, R. Ecologically noble Amerindians? Cattle ranching and cash cropping among Shuar and colonists in Ecuador. *Lat. Am. Res. Rev.* **2002**, *37*, 144–159.
119. Sierra, R.; Rodriguez, F.; Losos, E. Forest resource use change during early market integration in tropical rain forests: The Huaorani of upper Amazonia. *Ecol. Econ.* **1999**, *30*, 107–119. [CrossRef]
120. Pichón, F.J. Settler households and land-use patterns in the Amazon frontier: Farm-level evidence from Ecuador. *World Dev.* **1997**, *25*, 67–91. [CrossRef]
121. Murphy, L.L. Colonist Farm Income, Off-Farm Work, Cattle, Differentiation in Ecuador's Northern Amazon. *Hum. Org.* **2001**, *60*, 67–79. [CrossRef]
122. Murphy, L.; Bilsborrow, R.; Pichón, F. Poverty and prosperity among migrant settlers in the Amazon rainforest frontier of Ecuador. *J. Dev. Stud.* **1997**, *34*, 35–65. [CrossRef]
123. West, P. *Conservation is Our Government Now: The Politics of Ecology in Papua New Guinea*; Duke University Press: Durham, UK, 2006.
124. Sterling, E.J.; Betley, E.; Sigouin, A.; Gomez, A.; Toomey, A.; Cullman, G.; Malone, C.; Pekor, A.; Arengo, F.; Blair, M.; et al. Assessing the evidence for stakeholder engagement in biodiversity conservation. *Biol. Conserv.* **2017**, *209*, 159–171. [CrossRef]
125. Reimaan National Planning Team Reimaanlok. *Looking to the Future: National Conservation Area Plan for the Marshall Islands*; N. Baker: Melbourne, MEL, Australia, 2008.
126. Preuss, K.; Dixon, M. 'Looking after country two-ways': Insights into indigenous community-based conservation from the Southern Tanami. *Ecol. Manag. Restor.* **2012**, *13*, 2–15. [CrossRef]
127. Leenhardt, P.; Teneva, L.; Kininmonth, S.; Darling, E.; Cooley, S.; Claudet, J. Challenges, insights and perspectives associated with using social-ecological science for marine conservation. *Ocean Coast. Manag.* **2015**, *115*, 49–60. [CrossRef]
128. Ens, E.J. Vital Landscapes, Indigenous Futures. In *People on Country*; Altman, J., Kerin, S., Eds.; The Federation Press: Leichhardt, SYD, Australia, 2012; pp. 45–64.
129. Ens, E.J.; Daniels, C.; Nelson, E.; Roy, J.; Dixon, P. Creating multifunctional landscapes: Using exclusion fences to frame feral ungulate management preferences in remote Aboriginal-owned Northern Australia. *Biol. Conserv.* **2016**, *197*, 235–246. [CrossRef]
130. Boserup, E. *The Conditions of Agricultural Growth: The Economics of Agrarin Change under Population Pressure*; Transaction Publishers: New Brunswick, NJ, USA, 2011.
131. Schwartzman, S.; Moreira, A.; Nepstad, D. Rethinking tropical forest conservation: Perils in parks. *Conserv. Biol.* **2000**, *14*, 1351–1357. [CrossRef]
132. Hvalkof, S. Progress of the victims: Political ecology in the Peruvian Amazon. In *Reimagining Political Ecology*; Biersack, A., Greenberg, J.B., Eds.; Duke University Press: Durham, NC, USA, 2006.
133. Hicks, J.F.; Daly, H.E.; Davis, S.H.; de Freitas, M.D.L. *Ecuador's Amazon Region*; Development Issues and Options; World Bank Discussion Paper 75; World Bank: Washington, DC, USA, 1990.
134. Mainville, N.; Webb, J.; Lucotte, M.; Davidson, R.; Betancourt, O.; Cueva, E.; Mergler, D. Decrease of soil fertility and release of mercury following deforestation in the Andean Amazon, Napo River Valley, Ecuador. *Sci. Total Environ.* **2006**, *368*, 88–98. [CrossRef]
135. Torres, B.; Bilsborrow, R.; Barbieri, A.; Torres, A. Changes in income strategies of rural households in the Northern Ecuadorian Amazon. *Revista Amazónica Ciencia Tecnologia* **2014**, *3*, 221–257.
136. Hoffmann, U. Assuring Food Security in Developing Countries under the Challenges of Climate Change: Key Trade and Development Issues of a Fundamental Transformation of Agriculture. UNCTAD Discussion Papers. 2011, p. 50. Available online: http://www.unctad.org (accessed on 23 July 2020).

137. Nicholls, C. Enfoques agroecologicos para incrementar la resiliencia de los sistemas agricolas al cambio climático. In *Agroecología y Resiliencia Socioecológica Adaptándose al Cambio Climático*; Nicholls, C., Rios, L., Altieri, M., Eds.; Socla: Medellin, Colombia, 2013; pp. 18–29.
138. Montagnini, F.; Nair, P.K.R. Carbon sequestration: An underexploited environmental benefit of agroforestry systems. *Agrofor. Syst.* **2004**, *1*, 281–295.
139. Stückelberger, C. *Das Konzept der Nachhaltigen Entwicklung Um Zwei Dimensionen Erweitern*; Verlag Paul Haupt: Bern, Switzerland, 1999; pp. 103–122.
140. Loaiza, T.; Nehren, U.; Gerold, G. REDD+ and incentives: An analysis of income generation in forest-dependent communities of the Yasuní Biosphere Reserve, Ecuador. *Appl. Geogr.* **2015**, *62*, 225–236. [CrossRef]
141. Guzmán Gallegos, M.A. *Para Que la Yuca Beba Nuestra Sangre: Trabajo, Género y Parentesco en Una Comunidad Quichua de la Amazonía Ecuatoriana*; Abya-Yala: Quito, Ecuador, 1997.
142. Martínez Novo, R. Del tiempo insostenible y del sentido del tiempo en las comunidades kichwa canelos. *Desacatos* **2012**, *40*, 111–126.
143. Pérez, C.V.; Bilsborrow, R.; Torres, B. Income diversification of migrant colonists vs. indigenous populations: Contrasting strategies in the Amazon. *J. Rural Stud.* **2015**, *42*, 1–10. [CrossRef]
144. Mejía, E.; Pacheco, P.; Muzo, A.; Torres, B. Smallholders and timber extraction in the Ecuadorian Amazon: Amidst market opportunities and regulatory constraints. *Int. For. Rev.* **2015**, *17*, 38–50. [CrossRef]
145. EP. Gestión Social. Reporte Gerencial 2018, Petroamazonas EP. 2018. Available online: https://www.petroamazonas.gob.ec/wp-content/uploads/downloads/2019/04/INF-PRC-POR-PROVINCIA_2018.pdf (accessed on 20 August 2020).
146. Durango-Cordero, J.; Saqalli, M.; Laplanche, C.; Locquet, M.; Elger, A. Spatial Analysis of Accidental Oil Spills Using Heterogeneous Data: A Case Study from the North-Eastern Ecuadorian Amazon. *Sustainability* **2018**, *10*, 4719. [CrossRef]
147. Vargas, G.C.; Au, W.W.; Izzotti, A. Public health issues from crude-oil production in the Ecuadorian Amazon territories. *Sci. Total Environ.* **2020**, *719*, 134647. [CrossRef] [PubMed]
148. Webb, J.; Coomes, O.T.; Mergler, D.; Ross, N.A. Levels of 1-hydroxypyrene in urine of people living in an oil producing region of the Andean Amazon (Ecuador and Peru). *Int. Arch. Occup. Environ. Health* **2018**, *91*, 105–115. [CrossRef]
149. Burgherr, P.; Eckle, P.; Hirschberg, S. Comparative assessment of severe accident risks in the coal, oil and natural gas chains. *Reliab. Eng. Syst. Saf.* **2012**, *105*, 97–103. [CrossRef]
150. Burgherr, P.; Hirschberg, S. Comparative risk assessment of severe accidents in the energy sector. *Energy Policy* **2014**, *74*, S45–S56. [CrossRef]
151. UN. *Universal Declaration of Human Rights*; United Nations: New York, NY, USA, 1948; Available online: http://www.un.org/en/documents/udhr/index.shtml (accessed on 20 August 2020).
152. Suárez-Torres, J.; Suárez-López, J.R.; López-Paredes, D.; Morocho, H.; Cachiguango-Cachiguango, L.E.; Dellai, W. Agroecology and Health: Lessons from Indigenous Populations. *Curr. Envir. H* **2017**, *4*, 244–251. [CrossRef]
153. Mistry, J.; Berardi, A. Bridging indigenous and scientific knowledge. *Science* **2016**, *352*, 1274–1275. [CrossRef]
154. Tengö, M.; Hill, R.; Malmer, P.; Raymond, C.M.; Spierenburg, M.; Danielsen, F.; Elmqvist, T.; Folke, C. Weaving knowledge systems in IPBES, CBD and beyond-lessons learned for sustainability. *Curr. Opin. Environ. Sustain.* **2017**, *26*, 17–25. [CrossRef]
155. Gratani, M.; Butler, J.R.A.; Royee, F.; Valentine, P.; Burrows, D.; Canendo, W.I.; Anderson, A.S. Is validation of indigenous ecological knowledge a disrespectful process? A case study of traditional fishing poisons and invasive fish management from the Wet Tropics, Australia. *Ecol. Soc.* **2011**, *16*, 25. [CrossRef]
156. Berkes, F. *Sacred Ecology*, 4th ed.; Routledge: New York, NY, USA, 2018.
157. Donatuto, J.; Grossman, E.E.; Konovsky, J.; Grossman, S.; Campbell, L.W. Indigenous community health and climate change: Integrating biophysical and social science indicators. *Coast. Manag.* **2014**, *42*, 355–373. [CrossRef]
158. Vasco, C.; Bilsborrow, R. Contribution of off-farm employment to income of indigenous and mestizo households in the Ecuadorian Amazon. *Economía Agraria Recursos Natur. Agricult. Res. Econ.* **2016**, *16*, 5–18.

159. Ospina, V.; Torres, B.; Köthke, M.; Knapp, G.M.; Fisher, R.V.; Günter, S. Sistema socio-productivo y modelo de gobernanza en la comunidad kichwa "Shiwakucha", Pastaza, Ecuador. *Revista Amazónica Ciencia y Tecnología* **2017**, *6*, 126–149.
160. McNeely, J.A.; Schroth, G. Agroforestry and Biodiversity Conservation–Traditional Practices, Present Dynamics, and Lessons for the Future. *Biodivers. Conserv.* **2006**, *15*, 549–554. [CrossRef]
161. Sahoo, G.; Wani, A.M. Multifunctional agroforestry systems in India for livelihoods. *Ann. Horticult.* **2019**, *12*, 139–149. [CrossRef]
162. Almeida e Silva, J.V.; Gomes, F.A.; de Freitas, H.J.; Barbosa, M.; da Silva, J.C.T. Levels of cupuaçu seed by-product in feedstuff for free-range broilers reared in the Western Amazon. *Arquivos Ciências Veterinárias Zoologia UNIPAR* **2017**, *20*, 1–7.
163. Chauhan, B.S.; Johnson, D.E. The role of seed ecology in improving weed management strategies in the tropics. *Adv. Agron.* **2010**, *105*, 221–262.
164. Lazcano, C.; Gómez-Brandón, M.; Revilla, P.; Domínguez, J. Short-term effects of organic and inorganic fertilizers on soil microbial community structure and function. *Biol. Fertil. Soils* **2013**, *49*, 723–733. [CrossRef]
165. Alegre, J.; Lao, C.; Silva, C.; Schrevens, E. Recovering degraded lands in the Peruvian Amazon by cover crops and sustainable agroforestry systems. *Peruv. J. Agron.* **2017**, *1*, 1–7. [CrossRef]
166. Gordon, L.J.; Finlayson, C.M.; Falkenmark, M. Managing water in agriculture for food production and other ecosystem services. *Agricult. Water Manag.* **2010**, *97*, 512–519. [CrossRef]
167. León Alvear, V.; Torres, B.; Luna, M.; Torres, A.; Ramírez, P.; Andrade-Yucailla, V.; Muñoz-Rengifo, J.C.; Heredia, R.M. Percepción sobre cambio climático en cuatro comunidades orientadas a la ganadería bovina en la zona central de los Andes Ecuatorianos. *Livest. Res. Rural. Develop.* **2020**, *32*, 165.
168. Toulkeridis, T.; Tamayo, E.; Simón, D.; Merizalde, M.J.; Reyes, D.F.; Viera, M.; Heredia, M. Climate Change according to Ecuadorian academics—Perceptions ver-sus facts. *Granja Revista Ciencias Vida* **2020**, *31*, 21–49. [CrossRef]
169. Torres, B.; Starnfeld, F.; Vargas, J.C.; Ramm, G.; Chapalbay, R.; Rios, M.; Gómez, A.; Torricelli, Y.; Jurrius, I.; Tapia, A.; et al. *Gobernanza Participativa en la Amazonía del Ecuador: Recursos Naturales y Desarrollo Sostenible*; Universidad Estatal Amazónica: Puyo, Ecuador, 2013; 124p.
170. Lowan-Trudeau, G. Indigenous Environmental Education: The Case of Renewable Energy Projects. *Educ. Stud.* **2017**, *53*, 601–613. [CrossRef]
171. Heredia, M.; Bravo, C.; Torres, B.; Alemán, R. Innovación para el fortalecimiento de capacidades sobre sostenibilidad de los recursos naturales en poblaciones indígenas y mestizas—Colonas: Reserva de Biosfera Yasuní. *Revista Ibérica Sistemas Tecnologias Informação* **2020**, *25*, 103–116.
172. García, N.; Galeano, G.; Mesa, L.; Castaño, N.; Balslev, H.; Bernal, R. Management of the palm Astrocaryum chambira Burret (Arecaceae) in northwest Amazon. *Acta Bot. Brasilica* **2015**, *29*, 45–57. [CrossRef]
173. Innerhofer, S.; Bernhardt, K.G. Ethnobotanic garden design in the Ecuadorian Amazon. *Biodivers. Conserv.* **2011**, *20*, 429–439. [CrossRef]
174. Turner, N.J.; Łuczaj, Ł.J.; Migliorini, P.; Pieroni, A.; Dreon, A.L.; Sacchetti, L.E.; Paoletti, M.G. Edible and tended wild plants, traditional ecological knowledge and agroecology. *Crit. Rev. Plant Sci.* **2011**, *30*, 198–225. [CrossRef]
175. Leeuwis, C. *Communication for Rural Innovation: Rethinking Agricultural Extension*; John Wiley & Sons: Amsterdam, The Netherlands, 2013.
176. Hart, M.B.; Moore, M.J.; Laverty, M. Improving Indigenous health through education. *Med. J. Aust.* **2017**, *207*, 11–12. [CrossRef]
177. Cortina, R. *The Education of Indigenous Citizens in Latin America*; Multilingual Matters: Bristol, USA, 2013; Volume 95.
178. Heredia-R, M.; Falconí, A.K.; Barreto, D.; Amores, K.; Silva, J.H.; Torres, B. Conductas sustentables sobre el marco de evaluación SAFA–FAO: Un aporte para poblaciones rurales vulnerables de la Amazonía. *Revista Ibérica Sistemas Tecnologias Informação* **2020**, *33*, 312–326.
179. Lawlor, K.; Madeira, E.M.; Blockhus, J.; Ganz, D.J. Community participation and benefits in REDD+: A review of initial outcomes and lessons. *Forests* **2013**, *4*, 296–318. [CrossRef]
180. Chhatre, A.; Lakhanpal, S.; Larson, A.M.; Nelson, F.; Ojha, H.; Rao, J. Social safeguards and co-benefits in REDD+: A review of the adjacent possible. *Curr. Opin. Environ. Sustain.* **2012**, *4*, 654–660. [CrossRef]

181. Arnés, E.; Díaz-Ambrona, C.G.; Marín-González, O.; Astier, M. Farmer Field Schools (FFSs): A tool empowering sustainability and food security in peasant farming systems in the Nicaraguan Highlands. *Sustainability* **2018**, *10*, 3020.
182. Marinello, F.; Pezzuolo, A.; Chiumenti, A.; Sartori, L. Technical analysis of unmanned aerial vehicles (drones) for agricultural applications. In Proceedings of the Engineering for Rural Development, Jelgava, Latvia, 25–27 May 2016; Volume 15.
183. Aker, J.C. Dial "A" for agriculture: A review of information and communication technologies for agricultural extension in developing countries. *Agricult. Econ.* **2011**, *42*, 631–647. [CrossRef]
184. Mittal, S.; Gandhi, S.; Tripathi, G. *Socio-Economic Impact of Mobile Phones on Indian Agriculture*; Indian Council for Research on International Economic Relations (ICRIER): New Delhi, India, 2010.
185. Barreto-Álvarez, D.E.; Heredia-Rengifo, M.G.; Padilla-Almeida, O.; Toulkeridis, T. Multitemporal Evaluation of the Recent Land Use Change in Santa Cruz Island, Galapagos, Ecuador. In *Information and Communication Technologies, TICEC 2020*; Communications in Computer and Information Science; Springer: Cham, Switzerland, 2020; Volume 1307. [CrossRef]

3

Assessing Context-Specific Factors to Increase Tree Survival for Scaling Ecosystem Restoration Efforts in East Africa

Christine Magaju [1,*], Leigh Ann Winowiecki [1], Mary Crossland [2], Aymen Frija [3], Hassen Ouerghemmi [3], Niguse Hagazi [4], Phosiso Sola [1], Ibrahim Ochenje [1], Esther Kiura [1], Anne Kuria [1], Jonathan Muriuki [1], Sammy Carsan [1], Kiros Hadgu [4], Enrico Bonaiuti [5] and Fergus Sinclair [1,2]

1 World Agroforestry (ICRAF), UN Avenue, P.O. Box 30677, Nairobi 00100, Kenya; l.a.winowiecki@cgiar.org (L.A.W.); p.sola@cgiar.org (P.S.); ibraochenje@gmail.com (I.O.); muthoni.kiura@gmail.com (E.K.); a.kuria@cgiar.org (A.K.); j.muriuki@cgiar.org (J.M.); s.carsan@cgiar.org (S.C.); f.sinclair@cgiar.org (F.S.)
2 School of Natural Sciences, Bangor University, Bangor, Gwynedd LL57 2DG, UK; afp43d@bangor.ac.uk
3 International Center for Agricultural Research in the Dry Areas, ICARDA, Rue Hedi Karray, CP 2049 Ariana, Tunisia; a.frija@cgiar.org (A.F.); ouerghemmihassen@gmail.com (H.O.)
4 World Agroforestry (ICRAF), Gurd Shola, P.O. Box 5689 Addis Ababa, Ethiopia; n.hagazi@cgiar.org (N.H.); k.hadgu@cgiar.org (K.H.)
5 International Center for Agricultural Research in the Dry Areas, ICARDA, 6/106, Osiyo St, Tashkent 100084, Uzbekistan; E.Bonaiuti@cgiar.org
* Correspondence: c.magaju@cgiar.org

Abstract: Increasing tree cover in agricultural lands can contribute to achieving global and national restoration goals, more so in the drylands where trees play a key role in enhancing both ecosystem and livelihood resilience of the communities that depend on them. Despite this, drylands are characterized by low tree survival especially for tree species preferred by local communities. We conducted a study in arid and semi-arid areas of Kenya and Ethiopia with 1773 households to assess how different tree planting and management practices influence seedling survival. Using on-farm planned comparisons, farmers experimented and compared tree survival under different planting and management practices as well as under varying socioeconomic and biophysical contexts in the two countries. Seedling survival was monitored at least six months after planting. Results show that watering, manure application, seedling protection by fencing and planting in a small hole (30 cm diameter and 45 cm depth) had a significant effect on tree seedling survival in Kenya, while in Ethiopia, mulching, watering and planting niche were significant to tree survival. Household socioeconomics and farms' biophysical characteristics such as farm size, education level of the household head, land tenure, age of the household head had significant effects on seedling survival in both Ethiopia and Kenya while presence of soil erosion on the farm had a significant effect in Kenya. Soil quality ranking was positively correlated with tree survival in Ethiopia, regardless of species assessed. Current findings have confirmed effects of context specific variables some involving intrahousehold socioeconomic status such education level of the household head, and farm size that influence survival.

Keywords: trees on farm; options by context; on-farm planned comparison; tree seedling survival

1. Introduction

Restoration and avoiding further degradation through tree (re)establishment can be a key pathway towards achieving the UN's Sustainable Development Goals if successful restoration efforts can reach larger numbers of farmers and hectares over the coming decade. In fact, several global initiatives and commitments have come up in recognition of the need to intensify restoration efforts. This includes the UN decade on ecosystem restoration whose goal is to prevent, halt and reverse degradation of ecosystems by acting as an accelerator for ongoing restoration efforts[1]; the Bonn Challenge whose goal is to restore 150 million hectares of degraded and deforested landscapes by 2020 and 350 million hectares by 2030[2]; and the African Forest Landscape Restoration Initiative (AFR100) in which countries have committed to restoring 100 million hectares of degraded land in Africa by 2030[3]. Ethiopia and Kenya have each committed to restore 15 and 5.1 million hectares, respectively, as a contribution to AFR100 and the Bonn challenge. In addition, Kenya, through its national strategy, aims at achieving and maintaining over 10% tree cover by 2022 [1].

Achieving the restoration goals outlined in international and national commitments will require promotion of context-specific land restoration options across diverse social, economic, and biophysical realities [2–4]. These options will also need to be scaled up and out across the scaling domains [5]. For the most part, farmers targeted by restoration efforts work in complex, heterogenous and dynamic systems and, as such, no one single restoration option can suit all [3,4,6,7]. Many factors, both socioeconomic and biophysical, can affect the suitability and performance of the restoration options at different scales [8]. These include household characteristics such as farm size, age of the household head as well as education level and labor availability, farming practices, land degradation status, the policy environment [5,9] as well as farmer values and preferences [5,6,10,11].

Therefore, to understand which options are suitable, we must be cognizant of the variation within and between farms, landscapes and communities for all the context variables that might be important [3]. We must also take into account the goals, values and preferences of the people living on the land [6,10,11]. It is also imperative to consider the impact of restoration options such as tree planting on gender dynamics within a household if we are to ensure women, men and the youth benefit from restoration [10,12–14]. This includes understanding the roles and responsibilities of men and women in managing natural resources, decision-making within restoration options for example, who decides what restoration options to take part in and where on the farm these options are implemented.

Tree (re)establishment on agricultural land through natural regeneration or direct planting is often considered a key approach to restoration in the drylands [15–17]. This is largely driven by the recognition of the vital role that trees play in enhancing ecosystem and livelihood resilience [18–20]. Trees, for example, provide goods and services such as food and fuel, enhancing soil health, enhancing biodiversity, opportunities for generating income and contributing to climate change mitigation and adaptation [21–23]. However, low survival rates of planted trees, especially species highly valued by farmers, is a major limitation in the drylands [19,20]. This is partly due to unreliable rainfall, high levels of land degradation resulting in low soil productivity, planting of ecologically unsuitable tree species, and poor tree seedling management practices [18,19,24].

Understanding what determines tree seedling survival is fundamental to successful tree planting initiatives. For example, which trees species are suitable for which ecological context, which tree species farmers prefer, and what management practices increase tree survival in the different agroecological and socioeconomic conditions. In this paper, we evaluate how tree seedling planting and management practices influence tree survival across various agroecological conditions and the farmer circumstances in Ethiopia and Kenya, as well as the effect of socioeconomic and farm characteristics on tree survival.

---

1 https://www.decadeonrestoration.org/
2 https://www.bonnchallenge.org/
3 https://afr100.org/

2. Materials and Methods

The study was conducted within the context of a donor-funded project, 'Restoration of degraded land for food security and poverty reduction in East Africa and the Sahel: taking success in land restoration to scale'[4], (henceforth referred to as 'Restoration project'). The project's goal was to reduce food insecurity and improve livelihoods of poor people living in African drylands by restoring degraded land, and returning it to effective and sustainable tree, crop and livestock production, thereby increasing land profitability and landscape and livelihood resilience. To achieve this, the project employed a Research in Development (RinD) approach in which research activities are embedded within development activities [3]. Thus, the activities of the Restoration project were co-located with those of the Drylands Development Program[5], (henceforth referred to as 'DryDev'), an international development initiative.

A key aspect of RinD that was implemented by the restoration project, is the use of planned comparisons, an approach where farmers compare and test promising options, and variations thereof, in their fields across a varying range of ecological and socioeconomic conditions [3–5,24,25]. Ref. [26] defines planned comparisons as the systematic and deliberate comparison of options where options refer to what is being done differently to address a particular challenge [25]. The approach involves engagement of local communities including farmers, researchers, government extension agents, and development actors in identifying current challenges facing farmers, selecting and prioritizing the initial set of promising options to be compared, and in monitoring the performance of the options being compared. Involvement of local communities has an impact on the outcome on whether the options succeed or fail and, in most cases, local communities are responsible for long-term management of these options [27,28].

2.1. Site Description

The study was conducted across four woredas in Tigray and Oromia regions in Ethiopia and across six subcounties in Kitui, Machakos and Makueni counties in Kenya. Specifically, the study took place in Boset and Gursum woredas in the Oromia region of Ethiopia, Samre and Tsaeda Emba woredas in the Tigray region of Ethiopia. In Kenya, the study covered Kitui Rural and Mwingi East subcounties in Kitui, Mwala and Yatta subcounties in Machakos, and Mbooni East and Kibwezi East subcounties in Makueni counties in the Eastern drylands of Kenya (Figure 1).

All the sites in Ethiopia are classified as semi-arid with annual rainfall ranging from 400 mm to 800 mm which varies from year to year [29] and are located between 779 m and 1362 m above sea level. The sites are characterized by low vegetation cover, low soil fertility, high rates of soil erosion, and recurrent droughts [29,30]. The study area in Kenya is largely arid and semi-arid and is characterized by highly erratic and unreliable rainfall. Annual average rainfall varies across and within the three counties. For example, Machakos receives an average of 500 mm to 1250 mm per annum while Makueni receives 250–400 mm in the lower region and 800–900 mm in the higher regions. Annual average temperature also varies across the three counties. In Kitui, annual temperatures ranges from 14–32 °C, in Machakos from 14–32 °C and in Makueni from 20.2–24 °C [31–33].

Agriculture is the dominant land use in all the sites and is characterized by low input subsistence farming. Main crops grown across the sites in Ethiopia include *Eragrostis tef*, *Triticum*, *Zea mays* and *Sorghum bicolor*, while *Zea mays*, *Sorghum bicolor*, *Pennisetum glaucum*, and pulses such as *Phaseolus vulgaris* and *Vigna unguiculata* are commonly grown across the sites in Kenya. The majority of households also own livestock.

---

4 http://www.worldagroforestry.org/project/restoration-degraded-land-food-security-and-poverty-reduction-east-africa-and-sahel-taking

5 http://www.worldagroforestry.org/project/drylands-development-programe-drydev

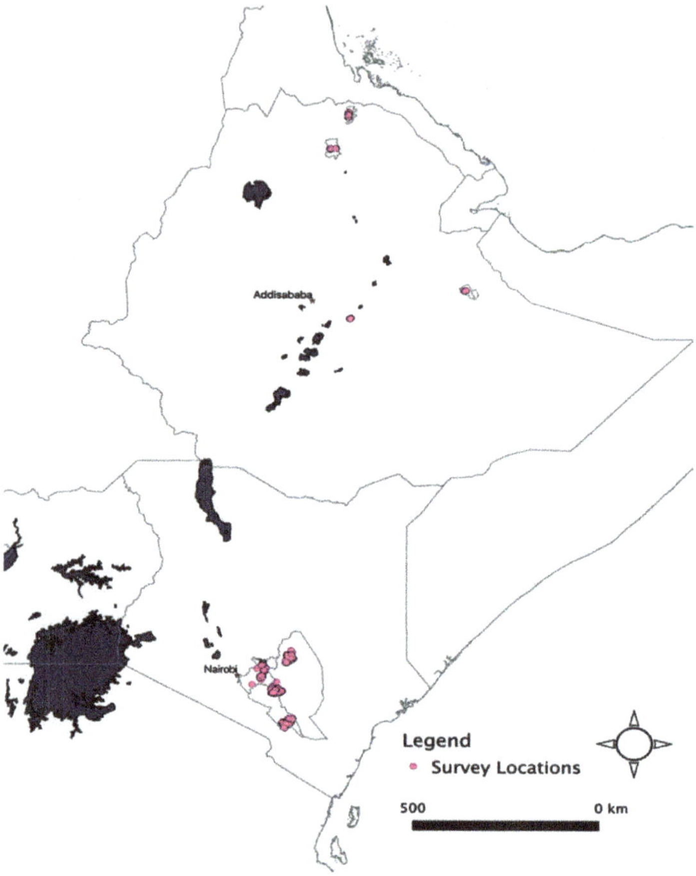

Figure 1. Map of the study locations in Ethiopia and Kenya.

2.2. *Tree Planting Planned Comparison*

The planned comparison on tree planting and management options was set up in response to a need by local communities in the study area to increase tree seedling survival. They identified tree planting and management as a learning priority during DryDev's community visioning and planning process [34]. Specifically, farmers identified low survival rates of planted trees as a key constraint to increasing tree cover and expressed interest in learning about planting and management methods that could increase establishment rates. Thus, the objective of the planned comparison on tree planting was to understand which planting and management practices can increase tree seedling survival rates for farmers across the study sites. More so, the practices that conferred the best chance of survival for the planted tree seedlings.

One thousand seven hundred and seventy-three households in both Kenya and Ethiopia volunteered to take part in the planned comparison during subsequent stakeholder and community engagement meetings. They compared the effect of different planting and management practices on seedling survival. Options compared included: tree species, planting hole sizes, planting with manure or without, physical protection of seedlings from livestock through spot fencing, and watering (Table 1). Water availability is a limiting factor for seedling survival in the study sites, thus the options compared were prioritized to increase the amount of water available to the seedlings after establishment. For example, digging a bigger planting hole to increase water capture thus increasing the rate of infiltration. This ensures that seedlings have sufficient water for initial growth and establishment [35]. Furthermore, browsing by livestock is one of the leading causes of seedling mortality.

Table 1. Options and contexts compared in the tree planting on-farm planned comparisons.

Options Compared	
Tree species	Categorical variable
Manure	Categorical variable with two levels, yes or no
Mulching	Categorical variable with two levels, yes or no (Mulch: a mix of grass, maize stalk and leaf litter is spread around the base of the seedling)
Hole size (diameter and depth)	Categorical variable with three levels (i) small hole (30 cm by 45 cm); (ii) big hole (60 cm by 45 cm); (iii) farmer usual hole size (diameter is not equal to 30 cm or 60 cm)
Watering	Categorical variable with two levels, yes or no
Watering frequency (Kenya)	Categorical variable with five levels (i) daily; (ii) every other day; (iii) weekly; (iv) bi-weekly; (v) monthly
Watering regime (Ethiopia)	Categorical variable with four levels (i) five litres every five days; (ii) five litres every ten days; (iii) three litres every five days; (iv) three litres every ten days
Fencing	Categorical variable with two levels, yes or no (physical protection of seedlings from livestock)
Pruning	Categorical variable with two levels, yes or no
Shading	Categorical variable with two levels, yes or no
Planting niche	Categorical variable with six levels, (i) external boundary; (ii) internal boundary; (iii) scattered in cropland; (iv) woodlot; (v) home compound; (vi) along terraces. (Planting niche: where on the farm the seedlings are planted)
Contexts Compared	
Age of household head	Continuous variable
Gender of household head	Categorical variable with two levels: (i) male; (ii) female
Education level of the household head	Categorical variable with four levels (i) no formal schooling; (ii) primary; (iii) secondary; (iv) tertiary (college/university/vocational)
Farm size (ha)	Continuous variable
Land tenure	Categorical variable with six levels: (i) private through purchase; (ii) private through customary inheritance; (iii) communal land; (iv) government; (v) settlement scheme; (vi) leasing
Erosion status	Categorical variable with two levels, yes or no. (Erosion status: Farmer experience with problems of soil erosion on the farm)
Soil quality ranking	Categorical variable with three levels (i) high (Good yields can be obtained without adding either chemical fertilizer or farmyard manure/compost); (ii) medium (Yields can be maximized with chemical fertilizer or farmyard manure/compost but fair yields can be obtained without); (iii) low (very little can grow without significant addition of chemical fertilizer or farmyard manure/compost). (Soil quality ranking: Farmer description of soil quality on the farm)

Farmers were trained on the various management options and on setting up the on-farm planned comparisons during several opportunities, during farmer workshops. This included practical training on how to set up the options being compared. Farmers had the choice of which option to implement and compare on their farms on condition of likeness. The number of options tested was subject to farmer interest and available resources for example if the farmer had access to manure or not. Farmers also had the choice of where on their farm to plant the seedlings (i.e., planting niche). Seedlings could be planted across all preferred niches provided variation in the farm characteristics was considered to ensure seedlings of a similar species were as homogenous as possible.

Implementing partners across the study sites distributed seedlings to farmers subsequent to training events. Tree seedlings were sourced from different nurseries across the study sites due to the large number required and the nature of the nursery enterprises in the study area. Seedlings were delivered potted in 10 cm by 15 cm potting bags and were on average, of good quality with variation depending on the nursery. The size of the seedlings varied depending on the nursery from which they were sourced. Farmers planted the seedlings within a week of receiving them.

In Kenya and Ethiopia, tree species selection was conducted through a consultative stakeholder engagement process. Twenty tree species were monitored in Ethiopia: *Acacia saligna, Azadirachta indica, Carica papaya, Casimiroa edulis, Citrus sinensis, Coffea arabica, Cordia africana, Faidherbia albida, Grevillea robusta, Jacaranda mimosifolia, Malus domestica, Mangifera indica, Melia volkensii, Moringa oleifera, Olea africana, Persea americana, Psidium guajava, Rhamnus prinodes and Vachellia seyal*. In Kenya, seven tree species were monitored: *Calliandra calothyrsus, Melia volkensii, Senna siamea, Mangifera indica, Carica papaya, Azadirachta indica* and *Moringa oleifera*. Farmers planted at least seven seedlings of each the selected species, applying the various management practices. Some species were planted exclusively in one woreda or county. For example, *Jacaranda mimosifolia* was only planted in Gursum Woreda while *Carica papaya* and *Melia volkensii* were exclusively planted in Boset Woreda, and in Kenya, *Calliandra calothyrsus* was only planted in Mwala, Machakos County.

2.3. Data Collection and Analysis

Data on tree seedling survival was collected from all households that received and planted tree seedlings through the project. Only tree seedlings planted as part of the on-farm planned comparisons were assessed. We assessed seedling survival under different planting and management practices at least six months after the seedlings were planted as it is widely accepted that the first six months after planting are most critical for survival as seedlings acclimatize and establish. This was done using a structured survey questionnaire administered through the Open Data Kit (ODK)[6] installed on mobile phones). Seedling survival was recorded as a categorical variable with two levels: yes or no represented by dummy variables 1 and 0. To understand men's and women's involvement in tree planting and management thereof, we collected information on the roles and responsibilities of men and women in decision-making within tree planting interventions.

We used a household survey to collect data on all farmers involved in the project. The survey included collection of basic socioeconomic and biophysical characteristics of the household and farm. Household demographic information collected included gender and age of household head, education level of the household head and household size. Farm characteristics such as farm size and ownership; soil erosion and control measures, trees on farm including the count, management and utility derived from the trees; crop and livestock production, climate change, including farmers' understanding, experience and response to climate change; and food security were also collected. We also conducted a tree inventory of the existing tree species within the household including data on species diversity within farms.

Statistical analysis on survival of planted trees under different planting and management options was conducted using the R software environment [36] including logistic regression to ascertain the probability of tree seedling survival under different management practices, and household socioeconomic and farm characteristics. Variation in the type and magnitude of the effect across the agroecological conditions represented in the study sites was also examined using a mixed-effect linear model. Since seedling survival varied highly across the tree species in Kenya, the effect of tree species was included as a constant in the model for Kenya. As not all practices were employed across the two countries, only those that were universally employed were compared across both countries in the results. For example, manure and variation in planting hole size was only assessed in study sites in Kenya while watering regime, disease, weed and pest control were only assessed in study sites in Ethiopia. Table 2 shows the number of observations in each option.

---

6 https://opendatakit.org/

Table 2. Number of tree seedlings under each option in Kenya and Ethiopia. Where available, a common name for each species is added in the brackets.

Country	Tree Species	Fencing	Manure	Mulch	Watering	Shade	Fertilizer
Ethiopia	*Acacia saligna*	14	-	4	12	-	-
	Azadiracta indica (Neem)	75	-	62	91	-	-
	Carica papaya (Papaya)	1	-			-	-
	Casimiroa edulis (White sapote)	5	-	13	19	-	-
	Citrus sinensis (Sweet orange)	172	-	102	155	-	-
	Coffea arabica (Arabica coffee)	36	-	35	55	-	-
	Cordia africana	1	-		1	-	-
	Faidherbia albida (Apple- ring acacia)	114	-	128	212	-	-
	Grevillea robusta (Silver oak)	450	-	222	346	-	-
	Jacaranda mimosifolia	6	-	12	11	-	-
	Malus domestica (Apple)	1	-		1	-	-
	Mangifera indica (Mango)	571	-	375	541	-	-
	Moringa oleifera (Moringa)	154	-	89	129	-	-
	Olea africana	2	-			-	-
	Persea americana (Avocado)	105	-	42	97	-	-
	Psidium guajava (Guava)	118	-	100	124	-	-
	Rhamnus prinoides	111	-	119	191	-	-
	Vachellia seyal (Shittah tree)	2	-	1	2	-	-
Kenya	*Azadirachta indica (Neem)*	899	1533	660	2436	82	8
	Calliandra calothyrsus	50	61	29	165	43	2
	Carica papaya (Pawpaw)	276	239	162	307	84	-
	Mangifera indica (Mango)	2157	4339	1658	6432	618	22
	Melia volkensii (Melia)	1113	1812	849	3221	335	-
	Moringa oleifera (Moringa)	167	489	113	883	41	9
	Senna siamea (Siamese senna)	662	1005	346	1808	59	25

3. Results

Data were collected from 173 households across the sites in Ethiopia in 2017 [37] and 1600 households in Kenya in 2018 [38]. Of these, 71% and 76% surveyed households in Kenya and Ethiopia were male-headed, respectively. Despite this, it was mostly women (i.e., their spouse) who registered to join the project and attended the training events. Median farm size was 2.02 Ha in Kenya and ranged from 0.05 Ha to 3.33 Ha while in Ethiopia, median farm size was 1 Ha and ranged from 0.1 Ha to 8 Ha. Household head age ranged from 18 years to 80 years in Ethiopia with a median of 40.5 years while in Kenya the age ranged from 23 years to 97 years with a median of 49 years.

3.1. Overview of Tree Seedling Survival

In Ethiopia, 4224 trees were monitored in 2017 and 17,520 trees were monitored in Kenya in 2018.

Overall, average seedling survival in the two countries varied across the species planted and agroecological conditions represented in the study sites. In Kenya, Kitui County had the highest average seedling survival at 53.4% while Machakos and Makueni counties had an average survival of 32.2% and 43.3%, respectively. The variation in average survival was also observed within each county. Highest variation was recorded in Makueni County with Mbooni East recording 66.7% survival compared to Kibwezi East which had survival rate of 33.8%. Comparatively, average seedling survival was high across all the woredas in Ethiopia. Tsaeda Emba and Boset had the highest average tree survival at 99% and 93% respectively while Gursum and Samre recorded 84% and 81% survival of all the trees planted respectively. Variation in average seedling survival was however observed within each woreda with highest variation recorded in Samre where Bara watershed recorded 95.8% seedling survival compared to Atami watershed which recorded 72.1% survival. We also found that different tree species performed better in some areas compared to others despite the broad similarities in agroecological conditions in Kenya (Figure 2) and Ethiopia (Table 3).

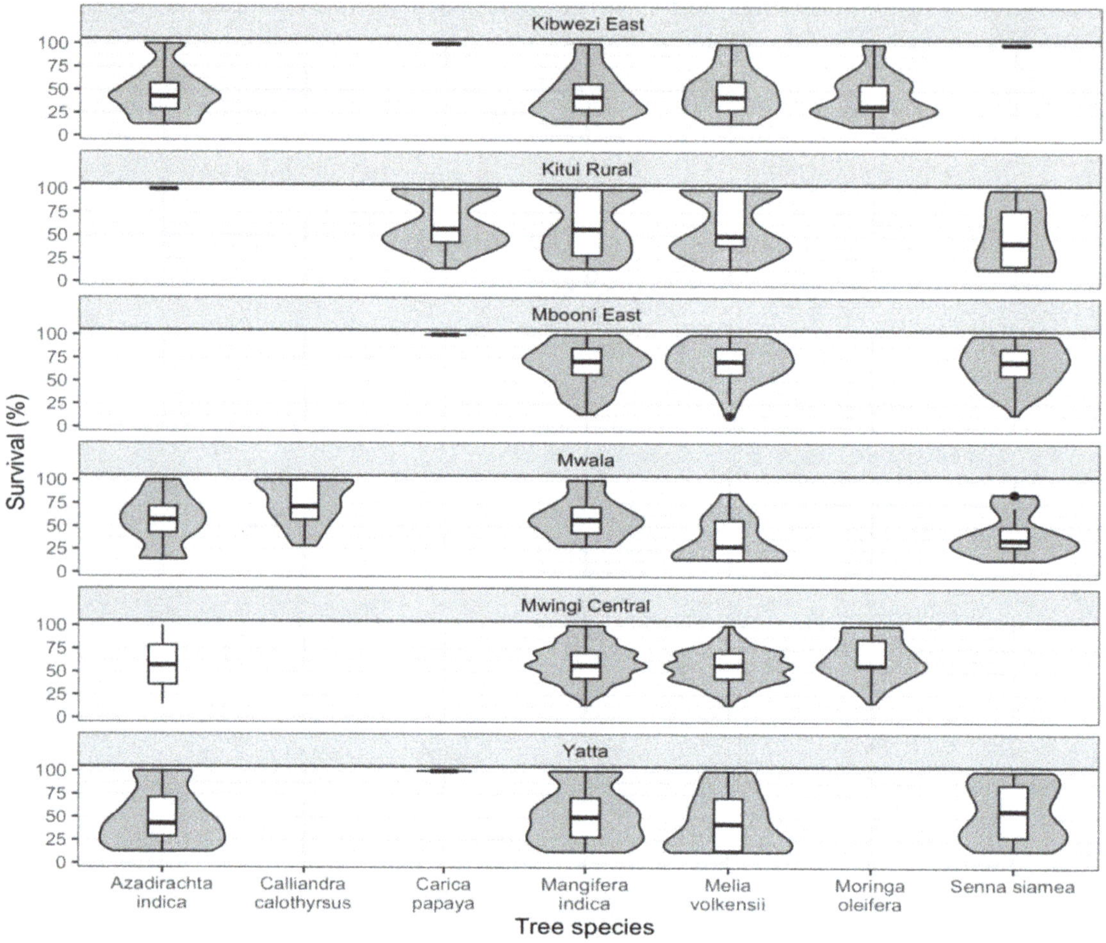

Figure 2. Boxplot of percentage tree seedling survival across the sites in Kenya (n = 7375 seedlings). The black horizontal line is the average percentage survival for each species and the gray shaded area shows the distribution of the data.

Table 3. Percentage seedling survival per species in each site in Ethiopia (number of trees monitored in each species is added in brackets).

Tree Species	Boset	Gursum	Samre	Tsaeda Emba
Acacia saligna	100 (1)	100 (5)		100 (8)
Azadiracta indica	66.7 (125)	85.5 (69)		
Carica papaya	100 (1)			
Casimiroa edulis				100 (19)
Citrus sinensis			93.8 (160)	
Coffea arabica				100 (59)
Cordia africana				100 (1)
Faidherbia albida	100 (1)			100 (215)
Grevillea robusta	93.6 (125)	71.3 (94)	87.7 (285)	
Jacaranda mimosifolia		91.7 (12)		
Malus domestica	100 (2)	100 (1)		
Mangifera indica	94.1 (169)	94.7 (95)	72.5 (280)	99.5 (204)
Melia volkensii	50 (2)			
Moringa oleifera			69.2 (156)	
Olea africana	100 (8)			
Persea americana	97.8 (45)	76.7 (103)		100 (1)
Psidium guajava		91.5 (94)		100 (60)
Rhamnus prinodes				99.5 (200)
Vachellia seyal				100 (2)

3.2. *Effect of Tree Planting and Management Practices on Seedling Survival*

Results show that despite influencing average survival, application of manure, watering regime, planting in a small hole relative to a big hole, and planting along the external boundary had a positive effect on seedling survival in Kenya (Table 4) while in Ethiopia, mulching, watering and physical protection were significant (Table 5). The type of effect (whether positive or negative) and magnitude of effect varied across the practices the sites where they were applied.

Overall, addition of manure increased seedling survival in Kenya by 12% with variation in the magnitude of effect across and within the counties. For example, we found that seedling survival in Kitui County increased by 8% with addition of manure while in Machakos and Makueni survival increased by 6% and 21% respectively. At subcounty level, Mbooni East in Makueni County had the greatest increase in seedling survival with addition of manure at 18.7%. There was no variation in the magnitude of the effect of manure at the species level although results of the model show that application of manure increased survival of all species by 10%. Addition of mulch had a positive effect on seedling survival and was significant in both Kenya and Ethiopia increasing survival by 5.8% and 5% respectively. Nevertheless, variation in the magnitude of the effect was observed across and within counties and woredas.

Watering the seedlings had a positive effect on tree survival and was significant in the two countries. In Kenya, survival increased by 35% when seedlings were watered with variation within and across the counties. Watering had the greatest effect on survival in Kitui (51% increase) and Makueni Counties (43% increase), which are considerably drier than Machakos County which recorded 7.5% increase in survival with watering. In Ethiopia, survival increased by 63% when the seedlings were watered. Farmers also compared four watering regimes in Ethiopia. They compared watering with five liters of water every five days, five liters every ten days, three liters every five days, and three liters every ten days. Our results show that the watering regime employed was not significant as long the seedlings were watered.

Planting in a small hole relative to a big hole had a negative effect on seedling survival in Kenya with variation across the counties. The greatest variation was observed in Machakos County where survival decreased by 21% when seedlings were planted in farmer hole size relative to a big hole. On the other hand, planting seedlings in a small hole relative to a big hole increased seedling survival by 12% in Makueni County and decreased survival by 2% and 8% in Machakos and Kitui Counties respectively. Variation in the effect of the size of planting hole was also observed across the tree species planted and was significant for *Calliandra calothyrsus*, *Mangifera indica* and *Melia volkensii*. For example, *Mangifera indica* had a 41.5% probability of survival when planted in a small hole compared to 45.6% when planted in a big hole and 39.3% when planted in normal-sized hole.

Physical protection of seedlings (fencing) had a significant effect on seedling survival in Kenya while in Ethiopia physical seedling protection was not significant. Overall, fencing increased seedling survival by 7.8% and 5% in Kenya and Ethiopia respectively. Variation in the type and magnitude of effect was observed across the counties and Woredas. For example, fencing increased survival by 9% and 15% in Machakos and Makeuni counties respectively and it decreased survival in Kitui County by 5%. On the other hand, survival increased by 8.5% in Boset, 5.3% in Gursum, 23.9% in Samre and by 0.4% in Tsaeda Emba when seedlings were fenced.

Finally, in Kenya, seedlings had a higher probability of survival when planted in woodlots and along the internal boundary and a lower probability of survival when planted along external boundaries, home compound and along terraces. In Ethiopia, seedlings had a higher probability of survival when planted along external boundaries and a lower probability of survival when planted scattered in cropland. All seedlings planted along terraces recorded 100% survival and, as such, were excluded from the regression model.

Table 4. Results of a logistic model on the probability of seedling survival under different planting and management practices on seedling survival in Kenya. Significance codes: 0 '***' 0.001 '**' 0.01 '*'.

Management Practice	n	Coefficient	Standard Error	Pr (>|z|)	Odds Ratio
Manure application	17,496				
No		−0.495	0.226	0.029 *	
Yes		0.466	0.033	<2 × 10^{-16} ***	1.641
Mulch application	17,496				
No		0.266	0.235	0.259	
Yes		0.233	0.041	1.34 × 10^{-8} ***	1.263
Watering	17,496				
No		−1.918	0.220	<2 × 10^{-16} ***	
Yes		1.838	0.069	<2 × 10^{-16} ***	6.284
Watering frequency	15,231				
Daily		0.249	0.239	0.296	1.283
Every other day		−0.063	0.069	0.363	0.939
Weekly		0.492	0.062	1.71 × 10^{-15} ***	0.611
Bi-weekly		−0.299	0.071	2.68 × 10^{-5} ***	0.741
Monthly		−0.416	0.143	0.004 **	0.66
Planting hole	17,496				
Big hole		−0.173	0.228	0.448	0.841
Small hole		−0.103	0.036	0.004 **	0.902
Farmer normal hole size		−0.046	0.071	0.52	0.955
Physical protection(fencing)	17,496				
No		−0.380	0.235	0.106	
Yes		0.405	0.037	<2 × 10^{-16} ***	1.5
Planting Niche	17,496				
Along terraces		−0.222	0.211	0.294	
External boundary		−0.271	0.076	0.000 ***	0.762
Home compound		−0.120	0.051	0.019 *	0.886
Internal boundary		−0.001	0.059	0.987	1.004
Scattered in cropland		−0.114	0.050	0.023 *	0.892
Woodlot		0.087	0.129	0.502	1.094

Table 5. Results of a logistic model on the probability of seedling survival under different planting and management practices on seedling survival in Ethiopia. Significance codes: 0 '***' 0.001 '**'.

Management Practice	n	Coefficient	Standard Error	Pr (>|z|)	Odds Ratio
Mulch application	2652				
No		2.414	0.812	0.295	
Yes		1.008	0.14	6.18 × 10^{-13} ***	2.739
Watering	2680				
No		−0.588	1.213	0.628	
Yes		5.021	0.366	<2 × 10^{-16} ***	151.608
Watering regime	2070				
Five liters every five days		3.447	1.028	0.001 ***	
Five liters every ten days		−0.004	0.277	0.990	0.996
Three liters every five days		−0.210	0.219	0.339	0.811
Three liters every ten days		−0.121	0.274	0.659	0.886
Physical protection (fencing)	2854				
No		2.202	0.842	0.009 **	
Yes		0.831	0.175	2 × 10^{-6} ***	2.296
Planting Niche	2695				
External boundary		3.065	0.844	0.000 ***	21.432
Home compound		0.591	0.391	0.131	1.806
Internal boundary		−0.538	0.359	0.134	1.713
Scattered in cropland		−0.623	0.422	0.140	1.864
Woodlot		−0.604	0.401	0.132	1.829

3.3. Options that Conferred the Best Chance of Survival for Planted Seedlings

To assess the planting and management options that conferred the best chance of survival, we calculated the survival rate of tree species planted using a combination of the options tested (Table 6).

Table 6. Description of options derived from a combination of those tested in Kenya.

Options	Planting Hole	Manure Quantity (kg)	Mulching	Watering	Observations
Option 1	Small*	2	No	No	32
Option 2	Small*	2	No	Yes	630
Option 3	Small*	2	Yes	No	26
Option 4	Small*	2	Yes	Yes	298
Option 5	Small*	0	Yes	No	34
Option 6	Small*	0	Yes	Yes	516
Option 7	Big**	4	No	No	19
Option 8	Big**	4	No	Yes	170
Option 9	Big**	4	Yes	No	7
Option 10	Big**	4	Yes	Yes	123
Option 11	Big**	0	Yes	No	20
Option 12	Big**	0	Yes	Yes	554
Option 0 (Farmer's practice)	Other***	Other	No/Yes	No/Yes	15,088
	Total				17,517

Small* = 30 cm diameter × 45 cm depth; Big** = 60 cm diameter × 45 cm depth; Other*** = Common hole sizes.

In Kenya, seedlings planted in a big hole with 4 kg of manure and mulch had the highest survival at 47% while those planted in a small hole with 2 kg of manure and mulch had a survival of 36.1%. Variation in percentage survival was observed across the species planted (Figure 3). For example, *Azadirachta indica* recorded higher survival when planted in a small hole with addition of mulch and two kilograms of manure while *Moringa oleifera* recorded higher survival when planted in a big hole with watering and addition of mulch. In Ethiopia, almost all combinations of the options conferred at least 90% chance of survival for the planted seedlings. However, seedlings that were not watered but were applied mulch and controlled for weed recorded the least survival at 44%. This is consistent with the results of the statistical analysis where watering and watering regime were found to have a positive effect on survival of seedlings in Ethiopia.

3.4. Effect of the Household Socioeconomic Characteristics on Tree Survival

We assessed, using the household survey data, the effect of age and education level of the household head, land tenure, soil quality ranking (farmers ranked the quality of soil on their farm as either low, medium or high) and long-term experience with farms' soil erosion problems on seedling survival. In Kenya, our results showed that that there was a significant relationship between age and education level of the household head and seedling survival as well as between farm size, the type of land tenure and whether the farmer experienced soil erosion on their farm (Table 7). Variation was observed across the sites, for instance, in Mwingi Central, seedlings had a 59.3% probability of survival when farmers described their soil quality as high compared to 47% when soil quality was ranked as low. In Kitui Rural however, seedlings had a higher probability of survival when farmers ranked their soil quality as low suggesting that the low soil quality rating was overcome by the management strategy.

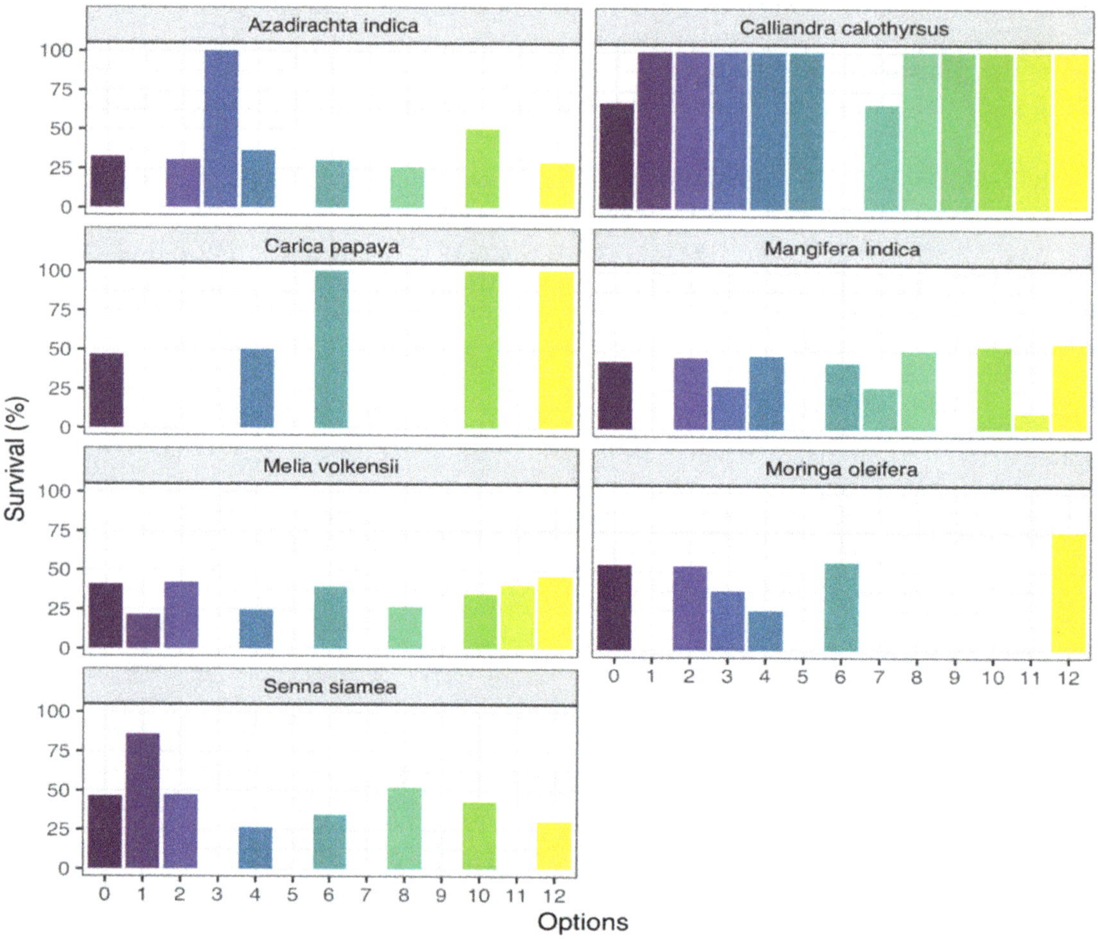

Figure 3. Performance of tree species against different combination of options that were tested in Kenya. Combinations presented here are described in Table 6.

Table 7. Results of the chi square test for a relationship between seedling survival and socioeconomic characteristics of a household and biophysical characteristics of the farm in Kenya. Significance levels: 1% '***' 5% '**'.

Variables	Seedling Survival (yes = 1)	n	Chi Square Value	*p* Value	Cramer's V	Relationship
Farm size		17,517	235.34 ***	0.000	0.12	weak
≤2 Ha	3145 (36%)					
>2 Ha	4239 (47%)					
Age category		17,517	50.95 ***	0.000	0.05	Very weak
<35 years	518 (40%)					
36–64 years	5833 (44%)					
>65 years	1033 (37%)					
Education level		15,723	59.84 ***	0.000	0.06	Very weak
No formal education	580 (40%)					
Primary education	4117 (44%)					
Secondary education	1484 (38%)					
Tertiary education	437 (49%)					
Land tenure		16,009	94.83 ***	0.000	0.08	Very weak
Title deed/	4841 (43%)					
Allotment/leasehold	1345 (35%)					
Others	501 (48%)					

Table 7. *Cont.*

Variables	Seedling Survival (yes = 1)	n	Chi Square Value	*p* Value	Cramer's V	Relationship
Soil quality		16,740	8.55 **	0.014	0.02	Very weak
High	1166 (44%)					
Low	1590 (42%)					
Medium	4262 (41%)					
Erosion		16,740	10.89 ***	0.001	0.03	Very weak
No	2372 (40%)					
Yes	4646 (43%)					

In Ethiopia, farm size and soil quality ranking had a significant relationship with seedling variation survival (Table 8). As soil quality ranking increased so did the likelihood of tree seedling survival (soil quality was ranked by the farmer as either low, medium or high).

Table 8. Results of the chi square test for a relationship between seedling survival and socioeconomic characteristics of a household and biophysical characteristics of the farm in Ethiopia. Significance levels: 1% '***' 5% '**' 10% '*'.

Variables	Seedling Survival (yes = 1)	n	Chi Square Value	*p* Value	Cramer's V	Relationship
Farm size		2854	22.72 ***	0.000	0.09	Very weak
<=2 Ha	980 (86%)					
>2 Ha	1562 (91%)					
Age category		2854	5.77 *	0.056	0.05	Very weak
<35 years	955 (88%)					
36–64 years	1234 (89%)					
>65 years	303 (93%)					
Education level		2710	9.58 **	0.022	0.06	Very weak
No formal education	1079 (87%)					
Primary education	1152 (91%)					
Secondary education	176 (86%)					
Tertiary education	7 (78%)					
Soil quality		2710	68.72 ***	0.000	0.16	weak
High	17 (44%)					
Low	516 (42%)					
Medium	1986 (41%)					

Variation in the effect of socioeconomic and biophysical variables was observed across the woredas. For example, increasing farm size decreased tree seedling survival in all woredas except Gursum which had increased tree survival with increased farm size.

3.5. *Gendered Management and Use of Trees*

In Kenya, analysis around intrahousehold decision-making over participation in the comparisons and division of labor in the management of trees was also conducted. The decision to get involved in the tree planting planned comparison was made by 94% of women respondents in female-headed households, 70% of women respondents in male-headed households and 70% of men respondents in female-headed and male-headed households respectively. A similar trend was observed in who was involved in deciding which tree species to plant and where on the farm to plant the trees.

We also wanted to understand the division of labor when managing the planted trees as, traditionally, men and women tend to have differing roles as well as areas of influence when it comes to tree planting [39,40]. Our results show that women-only labor was commonly used for adding mulch (44%), watering (42%), and adding manure (40%) compared to men-only labor, which was used for planting trees, fencing and pruning (Table 9).

Table 9. Gender of those involved in various management activities associated with tree planting.

	N *	Male Labor Only (%)	Male and Female Labor (%)	Female Labor Only (%)
Activities associated with tree planting				
Planting trees	1233	40	26	34
Adding manure	1064	34	26	40
Watering	1207	28	30	42
Fencing	668	50	16	33
Mulching	642	33	22	44
Pruning	613	52	16	32

* Differences in number of observations is due to "other" having been selected for those involved or households not being involved in certain activities

We also assessed if farmers were likely to invest in tree planting, which species they would consider for future planting among the surveyed households and how many seedlings they would plant. Results show that 74% of women respondents in female-headed households, 75% of women respondents in male-headed households, 77% of men respondents in female headed households and 75% of men respondents in male-headed households would consider planting additional trees in future. In terms of species consideration, *Mangifera indica* was the most considered species for future planting by 64% of all the surveyed households. Other species considered were *Melia volkensii* (24%), *Carica papaya* (21%), *Senna siamea* (20%) and *Azadirachta indica* (16%). Additional analysis revealed no major differences in the tree species desired for future planting between women and men respondents (Figure 4). We however found variation in the overall number of seedlings considered between female headed and male headed households. The maximum number of tree seedlings considered for future planting by respondents from male headed households (1000 seedlings) was twice that of respondents from female headed households (500 seedlings).

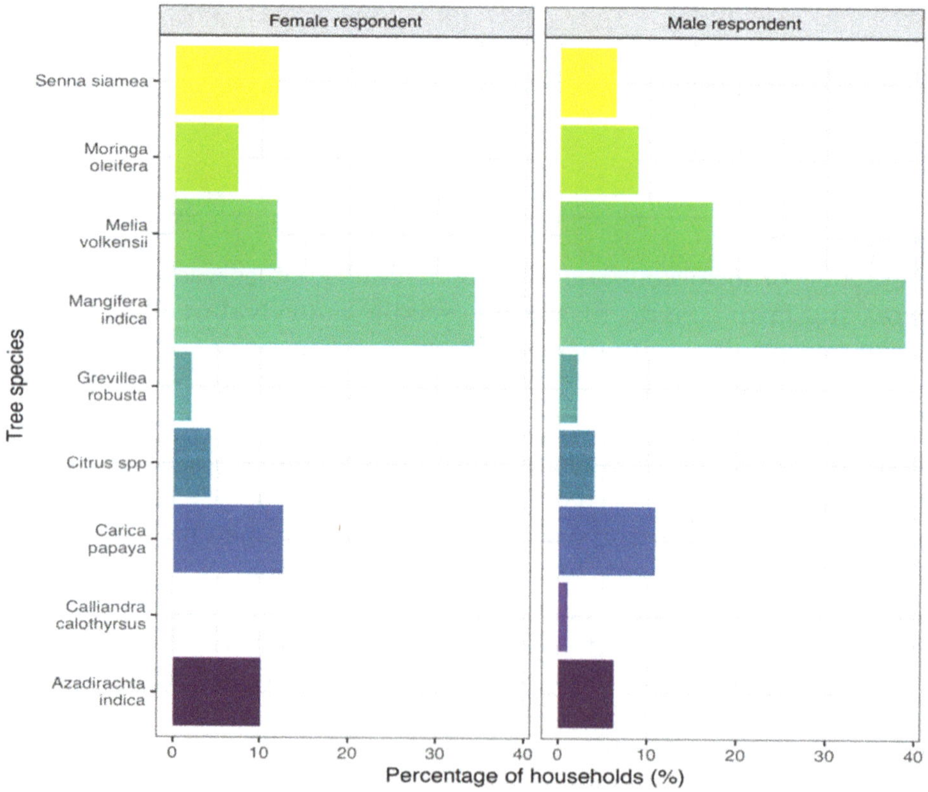

Figure 4. Species considered for future planting in female headed and male headed households by more than one percent of the surveyed households (n = 644 female respondents and 325 male respondents).

4. Discussion

The survival rate of tree seedlings in drylands is characteristically low partly due to unreliable rainfall, high levels of land degradation resulting in low soil productivity, planting of ecologically unsuitable tree species, and poor tree seedling management practices [18–20]. While planting and management practices employed can have a positive or negative effect on tree survival, the use of these practices is usually determined by knowledge, needs, perceptions and, availability and access to resources the communities implementing them [41]. The results show that planting and management practices such as watering and watering regime, manure application and seedling protection by fencing had a significant effect on tree seedling survival.

However, the type of effect (whether positive or negative) as well as the magnitude of this effect varied depending on the sites in which they were employed. Variation in average tree survival was also observed across tree species under the planting and management practices. For example, *Mangifera indica* recorded higher average survival when planted in a big hole in all the study sites in Kenya while *Melia volkensii* recorded higher survival when planted with smaller hole size (30 cm diameter by 45 cm depth). Water availability is a limiting factor to seedling survival especially in arid and semi-arid areas and the size of the planting hole can influence the water holding capacity. Bigger planting holes retain more water compared to smaller planting holes [35]. However, tree species have differing watering and management requirements. Additional discussion with farmers revealed that tree species such as *Melia volkensii* which are indigenous to arid and semi-arid areas are sensitive to waterlogging especially when young [42] and perform better when planted in smaller planting holes, while fruit trees such as *Mangifera indica* have a higher water requirement and perform better when planted in bigger planting holes.

That average seedling survival was higher when seedlings were physically protected from livestock across all the study sites in Kenya and Ethiopia was expected as browsing by livestock was reported as a key cause of seedling mortality across the study area. It is also consistent with findings by [43,44] who in their studies, found seedling survival rate was higher in the fenced plots compared to those not fenced. Farmers across the study sites used makeshift fences made up of twigs, old mosquito nets and old clothes. In some cases, seedlings whose tip had been previously browsed off, regenerated. The observed differences in the significance of seedling protection between Kenya and Ethiopia can partly be attributed to the varying grazing practices across the two countries. In Ethiopia, farmers practice communal grazing with designated grazing areas and as such is easier to protect seedlings planted. In contrast, grazing in Kenya is at the discretion of individual farmers with many letting their livestock roam in the homestead especially during the dry season when access to fodder is limited. This is not to mean that browsing by livestock is not a threat to seedling survival in Ethiopia. In fact, Ref. [45] found that browsing by livestock deterred farmers in Ethiopia from adopting high-value agroforestry.

Furthermore, seedling source plays a key role in the viability of planting material and is often overlooked in tree planting campaigns [28]. This is because seedling stage is considered the most sensitive stage in the lifecycle of trees and seedling care from early establishment is critical to survival [35,46]. Tree seedlings distributed to farmers were sourced from different nurseries across the study sites due to the large number of seedlings required and the structure of nursery enterprises in the area, that is, most nurseries are smallholder owned and run. As a result, the quality of seedlings varied depending on the nursery from which the seedlings were sourced.

Results also revealed a relationship between the household's socioeconomic characteristics and the survival of the planted seedlings in both Kenya and Ethiopia. The strength of that relationship was however either weak or very weak suggesting while farmer circumstances can influence the odds of survival of planted seedlings, they are not, on their own, determining factors for seedling survival. Instead, they interact with other factors for example the agroecological conditions prevalent in the locality. They also enhance or limit access to resources such as knowledge, manure and water, which were found to significantly increase the chances of survival for seedlings. Our study also found variation in the effect of both biophysical characteristics of the farm such as erosion status and soil quality and socioeconomic characteristic of the household at both national and local scale. For example, seedlings had a higher likelihood of survival when farmers described the soil quality as high across most of the study areas in the two countries. This was expected as high soil quality means that the soil has the requisite nutrients to support plant growth. However, in Kitui Rural within Kitui County, seedlings had a higher likelihood of survival when farmers described their soil quality as low. This suggests that farmers who considered their soil quality low were perhaps investing more in use of external inputs such as manure to improve their overall soil quality compared to farmers who considered the quality of their soil as high. This thus calls for quantitative analyses on seedling survival to be combined with in-depth qualitative work to understand the reason behind the variation across study sites.

Our results also revealed variation in survival among farmers within the same locality and farmer circumstances implying that some farmers have better experience growing and managing trees compared to their neighbors [22]. This can perhaps be explained by inherent local knowledge that has shaped how they grow and manage their trees necessitating the need for complementarity between scientific knowledge and local knowledge if good practices around tree growing that are suited to local agroecological conditions and farmer circumstance are to be identified and shared [22].

Our findings on the role of men and women in tree planting and management activities show that men-only labor was predominantly used for management activities such as pruning and fencing while women-only labor was mostly used for watering, adding manure and adding mulch. This was consistent with findings from [47], who found that traditionally women were discouraged from taking part in activities such as fencing and pruning as they were considered strenuous. The results were also consistent with previous studies that found that where both men and women were involved in tree management, women tended to be involved in the initial stages of tree establishment [13,39]. This suggests that existing customs and traditions influence the role that both men and women play in managing natural resources and as such tree planting.

We also found very little difference in tree species preferred for future planting between respondents from female-headed households and those from male-headed households in Kenya. However, there was variation in the total number of seedlings to be planted with respondents from male-headed households willing to plant twice the number of seedlings planted by respondents from female-headed households. This could partly be explained by differences in access to capital needed to purchase seedlings, labor availability and farm size between male-headed and female-headed households. These results are also consistent with the study by [48] who found that female-headed households tended to plant half the number of trees and/or shrubs planted by male-headed households and attributed the difference to variation in farm size. Tree species that are commonly found in households across the study sites were mostly preferred. Furthermore, species compared by farmers in the study were the top preferred species for future planting among the surveyed households. This suggests that farmers prioritize tree species whose survival they can assure based on their local knowledge and past experience with the tree species, species whose availability they are sure of, the value they attach to the tree species and species for which there is an already established market or demand [22].

5. Conclusions

Tree planting can have a positive effect on the environment as well as on the social-economic realities of farmers if seedlings are managed ensuring survival to maturity. However, to scale successful tree planting efforts, context-specific variables should be considered as they can impact the survival of planting seedlings either positively or negatively. Our study found that planting seedlings with manure, watering them and protecting them from physical harm significantly increased the probability of survival in Kenya while in Ethiopia, watering, the planting niche and protection from harm significantly increased the probability of survival. Our findings show that tree planting and management practices such as application of manure, mulch and fencing can significantly increase survival indicating the need to not only invest in such practices but also in continuous engagement and training with farmers and local community on how to implement the practices on their farms.

We also show that socioeconomic characteristics of the household as well as the biophysical characteristics of the farm can enhance or hinder tree seedling survival. That variation in both the type and magnitude of effect on seedling survival was recorded in most of the assessed socioeconomic variables across the study sites indicates the need to understand and be cognizant of the local agroecological conditions and the farmer circumstances in which these practices are employed. Results also show the need to consider the different role men and women play in tree planting including in the management and use of the planted trees as this can have an influence on practices that are feasible for different farmers. Finally, we show the need to consider the priorities and interests of farmers as well as the inherent local knowledge especially when it comes to the species preferred as well as the need for training on additional beneficial species for planting.

Author Contributions: Conceptualization of the work by F.S., L.A.W.; methodology by P.S., J.M., L.A.W., C.M., N.H., K.H., F.S.; formal analysis conducted by C.M., L.A.W., A.F., H.O., N.H., I.O., M.C.; data curation by A.F., H.O., E.B., E.K., C.M.; writing—original draft preparation, C.M., L.A.W., M.C., A.F., H.O., N.H., P.S., I.O., E.K., A.K., J.M., S.C., K.H., E.B., F.S., visualization, C.M., L.A.W., A.F., H.O., I.O.; supervision by L.A.W., F.S.; funding acquisition By F.S. All authors have read and agreed to the published version of the manuscript.

Acknowledgments: The authors acknowledge the collaboration of the Drylands Development Programme funded by the Ministry of Foreign Affairs (MoFA) of the Netherlands and its implementing partners in Kenya and Ethiopia, including World Vision Kenya, Netherlands Development Organization (SNV), Caritas-Kenya, Adventist Development and Relief Agency Kenya (ADRA), World Vision Ethiopia (WVE), Relief Society of Tigray (REST), Ethiopian Orthodox Church's Development and Inter-Church Aid Committee (EOC-DICAC). This project was also supported in part by the CGIAR Research Programme on Forests, Trees and Agroforestry (FTA), We would like to acknowledge the support of the enumerators and community facilitators including Angellinah Kimanzi, Caroline Mbuvi, Francisca Mutua, Felix Mbuvi, Mercy Mwea, Silas Muthuri, Stephen Maithya, and Sylvester Muendo.

References

1. Ministry of Environment and Forestry. *National Strategy for Achieving and Maintaining over 10% Tree Cover by 2022*; Ministry of Environment and Forestry: Nairobi, Kenya, 2019.
2. Reubens, B.; Moeremans, C.; Poesen, J.; Nyssen, J.; TewoldeBerhan, S.; Franzel, S.; Deckers, J.; Orwa, C.; Muys, B. Tree species selection for land rehabilitation in Ethiopia: From fragmented knowledge to an integrated multi-criteria decision approach. *Agrofor. Syst.* **2011**, 303–330. [CrossRef]
3. Coe, R.; Sinclair, F.; Barrios, E. Scaling up agroforestry requires research "in" rather than "for" development. *Curr. Opin. Environ. Sustain.* **2014**, *6*, 73–77. [CrossRef]
4. Sinclair, F.; Coe, R. The Options by Context Approach: A Paradigm shift in agronomy. *Exp. Agric.* **2019**, *55*, 1–13. [CrossRef]
5. Sinclair, F. System science at the scale of impact. In *Sustainable Intensification in Smallholder Agriculture: An Integrated Systems Research Approach*; Öborn, I., Vanlauwe, B., Phillips, M., Thomas, R., Brooijmans, W., Atta-krah, K., Eds.; Routledge: Abingdon, UK, 2017; pp. 43–57.
6. Kuria, A.; Pagella, T.; Winowiecki, L.A.; Mitiku, H.; Sinclair, F.L. Local knowledge promotes design of adaptive land restoration options that deliver multiple ecosystem services at scale. **2019**. Manuscript in Preparation.
7. Ojiem, J.O.; de Ridder, N.; Vanlauwe, B.; Giller, K.E. Socio-ecological niche: A conceptual framework for integration of legumes in smallholder farming systems. *Int. J. Agric. Sustain.* **2006**, *4*, 79–93. [CrossRef]
8. Kindt, R.; Van Damme, P.; Simons, A.J. Tree Diversity in Western Kenya: Using Profiles to Characterise Richness and Evenness. *Biodivers. Conserv.* **2006**, *15*, 1253–1270. [CrossRef]
9. Kindt, R.; Simons, A.J.; Van Damme, P. Do Farm Characteristics Explain Differences in Tree Species Diversity among Western Kenyan Farms? *Agrofor. Syst.* **2004**, *63*, 63–74. [CrossRef]
10. Crossland, M.; Winowiecki, L.A.; Pagella, T.; Hadgu, K.; Sinclair, F. Implications of variation in local perception of degradation and restoration processes for implementing land degradation neutrality. *Environ. Dev.* **2018**, *28*, 42–54. [CrossRef]

11. Coe, R.; Njoloma, J.; Sinclair, F. Loading the dice in favour of the farmer: Reducing the risk of adopting agronomic innovations. *Exp. Agric.* **2016**, *55*, 67–83. [CrossRef]
12. Kiptot, E.; Franzel, S. Gender and agroforestry in Africa: A review of women's participation. *Agrofor. Syst.* **2011**, *84*, 35–58. [CrossRef]
13. Naz, F.; Catacutan, D. Gender roles, decision-making and challenges to agroforestry adoption in Northwest Vietnam. *Int. For.* **2015**, 22–32. [CrossRef]
14. Pehou, C.; Djoudi, H.; Vinceti, B.; Elias, M. Intersecting and dynamic gender rights to néré, a food tree species in Burkina Faso. *J. Rural Stud.* **2020**, *76*, 230–239. [CrossRef]
15. Bastin, J.; Finegold, Y.; Garcia, C.; Mollicone, D. The global tree restoration potential. *Science* **2019**, *365*, 76–79. [CrossRef] [PubMed]
16. Lohbeck, M.; Albers, P.; Boels, L.E.; Bongers, F.; Morel, S.; Sinclair, F.; Takoutsing, B.; Vågen, T.G.; Winowiecki, L.A.; Smith-Dumont, E. Drivers of farmer-managed natural regeneration in the Sahel. Lessons for restoration. *Sci. Rep.* **2020**, *10*, 15038. [CrossRef]
17. Brancalion, P.H.; Niamir, A.; Broadbent, E.; Crouzeilles, R.; Barros, F.S.; Zambrano, A.M.; Baccini, A.; Aronson, J.; Goetz, S.; Reid, J.; et al. Global restoration opportunities in tropical rainforest landscapes. *Sci. Adv.* **2019**, *5*, 1–12. [CrossRef]
18. De Leeuw, J.; Njenga, M.; Wagner, B.; Liyama, M. *Treesilience: An Assessment of the Resilience Provided by Trees in the Drylands of Eastern Africa*; De Leeuw, J., Njenga, M., Wagner, B., Liyama, M., Eds.; World Agroforestry Centre: Nairobi, Kenya, 2014; ISBN 9789290593522.
19. Ndegwa, G.; Iiyama, M.; Anhuf, D.; Nehren, U.; Schlüter, S. Tree establishment and management on farms in the drylands: Evaluation of different systems adopted by small-scale farmers in Mutomo District, Kenya. *Agrofor. Syst.* **2017**, *91*, 1043–1055. [CrossRef]
20. Syano, N.M.; Wasonga, O.V.; Nyangito, M.; Kironchi, G.; Egeru, A. Ecological and Socio-Economic Evaluation of Dryland Agroforestry Systems in East Africa. 2016. Available online: http://repository.ruforum.org (accessed on 19 January 2020).
21. Kuyah, S.; Öborn, I.; Jonsson, M.; Dahlin, A.S.; Barrios, E.; Muthuri, C.; Malmer, A.; Nyaga, J.; Magaju, C.; Namirembe, S.; et al. Trees in agricultural landscapes enhance provision of ecosystem services in Sub-Saharan Africa. *Int. J. Biodivers. Sci. Ecosyst. Serv. Manag.* **2016**, *12*, 255–273. [CrossRef]
22. Derero, A.; Coe, R.; Muthuri, C.; Hadgu, K.M.; Sinclair, F. Farmer-led approaches to increasing tree diversity in fields and farmed landscapes in Ethiopia. *Agrofor. Syst.* **2020**. [CrossRef]
23. Iiyama, M.; Mukuralinda, A.; Ndayambaje, J.; Musana, B.; Ndoli, A.; Mowo, J.; Garrity, D.; Ling, S.; Ruganzu, V. Tree-Based Ecosystem Approaches (TBEAs) as multi-functional land management strategies-evidence from Rwanda. *Sustainability* **2018**, *10*, 1360. [CrossRef]
24. Mganga, K.Z.; Nyariki, D.M.; Musimba, N.K.R.; Mwang'ombe, A.W. Indigenous Grasses for Rehabilitating Degraded African Drylands. In *Agriculture and Ecosystem Resilience in Sub Saharan Africa. Climate Change Management*; Bamutaze, Y., Kyamanywa, S., Singh, B., Nabanoga, G., Lal, R., Eds.; Springer: Cham, Switzerland, 2019. [CrossRef]
25. Coe, R.; Hughes, K.; Sola, P.; Sinclair, F. *Planned Comparisons Demystified*; ICRAF Working Paper No 263; Nairobi World Agroforestry Centre: Nairobi, Kenya, 2017. [CrossRef]
26. Hughes, K.; Oduol, J.; Coe, R.; Sinclair, F.; Peralta, A. *Guidelines for Identifying and Designing Planned Comparisons*; ICRAF: Nairobi, Kenya, 2016.
27. Franzel, S.; Sinja, J.; Simpson, B. *Farmer-to-Farmer Extension in Kenya: The Perspectives of Organizations Using the Approach*; ICRAF Working Paper No. 181; World Agroforestry Centre: Nairobi, Kenya, 2014. [CrossRef]
28. Duguma, L.; Minang, P.; Aynekulu, E.; Carsan, S.; Nzyoka, J.; Bah, A.; Jamnadass, R. *From Tree Planting to Tree Growing: Rethinking Ecosystem Restoration Through Trees*; ICRAF Working Paper No 304; World Agroforestry: Nairobi, Kenya, 2020. [CrossRef]
29. Hadgu, K.M.; Kooistra, L.; Rossing, W.A.H.; van Bruggen, A.H.C. Assessing the effect of Faidherbia albida based land use systems on barley yield at field and regional scale in the highlands of Tigray, Northern Ethiopia. *Food Secur.* **2009**, *1*, 337–350. [CrossRef]
30. Lemma, B.; Kleja, D.B.; Olsson, M.; Nilsson, I. Factors controlling soil organic carbon sequestration under exotic tree plantations: A case study using the CO2Fix model in southwestern Ethiopia. *For. Ecol. Manag.* **2007**, *252*, 124–131. [CrossRef]

31. County Government of Kitui. *Kitui County Intergrated Plan 2018–2022*; County Government of Kitui: Kitui, Kenya, 2018. Available online: https://www.cog.go.ke/downloads/category/106-county-integrated-development-plans-2018-2022# (accessed on 22 April 2020).
32. County Government of Machakos. Machakos County Intergrated Development Plan II (2018–2022). 2018. Available online: https://www.cog.go.ke/downloads/category/106-county-integrated-development-plans-2018-2022# (accessed on 22 April 2020).
33. County Government of Makueni. Makueni County Integrated Development Plan (Cidp) 2018–2022. 2018. Available online: https://www.cog.go.ke/downloads/category/106-county-integrated-development-plans-2018-2022# (accessed on 14 June 2019).
34. Sola, P.; Zerfu, E.; Coe, R.; Hughes, K. *Community Visioning and Action Planning: Guidelines for Integrating the Options by Context Approach*; ICRAF: Nairobi, Kenya, 2017.
35. Mwamburi, A.; Musyoki, J. *Improving Tree Survival in the Drylands of Kenya: A Guide for Farmers and Tree Growers in the Drylands*; Kenya Forestry Research Institute: Nairobi, Kenya, 2010.
36. R Core Team. R: A Language and Environment for Statistical Computing. Vienna, Austria, 2020. Available online: https://www.r-project.org (accessed on 8 January 2020).
37. Hagazi, N.; Hadgu, K.; Sitotaw, A.; Tofu, A.; Lavoll, V.; Winowiecki, L.; Magaju, C.; Nyaga, J.; Carsan, S.; Muriuki, J.; et al. Tree Planting Data 2017—Ethiopia 2019. MELDATA, V1. 2019. Available online: ahttps://hdl.handle.net/20.500.11766.1/FK2/O9LOGI (accessed on 11 March 2020).
38. Magaju, C.; Winowiecki, L.; Nyaga, J.; Ochenje, I.; Makui, P.; Kiura, E.; Crossland, M.; Valencia, A.M.; Kuria, A.; Carsan, S.; et al. Tree Planting Data 2018—Kenya 2019. MELDATA, V1. 2019. Available online: https://hdl.handle.net/20.500.11766.1/FK2/BLHHPR (accessed on 19 March 2020).
39. Kiptot, E.; Franzel, S.; Degrande, A. Gender, agroforestry and food security in Africa. *Curr. Opin. Environ. Sustain.* **2014**, *6*. [CrossRef]
40. Meijer, S.S.; Catacutan, D.; Ajayi, O.C.; Sileshi, G.W.; Nieuwenhuis, M. The role of knowledge, attitudes and perceptions in the uptake of agricultural and agroforestry innovations among smallholder farmers in sub-Saharan Africa. *Int. J. Agric. Sustain.* **2015**, *13*, 40–54. [CrossRef]
41. Maluki, J.M.; Kimiti, J.M.; Nguluu, S.; Musyoki, J.K. Adoption levels of agroforestry tree types and practices by smallholders in the semi-arid areas of Kenya: A case of Makueni County. *J. Agric. Ext. Rural Dev.* **2016**, *8*, 187–196. [CrossRef]
42. Kenya Forest Service. *A Guide to Growing Melia Volkensii in the Dryland Areas of Kenya*; Kenya Forest Service: Nairobi, Kenya, 2018.
43. Wassie, A.; Sterck, F.J.; Teketay, D.; Bongers, F. Effects of livestock exclusion on tree regeneration in church forests of Ethiopia. *For. Ecol. Manag.* **2009**, *257*, 765–772. [CrossRef]
44. Love, B.; Bork, E.; Spaner, D. Tree seedling establishment in living fences: A low-cost agroforestry management practice for the tropics. *Agrofor. Syst.* **2009**, *77*, 1–8. [CrossRef]
45. Iiyama, M.; Derero, A.; Kelemu, K.; Muthuri, C.; Kinuthia, R.; Ayenkulu, E.; Kiptot, E.; Hadgu, K.; Mowo, J.; Sinclair, F.L. Understanding patterns of tree adoption on farms in semi-arid and sub-humid Ethiopia. *Agrofor. Syst.* **2017**, *91*, 271–293. [CrossRef]
46. Bhadouria, R.; Singh, R.; Srivastava, P.; Raghubanshi, A.S. Understanding the ecology of tree seedling growth in dry tropical environment: A management perspective. *Energy Ecol. Environ.* **2016**, *1*, 296–309. [CrossRef]
47. Crossland, M.; Valencia, A.M.P.; Pagella, T.; Magaju, C.; Kiura, E.; Winoweicki, L.; Sinclair, F. *Onto the Farm, into the Household: Intrahousehold Gender Relations Matter for Scaling Land Restoration Practices*, 2020; Manuscript Submitted for Publication.
48. Kiptot, E.; Franzel, S. Gender and agroforestry in Africa: Are women participating. *ICRAF Occas. Pap.* **2011**. [CrossRef]

4

Local Knowledge about Ecosystem Services Provided by Trees in Coffee Agroforestry Practices in Northwest Vietnam

Mai Phuong Nguyen [1,*], Philippe Vaast [1,2], Tim Pagella [3] and Fergus Sinclair [3,4]

1 World Agroforestry, Vietnam Office, Hanoi 100000, Vietnam; Philippe.vaast@cirad.fr
2 UMR Eco&Sols, CIRAD, Université Montpellier, 34090 Montpellier, France
3 School of Natural Sciences, Bangor University, Bangor, Gwynedd LL57 2DG, UK; t.pagella@bangor.ac.uk (T.P.); F.Sinclair@cgiar.org (F.S.)
4 World Agroforestry, Headquarter Office, P.O. Box 30677, Nairobi 00100, Kenya
* Correspondence: n.maiphuong@cgiar.org

Abstract: In recent decades in northwest Vietnam, Arabica coffee has been grown on sloping land in intensive, full sun monocultures that are not sustainable in the long term and have negative environmental impacts. There is an urgent need to reverse this negative trend by promoting good agricultural practices, including agroforestry, to prevent further deforestation and soil erosion on slopes. A survey of 124 farmers from three indigenous groups was conducted in northwest Vietnam to document coffee agroforestry practices and the ecosystem services associated with different tree species used in them. Trees were ranked according to the main ecosystem services and disservices considered to be locally relevant by rural communities. Our results show that tree species richness in agroforestry plots was much higher for coffee compared to non-coffee plots, including those with annual crops and tree plantations. Most farmers were aware of the benefits of trees for soil improvement, shelter (from wind and frost), and the provision of shade and mulch. In contrast, farmers had limited knowledge of the impact of trees on coffee quality and other interactions amongst trees and coffee. Farmers ranked the leguminous tree species *Leucaena leucocephala* as the best for incorporating in coffee plots because of the services it provides to coffee. Nonetheless, the farmers' selection of tree species to combine with coffee was highly influenced by economic benefits provided, especially by intercropped fruit trees, which was influenced by market access, determined by the proximity of farms to a main road. The findings from this research will help local extension institutions and farmers select appropriate tree species that suit the local context and that match household needs and constraints, thereby facilitating the transition to a more sustainable and climate-smart coffee production practice.

Keywords: agroforestry coffee; ecosystem services; shade tree species; pairwise ranking; Vietnam

---

1. Introduction

Coffee is the one of the most important commodities globally and contributes to the income of millions of smallholder farmers. Approximately twenty-five million farmers grow coffee in sixty developing countries [1]. There are approximately 10.5 million ha of coffee around the world consisting of both Arabica (*Coffea arabica*) and Robusta (*Coffea canephora*) varieties [2]. Vietnam is the second largest world producer after Brazil [3]. It is also the largest producer of Robusta coffee, accounting for 14.5% of global production and a share of about 40% of global trade [4].

Coffee was first introduced to Vietnam by the French in 1857. Between 1975 and 2010, the area of coffee planted in Vietnam increased from 134,000 ha to 513,000 ha. By 2016, the total area of coffee had reached 650,000 ha [5]. Vietnam exports over 90% of its total production but the value remains low mainly because low-quality beans rather than processed coffee are exported at a low price. The annual export volume was approximately 1.8 million tons with a value of USD 3.5 billion in 2018 [6].

Whilst Robusta is the most popular coffee variety grown in Vietnam, it is mainly Arabica that is cultivated in mountainous areas, generally at elevations above 600 m above sea level (masl). Arabica production accounts for approximately 7% of the total coffee area of Vietnam [7], including approximately 3995 ha in Dien Bien [8] and 17,128 ha in Son La [9] provinces in 2018.

Smallholder farmers have converted large areas of annual crops to coffee in northwest Vietnam, which has transformed their livelihoods from subsistence to commercial commodity dependent [10]. Most of the Arabica plantations in the northwest are full sun monoculture on sloping land, which is not sustainable in the long term because of soil degradation. The long-term sustainability of unshaded coffee is likely to be limited given climate change, resulting in higher temperatures and more variable weather patterns. Increasing temperatures, uneven precipitation and increased frequency and severity of extreme weather events such as storms, floods or frost may have negative impacts on Arabica coffee production.

Coffee was originally cultivated under moderate to heavy shade in Ethiopia, which was gradually changed over much of the area, grown under light or no shade to increase coffee yield, at least in the short term. However, reducing shade in coffee plots makes them more vulnerable to water run-off [11] and soil erosion [12] biodiversity loss [13] and climate change [14]. Shade coffee, which is a form of agroforestry where coffee is cultivated under a tree canopy [15], is generally considered to be more environmentally friendly. Studies have shown increased biodiversity, higher levels of natural pollination [16], greater erosion control, and higher carbon sequestration [13,17–19]. Boundary tree planting in coffee farms has potential for climate mitigation and adaptation [14,20].

Expanding the area of shaded coffee in Northwest Vietnam offers a potential mechanism to provide more sustainable cultivation, both economically and environmentally, when compared with either unshaded coffee or other monoculture crops such as maize. There has been an increased interest, globally, in shaded coffee given the sensitivity of coffee, particularly Arabica, to climate change [21,22]. Coffee cultivation, with between 20 to 40% shade and appropriate species and management, has been found to have a regulated microclimate which, in turn, leads to high coffee yield and quality [23]. However, poorly designed tree cover in coffee can lead to low coffee productivity [24]. Shade systems provide additional benefits beyond the direct impacts on coffee production in the shape of a more balanced supply of ecosystem services [25,26]. Shade trees in coffee can increase the sustainability of coffee production as well as the economic resilience of households to agricultural price volatility through production and revenue diversification. Furthermore, they facilitate the adaptation of rural communities to climate change via the adoption of more 'climate-smart' farming practices.

There have been a number of studies looking at farmers' knowledge about coffee agroforestry [27–29] however, to date, there has been no study assessing farmers' attitudes and knowledge associated with coffee agroforestry in northwest Vietnam. Understanding farmers' knowledge of the benefits that trees potentially provide alongside any trade-offs or disservices is critical for developing interventions that enable more resilient landscapes [30], whilst reconciling production and conservation objectives. This is particularly important in the context of northwest Vietnam given the broad spectrum of ethnicities at play and the way in which factors such as gender might influence tree species selection.

To address this knowledge gap, the principal aim of this study was to explore farmers' perceptions of the range of benefits associated with shade trees in coffee systems and then to use this information to customize a decision-support tool (the 'Shade Tree Advice tool' developed by researchers from World Agroforestry, French Agricultural Research Centre for International Development (CIRAD) and International Institute of Tropical Agriculture (IITA)) to local conditions in Vietnam. The original tool was developed using data from Uganda, Ghana and China; http://www.shadetreeadvice.org [31,32]. The 'Shade Tree Advice' tool is aimed primarily at extension services, members of farmers' cooperatives and non-governmental organizations working in northwest Vietnam. The tool supports farmers in selecting the most appropriate tree species that are both adapted to local ecological conditions and match household needs and constraints.

2. Materials and Methods

2.1. Study Sites

Son La and Dien Bien are the only two provinces in northwest Vietnam where coffee is planted, and Son La is the second largest Arabica coffee producer in Vietnam. This study was conducted in seven communes of Son La and Dien Bien (Figure 1). The criteria for commune selection were coffee planting area, ethnicity, proximity to main road to compare the different characteristics of coffee agroforestry practices, and willingness to share perceptions about tree services and tree species ranking. Information on these criteria were provided by district extension workers. The communes were selected from the most well known coffee planting areas in two provinces. Coffee agroforestry households were suggested by commune extension workers. As Thai, H'mong and Kinh are the major ethnic groups in the region, households from these three ethnicities that were actively involved in coffee agroforestry were selected for the surveys with specific care taken to ensure gender balance amongst interviewees. The Kinh are the largest ethnic group in Vietnam but account for less than 30% of the population in the northwest. Kinh people generally live at lower altitudes (below 600 m) while Thai people are generally found at altitudes of 500 to 800 m and H'mong people generally live in areas above 800 m. The proximity to main roads were categorized into three levels: 0–2, 2–5 km or further than 5 km. As farmers often sell their fruit products, such as mango or plum, along the main road, this represents an indicator of their proximity to markets.

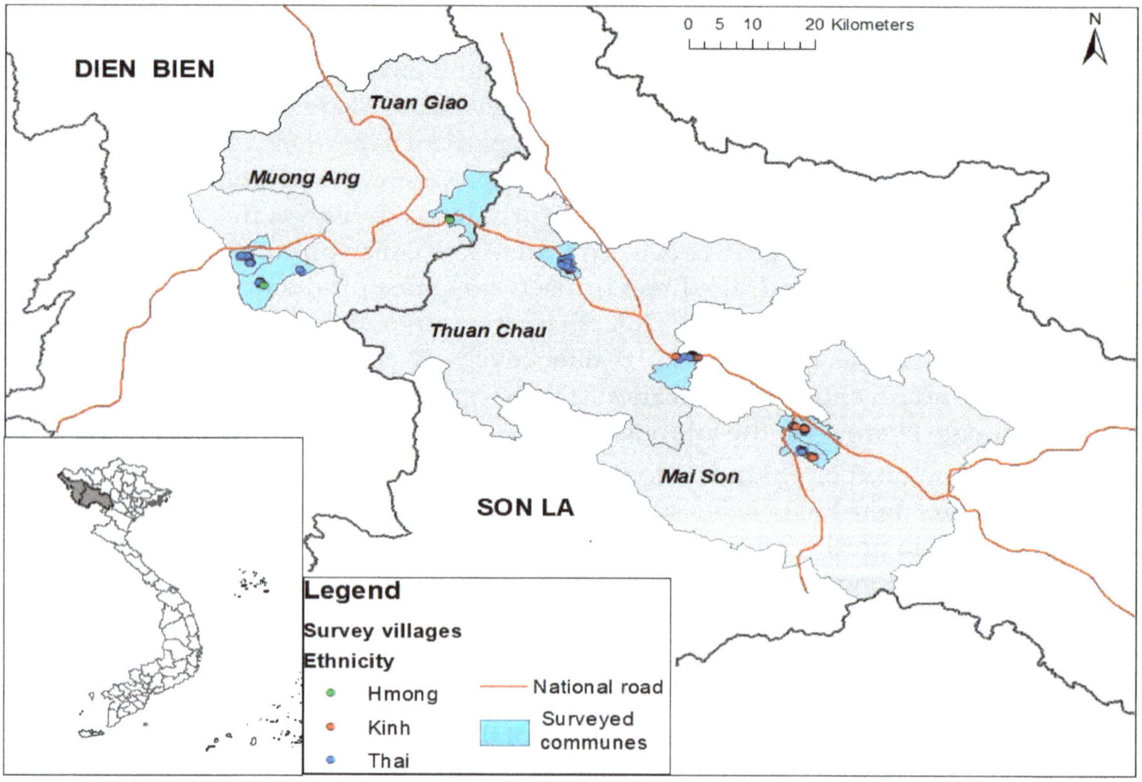

Figure 1. Map of the study sites and the communes surveyed in the Son La and Dien Bien provinces of northwest Vietnam.

The elevation of the study sites ranges from 300 to 2000 masl on 5 to 50% slopes. Annual temperature ranges from 21 to 24 °C, and annual precipitation ranges from 1500 to 2500 mm. Rainy season is from

April to September. The main soil type is Ferrasols, and average soil layer thickness ranges from 50 cm to 1 m. These natural conditions are favorable for Arabica coffee. The upper limit frost in winter is the most serious constraint for Arabica coffee cultivation at high elevations. In addition to coffee, the main agricultural land uses are annual crops (upland rice (*Oryza sativa*), maize (*Zea mays*), fruit tree plantations such as longan (*Dimocarpus longan*), plum (*Prunus salicina*), and mango (*Mangifera indica*). There are also limited areas of planted forest and secondary natural forest.

2.2. Data Collection

Data were collected through two rounds of surveys between March to May 2018, following a well established tree attribute ranking methodology [31,33]. The first survey consisted of a household interview and tree inventory in coffee agroforestry farms in March 2018. A total of 124 households were surveyed; consisting of 16 H'mong farmers, 25 Kinh farmers and 83 Thai farmers (with those proportions largely representative of the ethnic distributions at the study sites). Sicty-eight men and 56 women participated in the interview process. A sub-section of 50 farms were inventoried to provide a baseline of potential shade tree species found in existing coffee agroforestry plots. The survey questions captured social and economic characteristics, coffee plot descriptions, including information about trees in agroforestry plots, including those with coffee as the major crop and non-coffee plots (where annual crops or tree plantations were the main productive component), the benefits that trees provide to coffee, and coffee management. Additional information on farmer perceptions of ecosystem services and disservices was captured via focus group discussions. These were also used to explore how perceptions might vary in relation to farmers' age, ethnicity and gender.

From the first survey, all the tree species on coffee farms were documented and the most dominant 25 species were selected for the second survey in May 2018. Based on the initial discussions with farmers, the perceived benefits of shade trees were grouped into 11 broad topic areas (which effectively acted as proxies for ecosystem services). These topics were tree effects on: coffee production; soil moisture, soil fertility, soil erosion, biodiversity, micro-climate regulation, wind control, frost control, effects of and on shade provision, the value of mulch provision and effects on the use of fertilizer (which relates to soil fertility and greenhouse gas emission). The same farmers involved in the previous survey were invited for the second set of interviews. There were 118 farmers from three ethnic groups Kinh, Thai and H'mong, of which were 64 men and 54 women, that were involved in the second survey. The second survey interview focused on ranking exercises [33]. Interviewees were asked to select up to 10 preferred species to grow with coffee. They must have been managing, or at least been familiar with these selected species. Pictures of both tree species and tree services were printed on cards with which farmers then ranked the performance of their selected trees for each topic area (stating whether their effect was positive or negative). Results of from individual farmer ranking of trees were recorded in the ranking sheet for later analysis. During this exercise, farmers also provided the explanation for their rankings.

A final workshop was conducted in July 2018 in Mai Son district (Son La) and Muong Ang district (Dien Bien) with 25 farmers to feedback the analysis results.

2.3. Data Analysis

The results of tree species inventory were analyzed using the principles of BiodiversityR packages in R studio and combined with Excel for visualization. Species accumulation was presented with the trendline of a data plot using logarithm function and first order Jackknife asymptote [34]. Several species accumulation analyses were made for three ethnic groups, coffee farms and non-coffee farms (for example, maize, rice, orchard plantation) to compare the biodiversity among Thai, H'mong, Kinh farms, coffee and non-coffee agroforestry plots. Coffee monoculture plots were excluded from analyses, as there were no tree species in those plots.

Following the ranking in the second survey, an analysis was undertaken using the Bradley–Terry model in R studio [35]. Ranking was converted into pairwise comparison as input data for the model.

The model was run for eleven tree services for three ethnic groups, three distances to road (near, medium and far), and gender (male and female). Species, ranked less than 10 times, were excluded from the results. The scores reflect the comparison of performance rather than the absolute values. Scores were normalized to values between 0 and 1, where 1 is the maximum value for the best tree species. Ranking data was uploaded to the online database at www.shadetreeadvice.org. This study followed the same methods as those used for other databases from Ghana [31] and China [32] to ensure that the results were comparable.

3. Results

3.1. Farm Characteristics of Coffee Agroforestry Systems

All the interviewed farmers were between 25 and 50 years old with a mean age of 39 years. Among the three ethnic groups, Kinh people had the highest mean annual income while Thai farmers had the next highest and H'mong the lowest at approximately USD 1762 per year. Mean farm size was about 1.5 ha; Kinh farmers have smaller cultivation areas compared to those of the H'mong group. Coffee was cultivated on about two thirds of their lands (approximately 1 ha on average) and the remaining areas were dedicated to annual crops for family and livestock consumption. Almost all their coffee areas were 'shaded' coffee, i.e., coffee intercropped with trees such as fruit trees, timber trees, nut trees or *L. leucocephala* (Table 1). These were considered coffee agroforestry plots. Fruits were often sold at home or on the side of the main road. Timber and coffee were collected by middlemen at home or sometimes at the local market. All the Kinh villages were near the main road, at a distance of 0–2 km. Thai farmers typically lived at distances of 0–5 km from a main road. H'mong farmers mostly lived far from main roads (5–7 km) except for the H'mong group in Toa Tinh commune, Dien Bien province, because of the newly built highway near their village.

Table 1. Social and plot characteristics of the coffee farms; data are the mean values with standard errors.

Social and Farm Characteristics	Total Population	Kinh	Thai	H'mong
Total number of respondents	124	25	83	16
Mean agricultural income (USD/year/household)	4571 ± 390	8762 ± 1286	3810 ± 324	1762 ± 276
Mean farm area (ha)	1.50 ± 0.09	1.60 ± 0.26	1.30 ± 0.10	1.90 ± 0.30
Mean number of coffee plots	2-3	2-3	2-3	2-3
Mean area of coffee plots (ha)	1.00 ± 0.09	1.10 ± 0.16	1.00 ± 0.08	1.20 ± 0.20
Mean area of agroforestry coffee plots (ha)	0.80 ± 0.06	1.00 ± 0.12	0.80 ± 0.07	1.00 ± 0.21
Mean distance to main road		0–2 km	0–5 km	0–3 km (Tuan Giao district) or 5–7 km (Muong Ang district)

3.2. Tree Species Inventory in Coffee Agroforestry Systems

Forty-seven tree species were identified during the farm inventories. These included fruit trees, timber trees, nut trees, and shade trees. All inventoried farms were located from 580 to 1000 masl. Two native timber tree species could not be botanically identified and are only referred to by their local name. Nearly half of the tree species were native (Table 2). Frequencies of tree species were calculated by dividing the number of farms a tree species was observed on by the total number of farms inventoried and expressed as a percentage. Many species were quite rare (37% were encountered on only 1% of the coffee farms while a further 37% were found on 2–10% of coffee farms). The most abundant species, found on 13–46% of farms, constituted 26% of the total species encountered. The majority of the most abundant species were commercial fruit and nut trees except for the leguminous *L. leucocephala* and *Melia azedarach* (a valuable timber tree). There were 25 exotic species out of the 48 species recorded. Longan and mango are native species, however, only grafted hybrid varieties are being planted on farms.

Table 2. Characterization of the tree species mentioned by farmers as growing on their coffee farms and observed during the farm inventory in the provinces of Son La and Dien Bien, northwest Vietnam.

No.	English Name	Scientific Name	Dominant Function	Exotic/Native	Frequencies of Species (%)
1	Longan	*Dimocarpus longan*	Commercial fruit	Exotic	46
2	Plum	*Prunus salicina*	Commercial fruit	Native	43
3	Mango	*Mangifera indica*	Commercial fruit	Exotic	42
4	Leucaena	*Leucaena leucocephala*	Shade	Exotic	23
5	Jackfruit	*Artocarpus heterophyllus*	Family consumption/ Commercial fruit	Native	22
6	Pomelo	*Citrus grandis*	Commercial fruit	Exotic	22
7	Melia	*Melia azedarach*	Commercial timber	Exotic	20
8	Peach	*Prunus persica*	Commercial fruit	Native	15
9	Macadamia	*Macadamia* spp.	Commercial nut	Exotic	15
10	Avocado	*Persea americana*	Commercial fruit	Exotic	13
11	Docynia indica	*Docynia indica*	Commercial fruit	Native	9
12	Guava	*Psidium guajava*	Commercial fruit	Native	8
13	Orange	*Citrus sinensis*	Commercial fruit	Exotic	6
14	Lime	*Citrus aurantifolia*	Family consumption	Native	6
15	Litchi	*Litchi chinensis*	Commercial fruit	Exotic	6
16	Eucalyptus	*Eucalyptus* spp.	Commercial timber	Exotic	6
17	Vernicia montana	*Vernicia montana*	Commercial timber	Native	6
18	Apricot	*Prunus mume*	Commercial fruit	Native	5
19	Tamarind	*Tamarindus indica*	Family consumption	Native	4
20	Chukrasia	*Chukrasia tabularis*	Commercial timber	Exotic	4
21	Dalbergia	*Dalbergia tonkinensis*	Commercial timber	Exotic	4
23	Manglietia	*Manglietia conifera*	Commercial timber	Exotic	3
22	Michelia	*Michelia mediocris*	Commercial timber	Exotic	3
24	Local pear	*Pyrus granulosa*	Family consumption	Native	2
25	Oroxylum indicum	*Oroxylum indicum*	Timber/flowers	Native	2
26	Lucuma	*Pouteria lucuma*	Family consumption	Native	1
27	Fig	*Ficus auriculata*	Timber/fruit	Native	1
28	Baccaurea sapida	*Baccaurea sapida*	Timber/fruit	Native	1
29	Bischofia javanica	*Bischofia javanica*	Timber	Native	1
30	Papaya	*Carica papaya*	Family consumption	Native	1
31	Star apple	*Chrysophyllum cainito*	Family consumption	Exotic	1
32	Star fruit	*Averrhoa carambola L.*	Family consumption	Exotic	1
33	Pomegranate	*Punica granatum*	Family consumption	Exotic	1
34	Indian Jujube	*Indian Jujube*	Family consumption	Native	1
35	Pine	*Pinus latteri*	Commercial timber/resin	Exotic	1
36	Styphnolobium japonicum	*Styphnolobium japonicum*	Timber	Native	1
37	Teak	*Tectona grandis*	Timber	Exotic	1
38	Alstonia scholaris	*Alstonia scholaris*	Timber	Exotic	1
39	Syzygium nervosum	*Syzygium nervosum*	Timber/leaf/flower	Exotic	1
40	Khaya senegalensis	*Khaya senegalensis*	Timber	Exotic	1
42	Dillenia Indica	*Dillenia Indica*	Timber	Exotic	1
43	Tea	*Camellia sinensis*	Leaf	Exotic	1

Table 2. *Cont.*

No.	English Name	Scientific Name	Dominant Function	Exotic/Native	Frequencies of Species (%)
44	Zanthoxylum rhetsa	*Zanthoxylum rhetsa*	Seed/timber	Native	1
45	Agarwood	*Aquilaria malaccensis*	Timber/resin	Exotic	1
46	Schima wallichii	*Schima wallichii*	Timber	Native	1
47	Local timber tree (mý) *		Timber	Native	1
48	Local timber tree (thro) *		Timber	Native	1

* Scientific names of those species that have not been identified.

Trees were planted in different settings such as in rows or scattered within the coffee plots or along plot boundaries. Coffee–tree intercropping practices were characterized as four common types including: coffee–fruit trees, coffee–timber trees, coffee–nut trees, and coffee–*L. leucocephala* that could be combined on one farm. In all instances, trees were planted at regular spacing in rows if there were one to two species being intercropped with coffee. Trees were planted at 5 m × 5 m for plum or mango, macadamia, *L. leucocephala*, and up to 10 m × 10 m for avocado. If more than three species were intercropped with coffee, farmers often planted them scattered between coffee rows or along the boundaries of the coffee plots. Most of the coffee-multiple fruit tree systems were found in home gardens and coffee—native timber trees were more frequently planted on plots far from the homestead. Tree products in both cases were for family consumption. Coffee with commercial trees were planted at medium to close distance to home for the ease of management, harvest and transport.

3.3. Tree Species Richness in Coffee Farms

There were significant differences in the species and area accumulations between coffee agroforestry plots and non-coffee agroforestry plots ($n = 124$, Figure 2). Non-coffee plots were fields with annual food crops or tree plantations as the main productive component. First order Jackknife asymptote showed that the extrapolated value of the total tree species richness of agroforestry coffee plots was 48. All tree species encountered during the inventories also appeared in coffee agroforestry plots. Species richness and the accumulated area of coffee agroforestry plots were almost double those of non-coffee plots as was the species richness per ha.

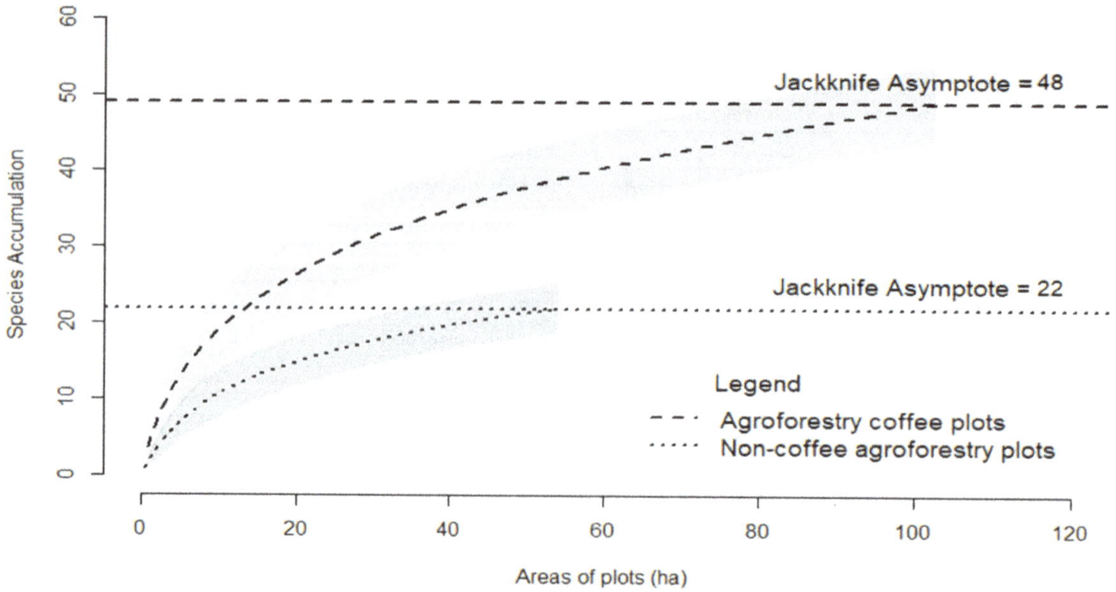

Figure 2. Species accumulations and first order Jackknife asymptotes by area from coffee agroforestry plots and non-coffee plots (orchards, annual crops, timbers).

Higher species richness was found on Thai farmers' coffee agroforestry plots compared to those of the Kinh and H'mong ethnic groups. First order Jackknife asymptotes from Kinh coffee plots was the same as that of H'mong coffee plots (Figure 3). This reflects the higher number of tree species in coffee plots of the Thai group in comparison to other groups as the Thai people live at mid-elevation which is suitable for most of the species encountered at all elevations (Table 2).

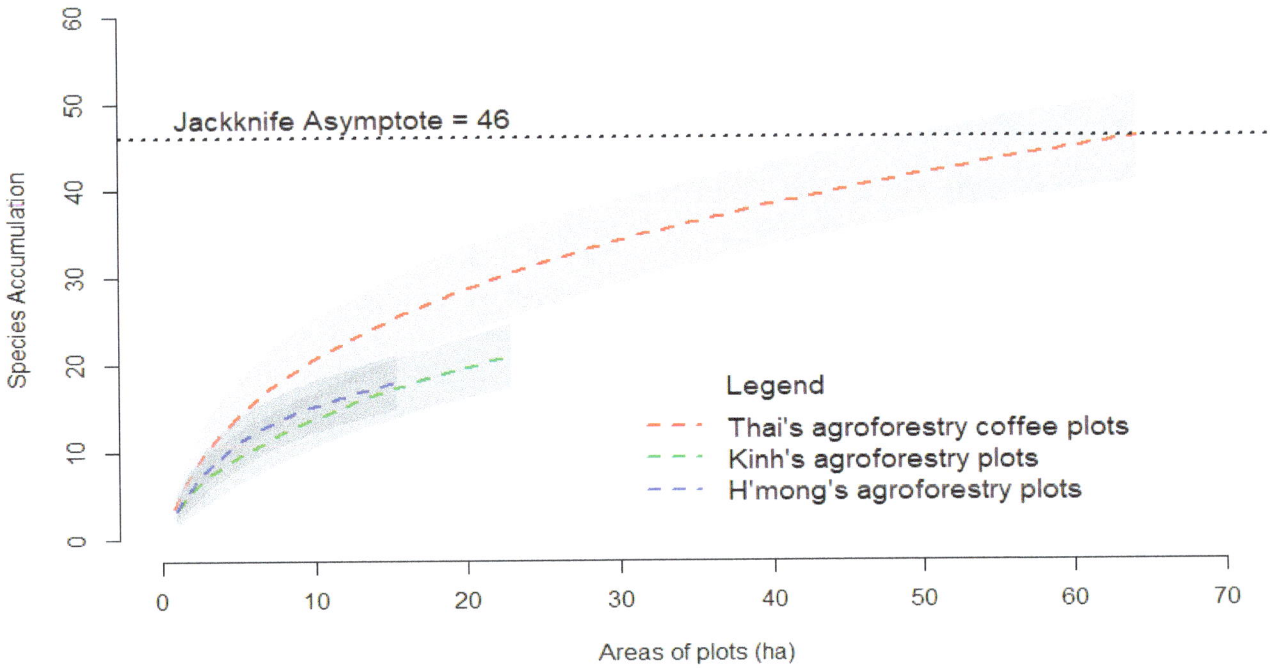

Figure 3. Species accumulations and first order Jackknife asymptotes by area from all coffee agroforestry plots of Thai, H'mong and Kinh ethnic groups.

3.4. Variation in Shade Tree Species Composition of Coffee Farms in Relation to Proximity to Main Roads

When farmers selected tree species to be intercropped with coffee in agroforestry plots, the primary factor that influenced their selection was their commercial value, i.e., their value for the provision of fruits or timbers that could be sold. There was a number of variables related to this factor including market price, proximity to market (indicated in the present study, by proximity to a main road because farmers usually sold fruits on the side of the main road). Among the three ethnic groups, the H'mong were generally at the furthest distance from the road compared with the other two groups. The Kinh people had good road networks because they live in the lower altitudes. The proximity to the road was identified as a critical factor for determining tree species selection by farmers. Fifty-two of the sampled farms were in communes near the main road (0–2 km), 54 farms were at medium distance to the main road (2–5 km) and 18 farms were were far from the main road (5–7 km) (Figure 4). High-value commercial species (fruits, timbers and nuts) were commonly integrated with coffee in the close and medium distance farms (accounting for 81 and 66% of the total trees species, respectively). Commercial species were less commonly (54%) far from the main road. In these farms, more native timbers were intercropped in the coffee plots. The leguminous species, *L. leucocephala*, was more common in the farms far from the main road.

3.5. Farmer Perspectives on Ecosystem Services Associated with Trees in Coffee Systems

Farmers had in-depth knowledge of the benefits trees provided to coffee. Most of them were aware of observable ecosystem services such as reducing soil erosion, improving soil fertility, enhancing biodiversity, preventing damage from wind and frost, and providing shade and mulch.

Some ecosystem services were linked to one another, such as shade provision, mulch provision and soil moisture or frost control and wind control. Trees with big leaves and wide crowns were most generally associated with these benefits. Interestingly, farmers had limited experience or knowledge on the effects of tree species on coffee quality. In addition, their responses suggest that they were not highly knowledgeable about the light and nutrient interactions compared to other services with more than half of their answers consisting of 'don't know', indicating knowledge gaps (Figure 5).

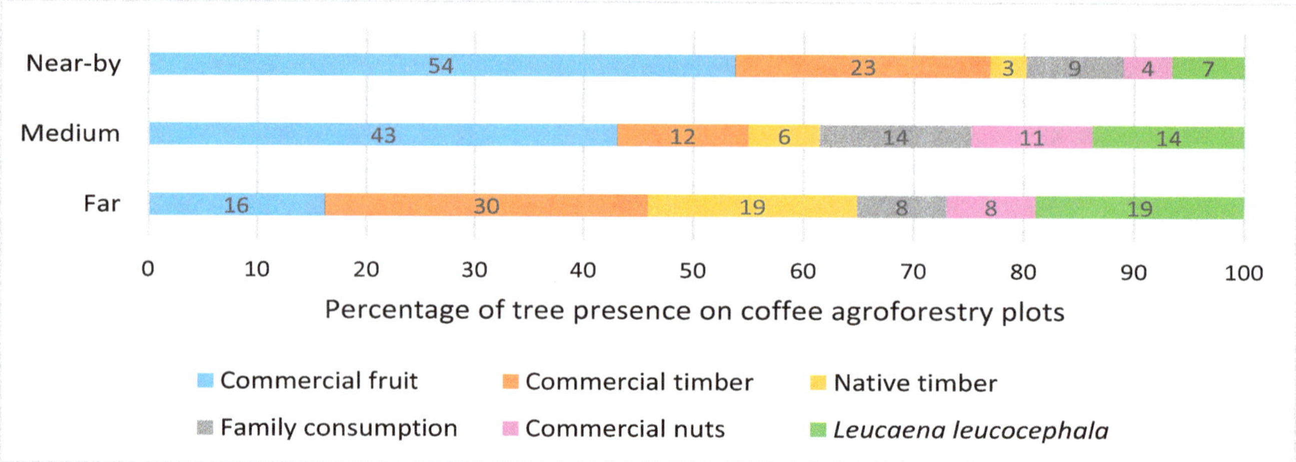

Figure 4. Percentage of tree species present in coffee agroforestry plots with respect to proximity to the road (Note: number of farmers near the road was 52, medium to road was 54, far from road is 18).

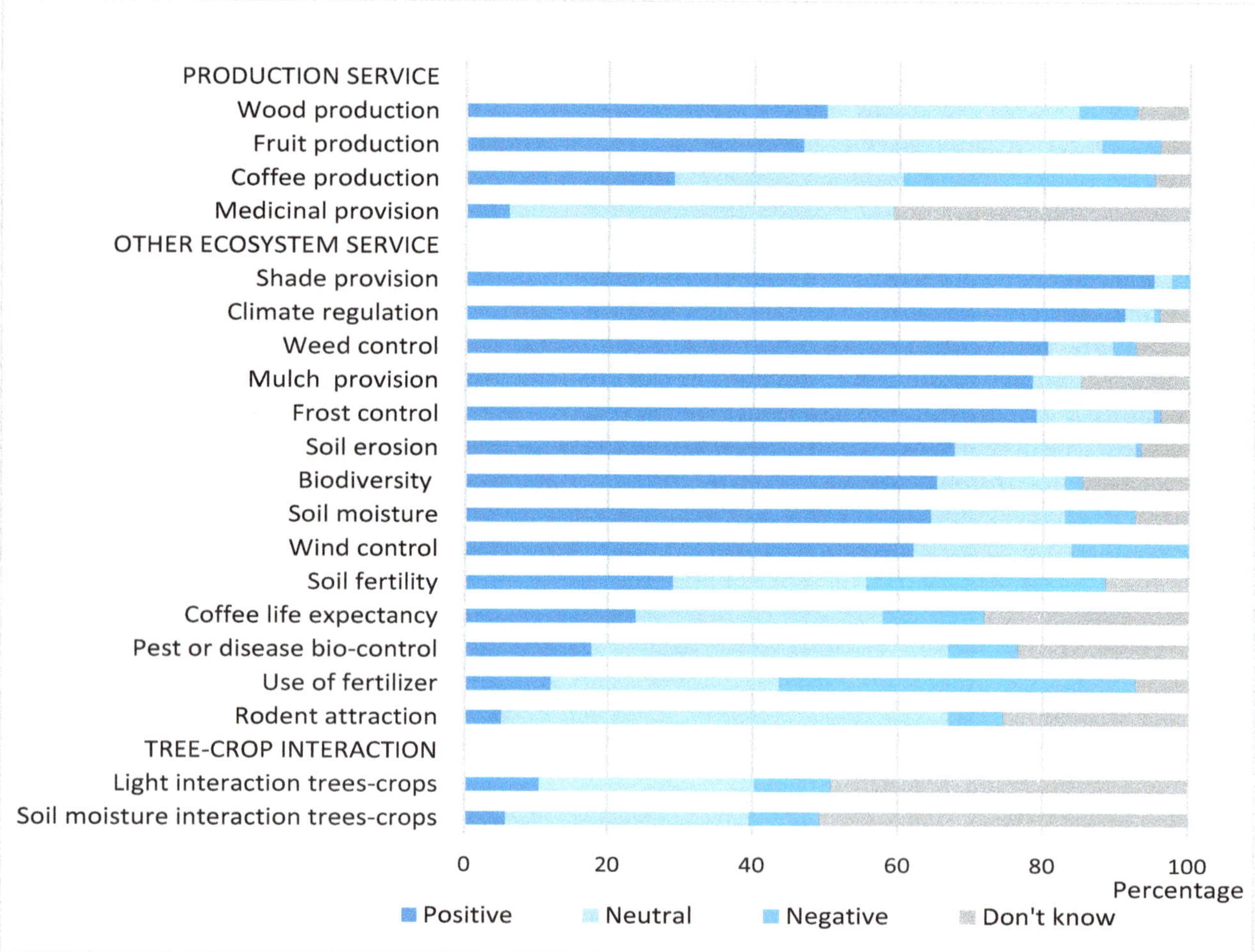

Figure 5. Farmers' perspectives on tree services to coffee in agroforestry practices in the two provinces of Son La and Dien Bien, northwest Vietnam.

When taking ethnicity into account in the analyses, the proportion of four answers were quite similar among the three ethnic groups, except for coffee production, soil fertility, use of fertilizers (Kinh farmers were more negative about these), wood production (H'mong farmers were more positive about this), biodiversity (H'mong farmers were more negative about this), and climate regulation (Thai and Kinh farmers were more positive about this) (Appendix A).

3.6. Tree Species Pairwise Ranking

Among 48 tree species encountered in coffee plots, only 25 species were intercropped in more than 2% of coffee farms. These species were selected for tree pairwise ranking exercises. Accumulated values for each species are shown in Figure 6. The leguminous shade tree species (*L. leucocephala*) was the highest valued species based on its services provided to coffee. *Dimocarpus longan* was the second highest valued species but was the most common species intercropped across coffee farms, followed by mango due to the high commercial value of their fruit. Timber trees were not preferred as farmers stated that timber trees compete for nutrients and water with coffee. Details on the ranking for each tree service are presented in Appendix B.

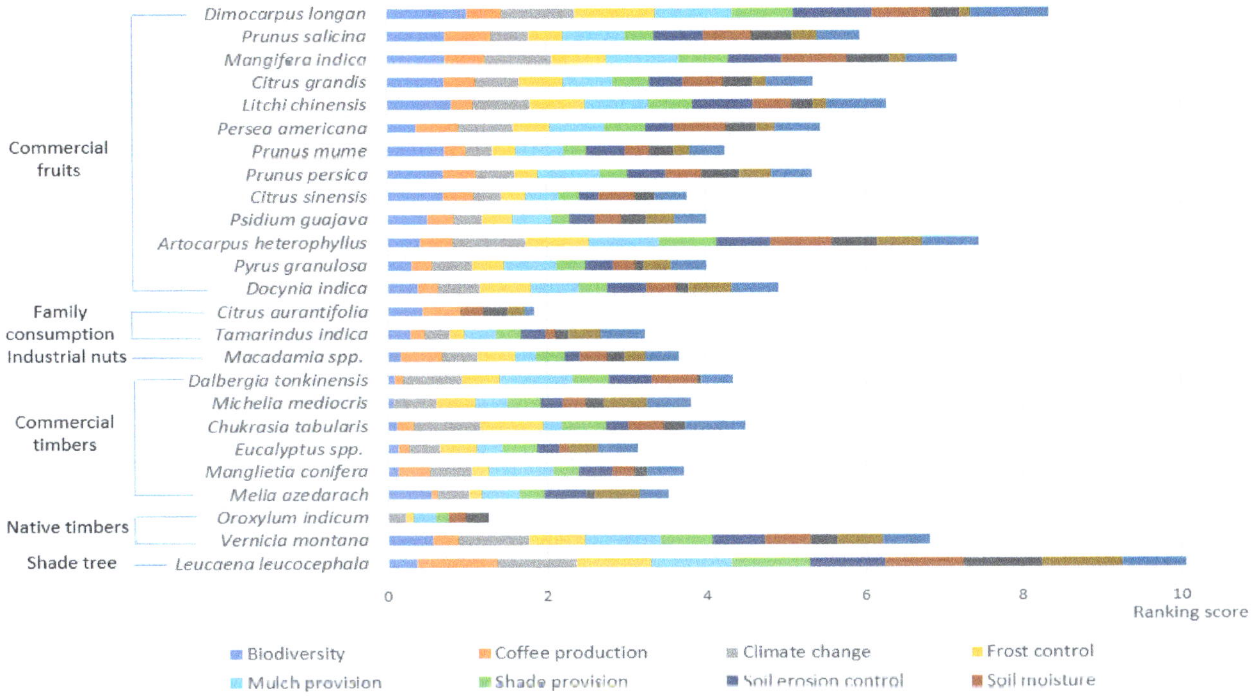

Figure 6. Accumulated tree species ranking scores for various ecosystem services based on the whole group of interviewed male and female farmers in the provinces of Son La and Dien Bien, northwest Vietnam.

Variation in how ranking changed in relation to distance to road was also analyzed and showed differences in tree ranking for coffee production (Table 3). The farmer group living near the main road ranked fruit trees such as plum and longan highly because the fruit selling prices were high, and it was easy to sell them by the roadside. They also liked the shade provided by these trees. They explained that the economic value of *L. leucocephala* was much lower than commercial fruit trees, although this tree species was good for coffee. Therefore, when their coffee farms were far from the main road, where market forces held less sway, they generally grew more *L. leucocephala* with coffee.

Table 3. Best tree species ranked by farmers with respect to the tree services provision to coffee in agroforestry practices (grouped by ethnicity and proximity to main road).

Tree Services	By Ethnic Group			By Proximity to Road			Overall Ranking
	Kinh	Thai	H'mong	Nearby	Medium	Far	
Coffee production	Leucaena *	Leucaena	Leucaena	Plum	Leucaena	Leucaena	Leucaena
Soil fertility	Leucaena *	Leucaena	Leucaena	Leucaena	Leucaena	Leucaena	Leucaena
Shade provision	Longan	Leucaena	Leucaena	Longan	Leucaena	Leucaena	Leucaena
Climate regulation	Longan	Leucaena	Mango	Longan	Leucaena	Leucaena	Leucaena
Soil moisture	Longan	Leucaena	Leucaena	Jackfruit	Leucaena	Leucaena	Leucaena
Soil erosion	Longan	Longan	Longan	Longan	Longan	Longan	Longan
Wind control	Longan	Longan	Longan	Longan	Longan	Longan	Longan
Frost control	Longan	Leucaena	Longan	Longan	Leucaena	Longan	Longan
Mulch provision	Plum	Leucaena	Longan	Longan	Jackfruit	Leucaena	Longan
Biodiversity	Longan	Longan	Mango	Longan	Longan	Longan	Longan
Use of fertilizer	Jackfruit	Leucaena	Leucaena	Leucaena	Leucaena	Peach	Leucaena

(Note: Leucaena: *L. leucocephala*, Longan: *D. longan*, Jackfruit: *A. heterophyllus*, Plum: *P. salicina*, Peach: *P. percia*; Mango: *M. indica*) (*) indicates low level of confidence in ranking results.

The leguminous tree, *L. leucocephala*, was introduced in coffee plantations first by the French since they introduced coffee cultivation in Vietnam in the 19th century. In the 1980s, *L. leucocephala* was promoted to be intercropped with coffee by Son La coffee company in Son La and Dien Bien provinces. After the company went bankrupt, farmers continued planting shade trees by themselves. More recently, some of them still keep *L. leucocephala* in their coffee plantations while an increasing number of them replaced it with commercial fruit trees. Farmers in Son La replaced *L. leucocephala* with fruit trees such as plum and longan which provide more immediate and constant income.

Ninety percent of *L. leucocephala* was found in Dien Bien province and grown by Thai farmers ($n = 29$). A few H'mong farmers also planted *L. leucocephala* intercropped with coffee and other trees. Some H'mong farmers stated that they learned the technique and took the seedlings from fellow farmers in Thai villages. Son La province is famous for fruit production while fruit tree cultivation has not been well developed in some areas of Dien Bien province despite some fruit tree support programs from the local government. Farmers in Dien Bien said that it was hard to manage fruit trees as young seedlings and fruits during harvesting season were often stolen. Harsh weather like drought and frost were also mentioned as major constraints in planting fruit trees. Most fruits for domestic consumption in Dien Bien were imported from Son La and other provinces.

Results of pairwise comparisons between men and women did not show any significant difference with the overall ranking. Men and women had quite similar average ranking for the best tree species with respect to all tree services. The main differences in their ranking were related to timber trees with respect to soil erosion, mulch provision, use of fertilizer and coffee production (Appendix C). Rankings from the three ethnic groups showed remarkable differences, mainly for timber trees and the local fruit tree son tra (*D. indica*) (Appendix D).

Longan (*D. longan)* was a highly popular fruit tree species in the northwest. From the tree inventory, *D. longan* was the most abundant species, appearing in 46% of surveyed farms. Most of the *D. longan* trees were planted by Kinh and Thai farmers. *D. longan* was highly ranked for coffee agroforestry practices because of its good services to coffee such as providing shade, litterfall to keep the soil humidity, reducing wind and frost impact.

All ranking data were uploaded to the online tool supporting decision-making process for selecting shade tree species on coffee farms together with data from other countries. The tool is available at https://www.shadetreeadvice.org/. Vietnamese data can be found under the Vietnam country category, northwest region, Arabica variety and medium zone. After the users select the ecosystem services that they wish to prioritize, the tool produces outputs suggesting appropriate shade tree species to be considered to be integrated into their coffee practices.

4. Discussion

4.1. *Tree Diversity in Coffee Agroforestry Systems*

Shaded coffee systems were well recognized for biodiversity enhancement according to the farmers' perception. Tree species richness from agroforestry coffee plots was almost double that of non-coffee plots, supporting the notion that shaded coffee systems have higher tree diversity compared to other agricultural land uses [23]. The highest diversity of trees found on coffee plots were associated with Thai farmers (Figure 3), which can be explained by the favorable conditions of their mid-elevation location for various tree species compared to other two groups. Thai people generally live at a medium elevation from 600 to 800 m and are near or at medium distance to main roads.

Among 48 recorded species from the tree inventory, 10 species are the most abundant as being planted in more than 10% of total inventoried coffee plots. This is relatively low compared to the 162 tree species found in the Yunnan, China [32], 165 tree species in Mount Kenya [36] and 100 species associated with coffee in Muranga, Kenya, 29 of which were sufficiently abundant to be ranked [28]. However, this finding was similar to the 36 species that farmers had knowledge of in coffee farms in Costa Rica [27] and the number of species found by Nyaga et al. [37] in agroforestry in the Rift Valley in Kenya with 44 species (55% native). However, the primary functions of trees in Vietnam were commercial fruits with a ready market while the most abundant species in the Rift Valley in Kenya were firewood and fast growing fodder or fertilizer species [37].

4.2. *Tree Ranking and Farmer's Perception on Tree Services or Disservices*

In northwest Vietnam, farmers did not like planting timber trees with coffee. They only planted timber trees when they had no other choices for fruits (either because of unsuitable conditions or distance from the market). Farmers' ranked timber trees lowest indicating competitive rather than complementary relationships with coffee. This was consistent with the tree frequencies (Table 2) where timber species appeared in less than 6% of surveyed farms. Farmers were knowledgeable about the benefits that trees can provide to coffee as documented in other continents [33,38]. In this study, farmers could confidently describe the effects of trees on climate regulation, soil erosion, shade provision, mulch provision, frost and wind control.

Unlike in the Central Highlands of Vietnam where coffee was blamed for deforestation [39], in the northwest, most coffee farms were established on annual crop land or fallow land [10]. Therefore, farmers could clearly explain the benefits, based on direct observation, of integrating trees on their coffee farms on steep slopes, such as soil becoming soft and moist, and having darker brown color compared to a more yellow color in the past. Farmers were much less clear about tree shade interactions with the coffee plants, particularly with regard to the competition between coffee and trees for light and water (Figure 5), although they were aware that 'too much' shade made coffee beans take too long to ripen, and that the coffee produced fewer cherries but the size of the cherries was bigger. Coffee quality remains a big knowledge gap because farmers do not drink the coffee they produce. These are areas where better information about the effects of different shade species on the quality of the coffee may lead to the greater diversification of shade trees—provided that quality influences the price farmers receive for their coffee.

4.3. *The Relevance of Tree Knowledge and Tree Biophysical Suitability*

Nitrogen-fixing species are common in many coffee regions of the world such as *Erythrina* spp. and *Inga* spp. in Central America, although they have no timber values [40]; and *L. leucocephala* was ranked as best species for most ecosystem services, but was only planted in 23% of coffee fields, mainly in Dien Bien. As observed by farmers, *L. leucocephala* grew well with coffee and provided good shade as its leaf size was not too small or too large. Furthermore, *L. leucocephala* leaves were soft, which was good for soil fertility because they decompose quickly. Although legume species are good for coffee, this also depends on the management of legume trees on farms [12]. Farmers in Dien Bien expressed their concern about too much shade from *L. leucocephala* trees. Coffee density under shade was not discussed with farmers while it appeared as a factor significantly influencing performance. Medium intensity coffee management under shade is recommended in Central America [38].

Fruit tree cultivation in coffee farms is also popular in other countries thanks to their high profit. Farmers in Yunnan ranked fruit trees including *D. longan* quite high [30]. *Dimocarpus longan* is restricted to elevations lower than 600 m whereas Arabica coffee is suitable at elevations of 500 masl and above. Thai and H'mong farmers ranked *D. longan* as a good species for some ecosystem services and they expressed the wish to expand coffee–*D. longan* systems if they had available land. This raises a concern about the suitability of *D. longan* at higher elevations, as it may lead to the low productivity of the trees and integrated coffee–*D. longan* cultivation—although farmers stated that *D. longan* has a large crown and wide root system which negatively interact with coffee. The farmers' selection of tree species in coffee agroforestry systems was highly influenced by the economic benefits of intercropped trees and market access, particularly the proximity of farms to a main road. The oversupply of *D. longan* could lead to market price reduction in the near future. Moreover, the farmers' choice of trees were just a few species (*L. leucocephala*, longan, peach, plum, jackfruit), and hence the low biodiversity of the coffee landscape can be foreseen.

Farmers ranked timber species as the worst species to integrate in coffee plots. They explained that timber trees often compete with coffee for water and nutrition, and provide heavy shade and hence were not good for coffee. Poor quality road networks make farmers reluctant to grow timber trees as the transportation costs are high. This is quite similar to Central America where timber trees are also less common in coffee farms although they can provide additional values for farmers [41]. Compared to fruit trees, timber trees provide long-term benefit and could be considered as analogous to a savings bank account for the next generation. This suggests a need for further studies to focus on the economic benefits and trade-offs associated with integrating various types of trees in coffee farms, such as legume trees for shade and fertility versus timber or fruit trees for their commercial value.

4.4. Gender, Ethnicity and Tree Selection

Both men and women farmers from the three ethnic groups: Kinh, Thai, H'mong, ranked *L. leucocephala* as the best tree species providing multiple services to coffee. Overall ranking shows that the most preferred options were *D. longan* or plum and jackfruit if the farms were close to the main road. *Leucaena leucocephala* was preferred if the farms were further away from the road. Thus, market access appears to be a main driver of tree species selection. Moreover, coffee was produced only for trade, as the farmers did not consume it themselves and hence had no knowledge about tree species or the management effects on the quality of the final product; that is, the cup of coffee is mostly consumed in consuming countries. In this commercial system, market factors seemed to be the most important consideration rather than ethnicity, gender or the local knowledge about different tree species.

5. Conclusions

This study suggests that integrating trees into monoculture coffee systems can bring multiple benefits to the environment and rural livelihoods in northwest Vietnam. From the field surveys, it was observed that most of the smallholder coffee area was under agroforestry, but highly dependent on plot location, ecological suitability and market access. The tree diversity of coffee agroforestry plots were double those of agroforestry with annual crops, orchards or timber plantations. There were observable differences between the three ethnic groups. Thai farms had higher tree species richness in comparison with Kinh and H'mong groups—but it was not clear the degree to which this was driven by cultural rather than other contextual factors (such as proximity to roads and elevation). Commercial fruit trees were more common when farmers lived near or at a medium distance to main roads. Timber trees and a legume tree, *L. leucocephala*, were more common on coffee farms which were far from the road.

Farmers had detailed knowledge about observable tree services such as soil erosion control and mulch provision but they appeared to lack knowledge about the effects of shade trees on soil fertility, pest and disease control, and interestingly, on the competition or complementarity between coffee and trees in respect of light and water capture. This is where integrating local knowledge and scientific knowledge can provide richer advice to farmers on selecting appropriate tree species for their coffee farms.

By uploading the data from this research to the online tool http://shadetreeadvice.org, the information is available to help farmers, researchers and policy makers to choose suitable tree species according to the services that they can provide. Further improvement of the tool could include a focus on management aspects with farmers ranking additional information, such as the labor requirement for species associated with the timing and severity of tree pruning. This could be used to create a more comprehensive knowledge management system that goes beyond tree species selection.

Author Contributions: Conceptualization, P.V., T.P., M.P.N. and F.S.; methodology, M.P.N., P.V. and T.P., data collection and analysis, M.P.N.; original draft preparation, M.P.N.; review and editing, T.P., P.V., F.S.; funding acquisition, P.V. and F.S. All authors have read and agreed to the published version of the manuscript.

Acknowledgments: This research was supported by AFLi II project (Developing and Promoting market-based agroforestry and forest rehabilitation options for northwest Vietnam) and CCAFS (CGIAR Research Program on Climate Change, Agriculture and Food Security). We thank our colleagues Nguyen Thi Thanh Hai, Nguyen Van Huan from Northern Mountainous Agriculture and Forestry Science Institute (NOMAFSI) CANSEA, Pham Thanh Van, Nguyen Van Cuong, Duong Minh Tuan (World Agroforestry—Vietnam office), Master student Mathilde Lépine for their contribution in farmer interview, farmer ranking, focus group discussion and Clement Rigal for guidance in data analysis.

Appendix A

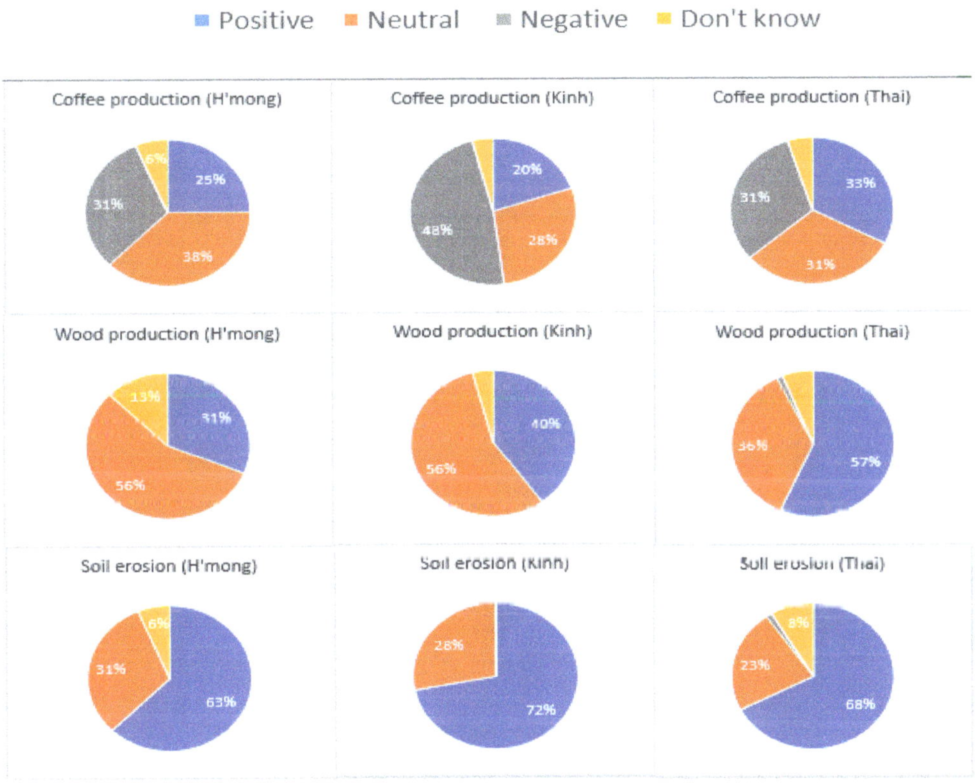

Figure A1. *Cont.*

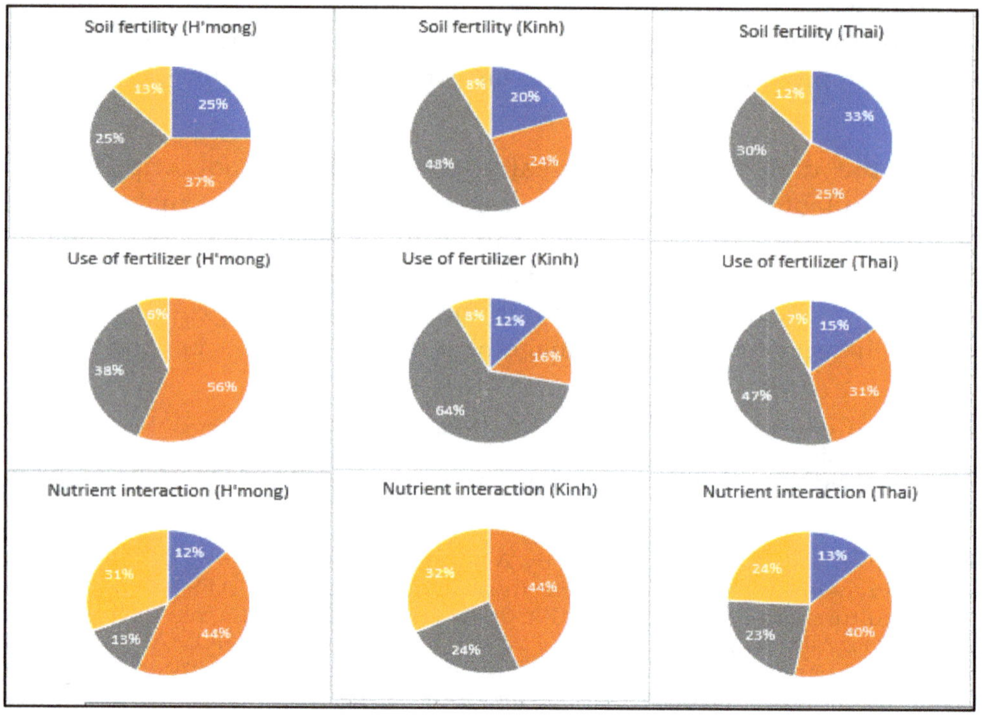

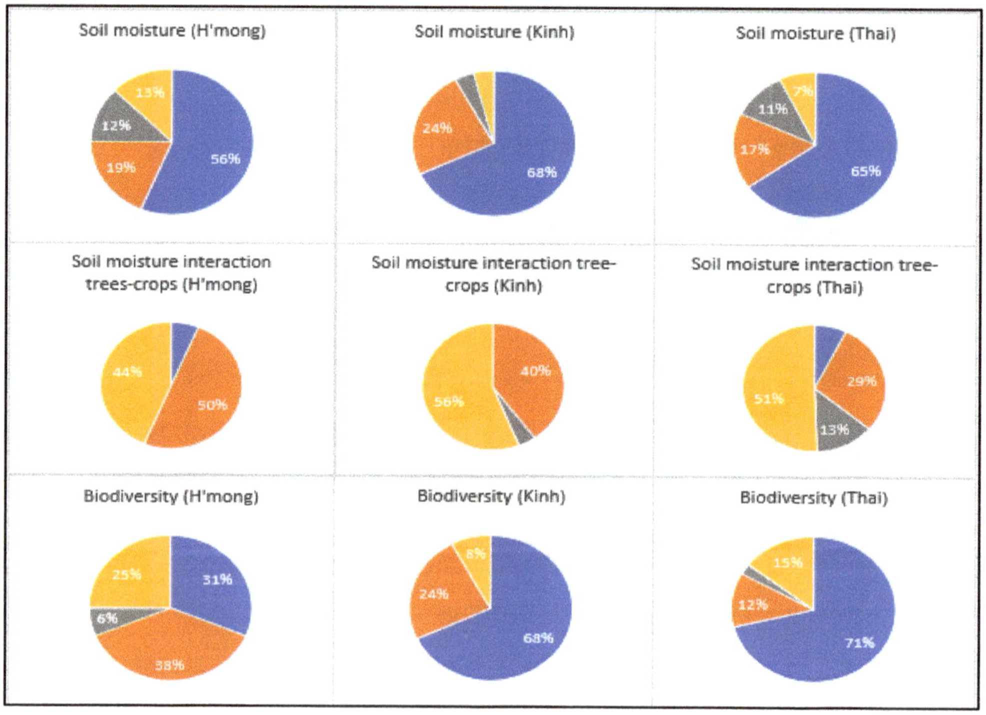

Figure A1. *Cont.*

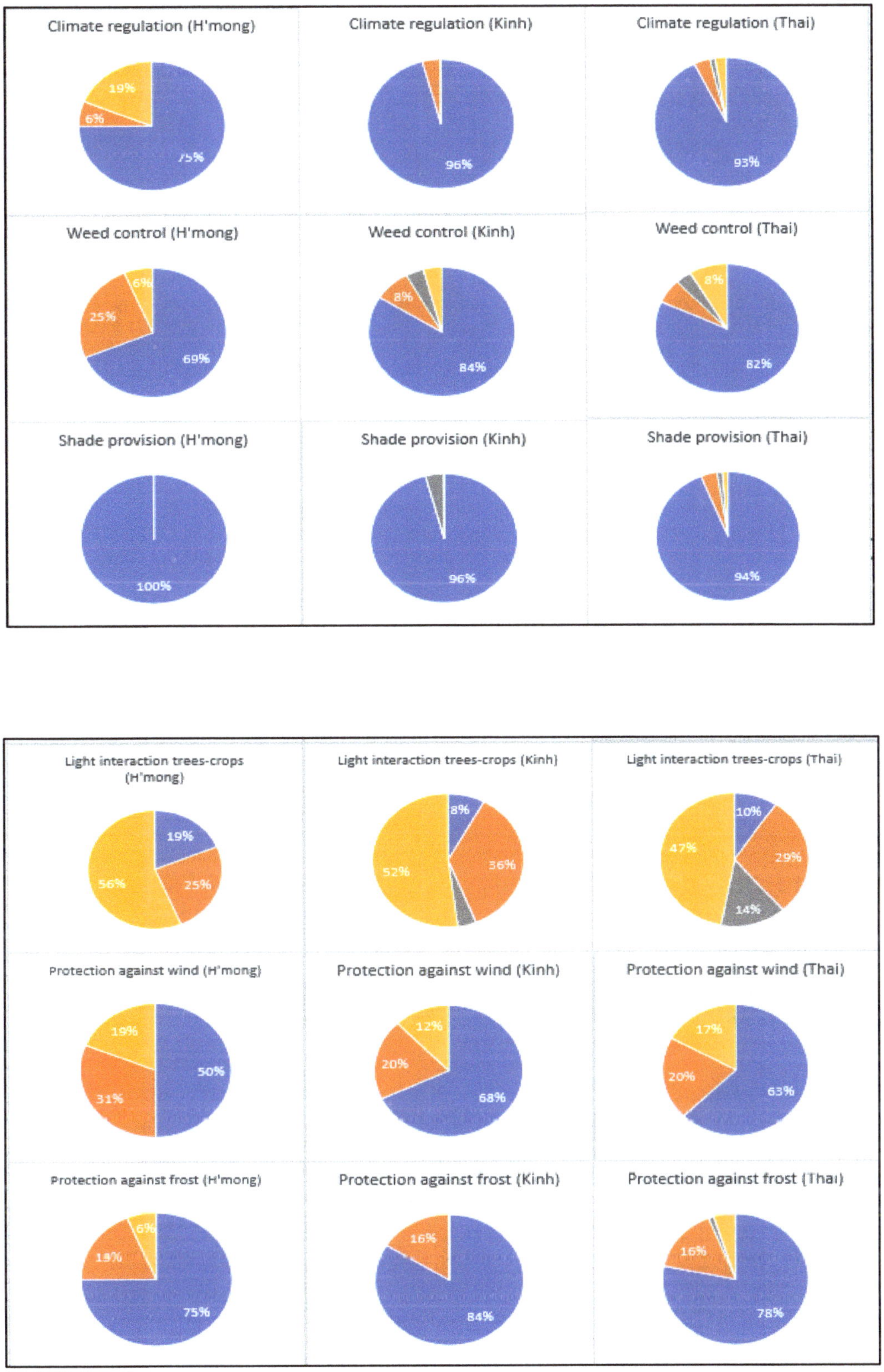

Figure A1. *Cont.*

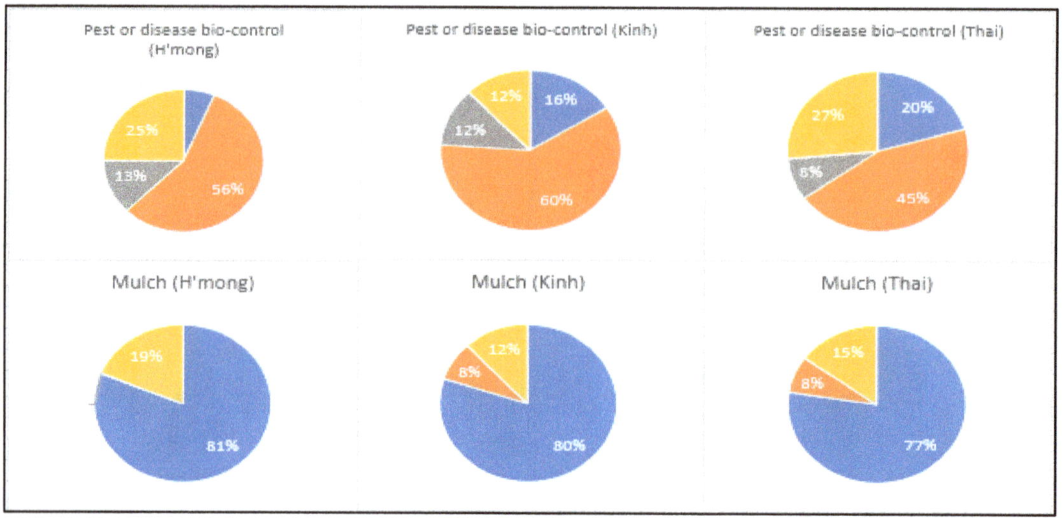

Figure A1. Farmers' perception of tree services for coffee in coffee agroforestry systems by ethnicity.

Appendix B

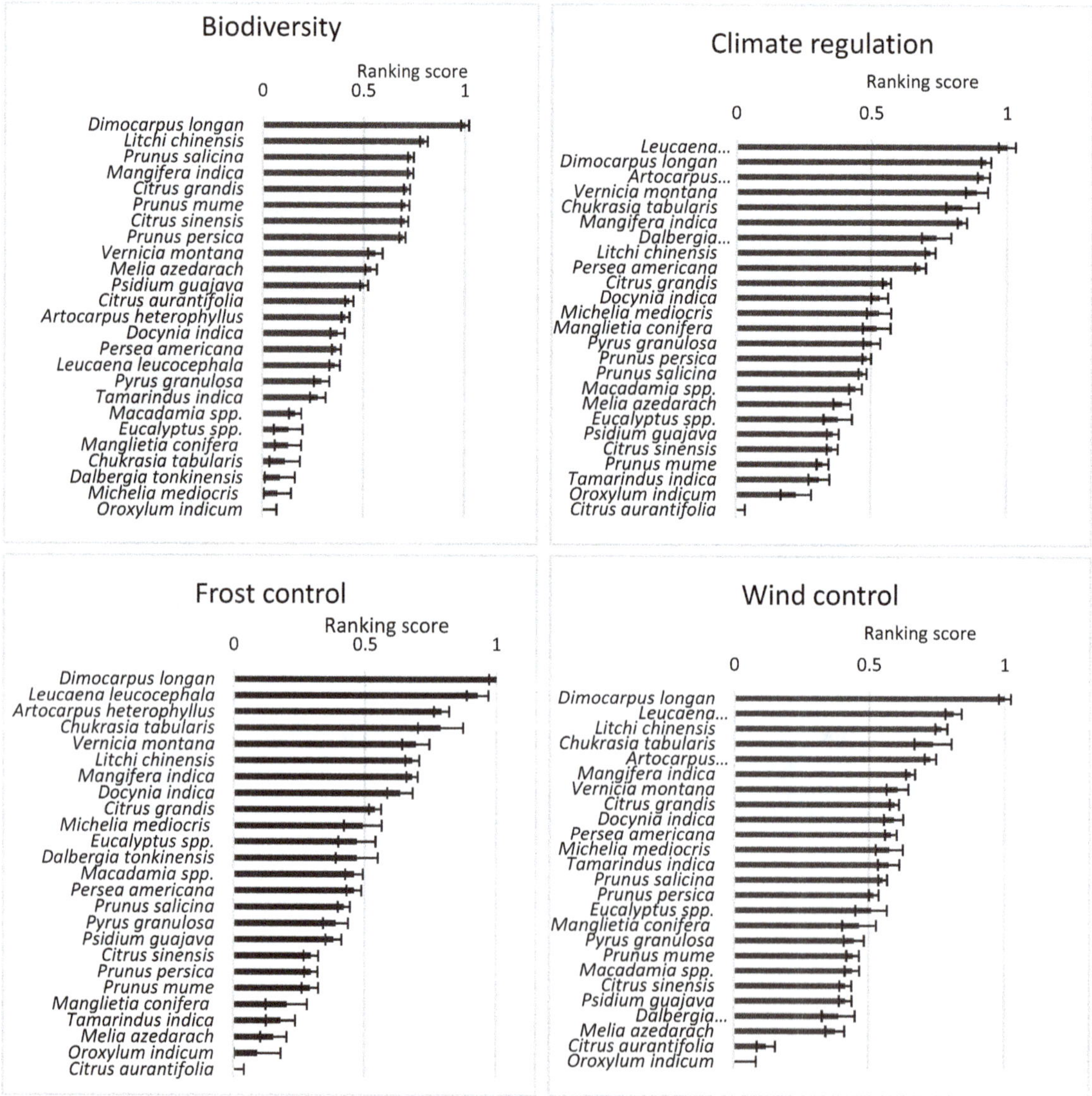

Figure A2. *Cont*

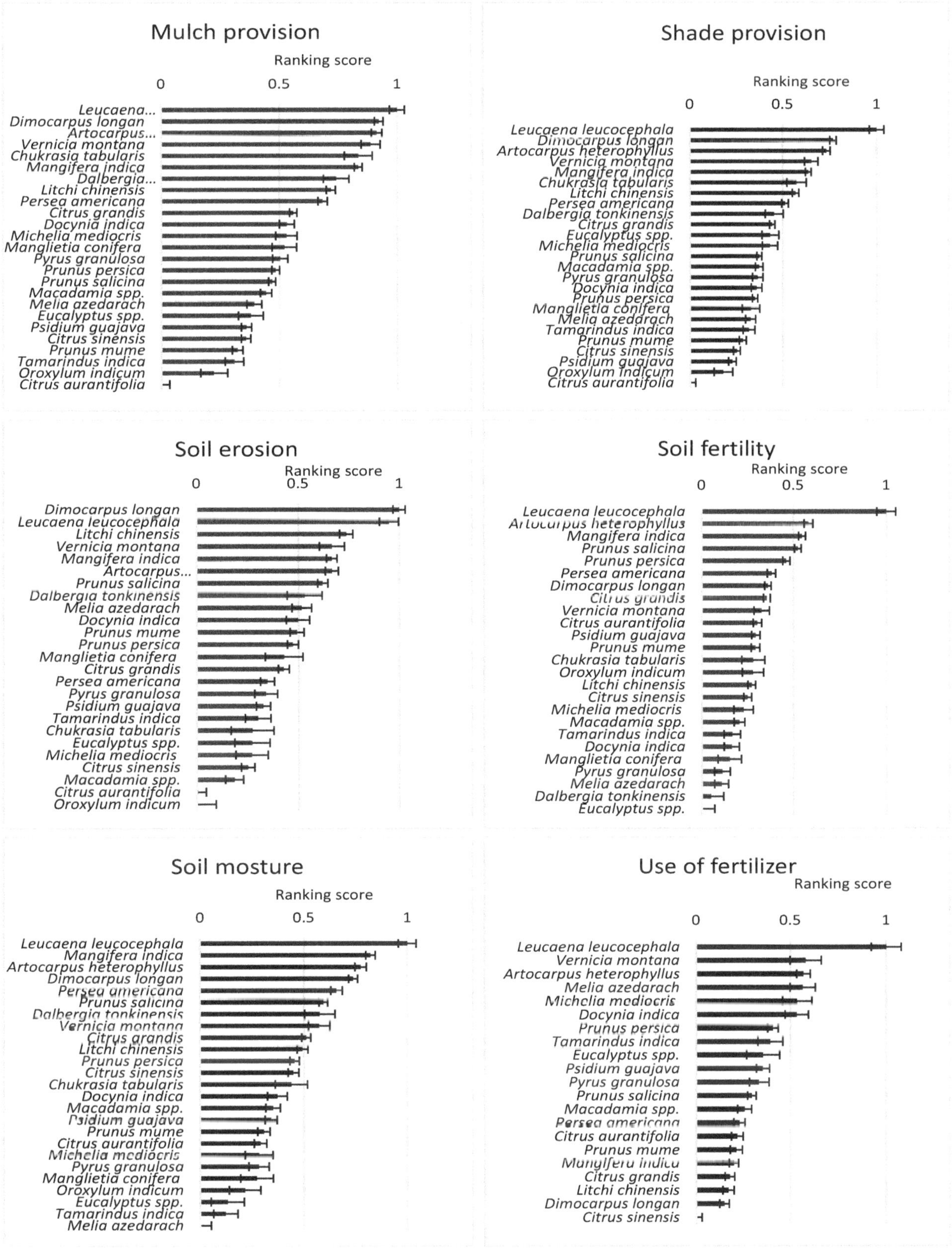

Figure A2. *Cont.*

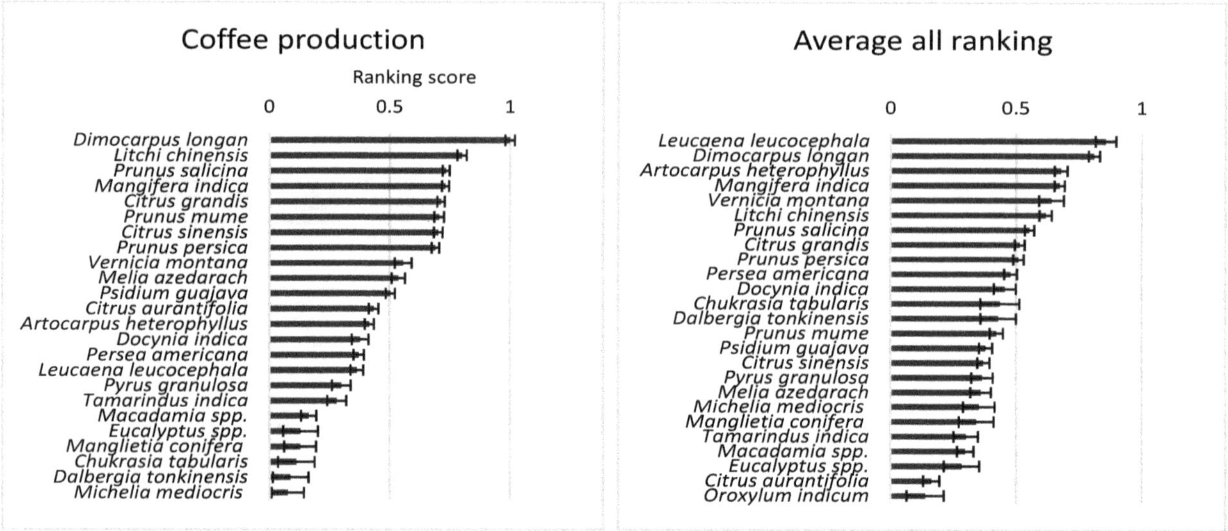

Figure A2. Pairwise ranking of tree species contributing to various tree services in coffee agroforestry systems with all groups combined.

Appendix C

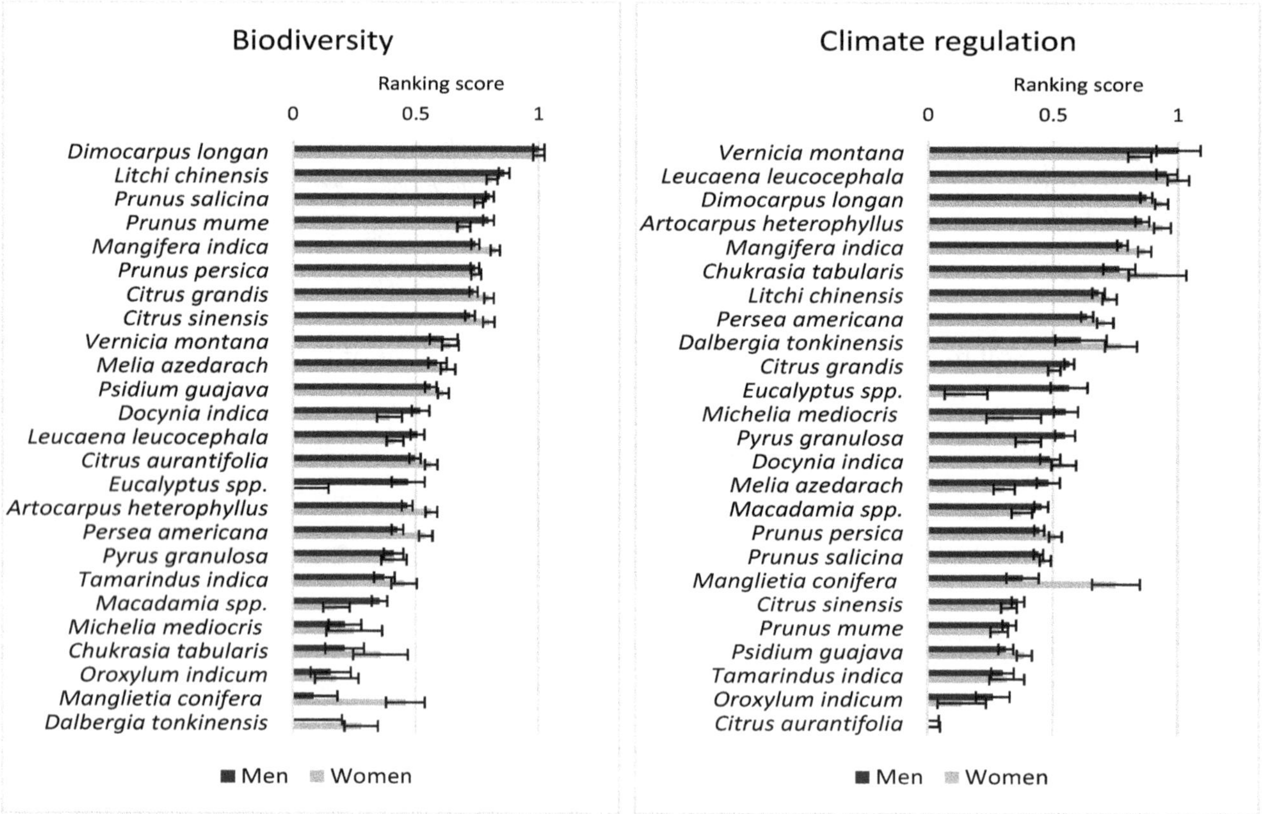

Figure A3. *Cont.*

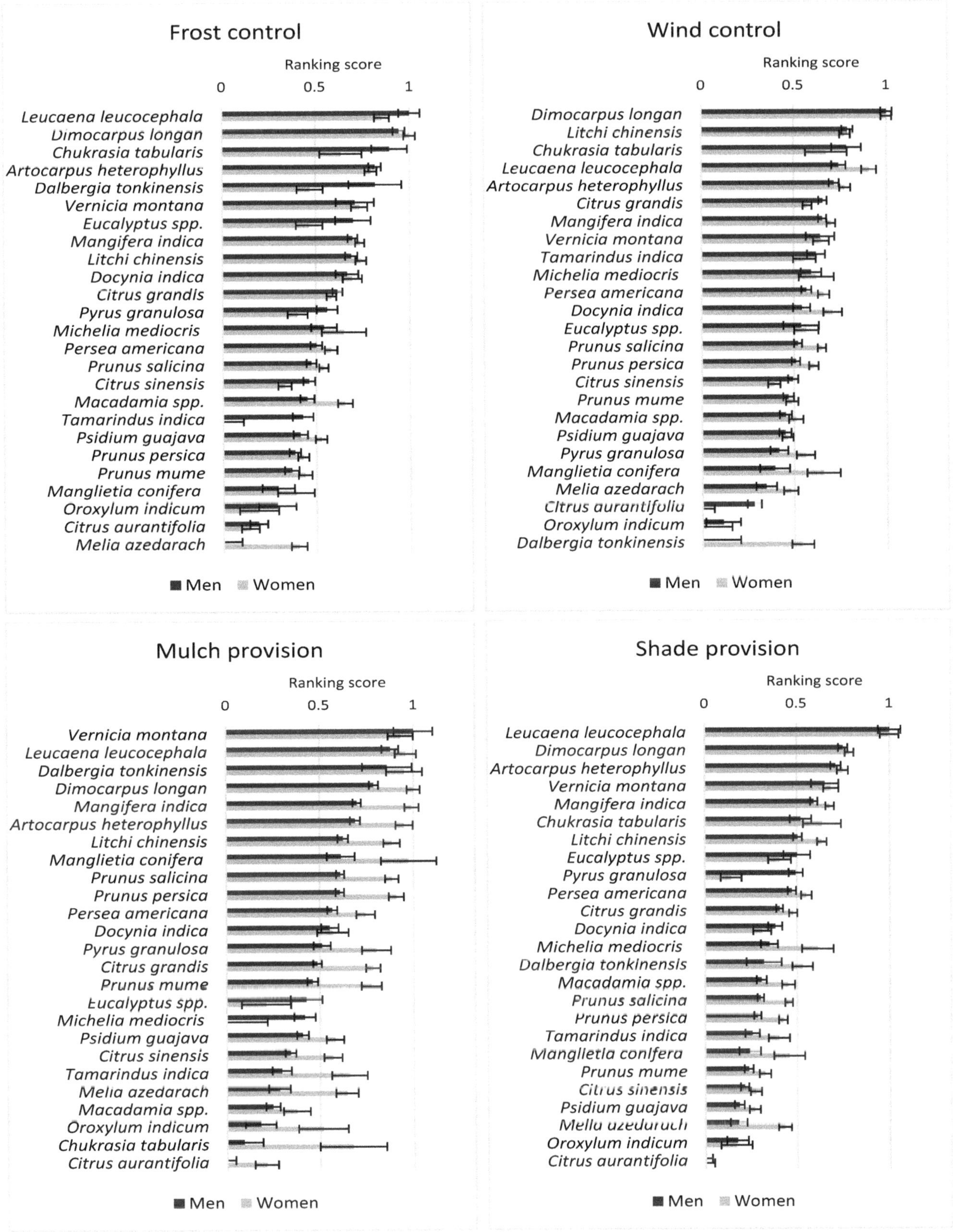

Figure A3. *Cont.*

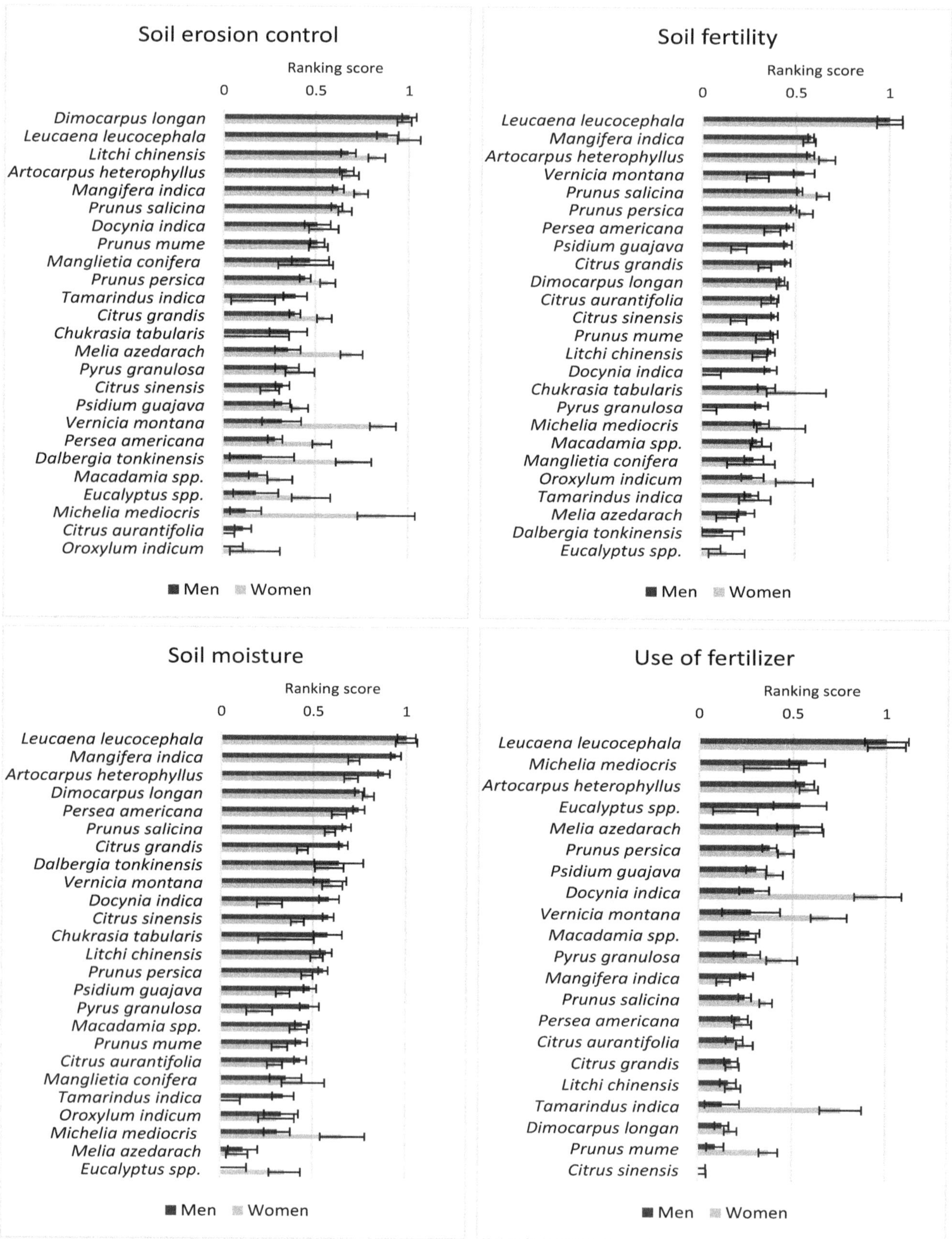

Figure A3. *Cont.*

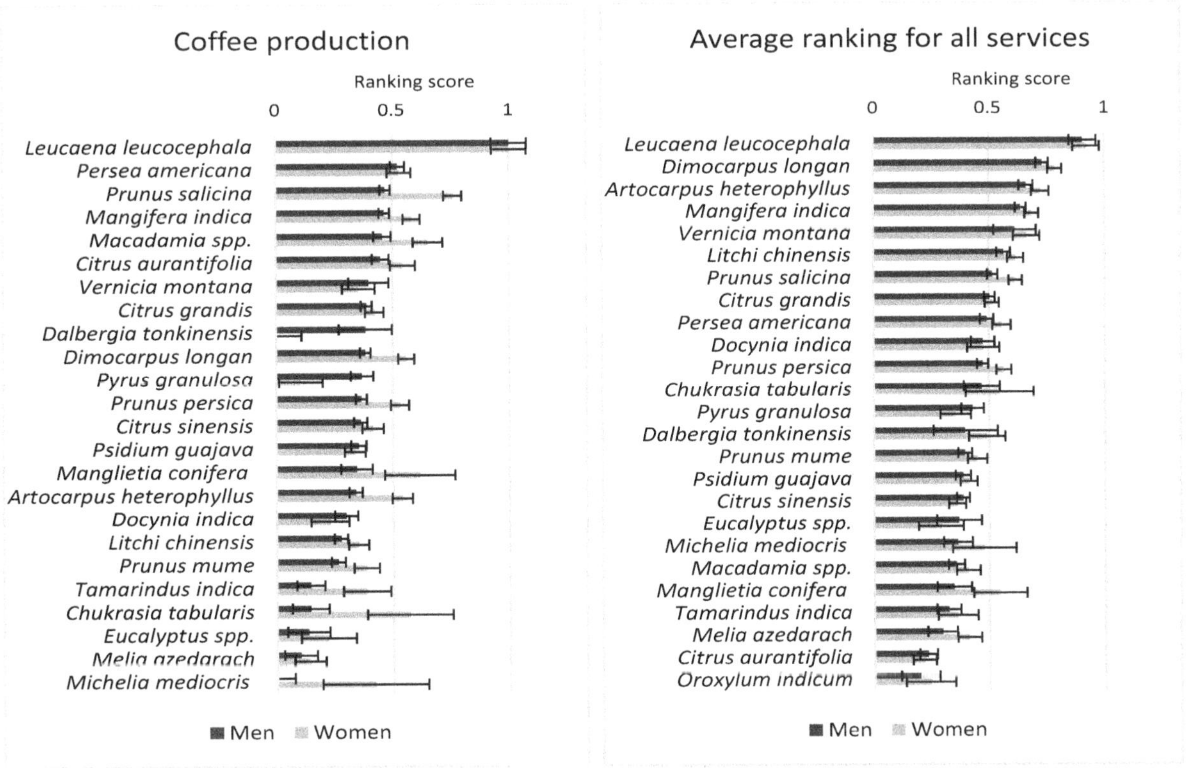

Figure A3. Pairwise ranking of tree species contributing to different tree services in coffee agroforestry systems by gender.

Appendix D

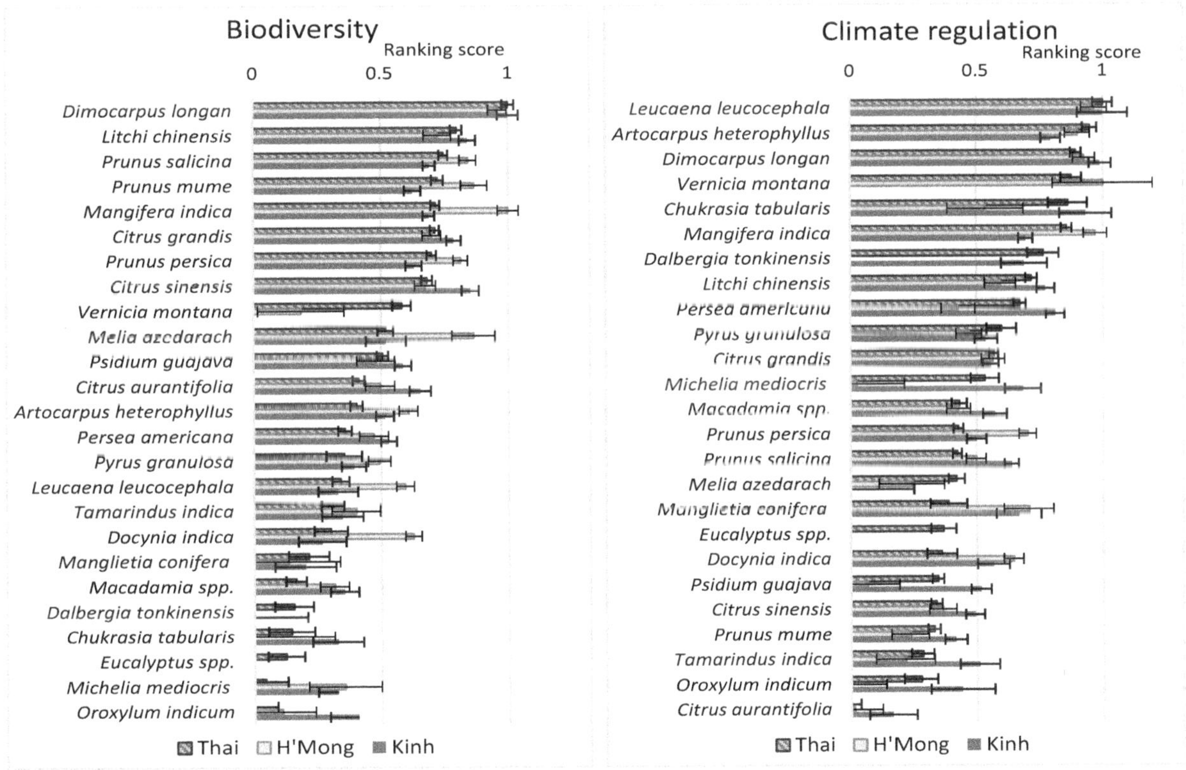

Figure A4. *Cont.*

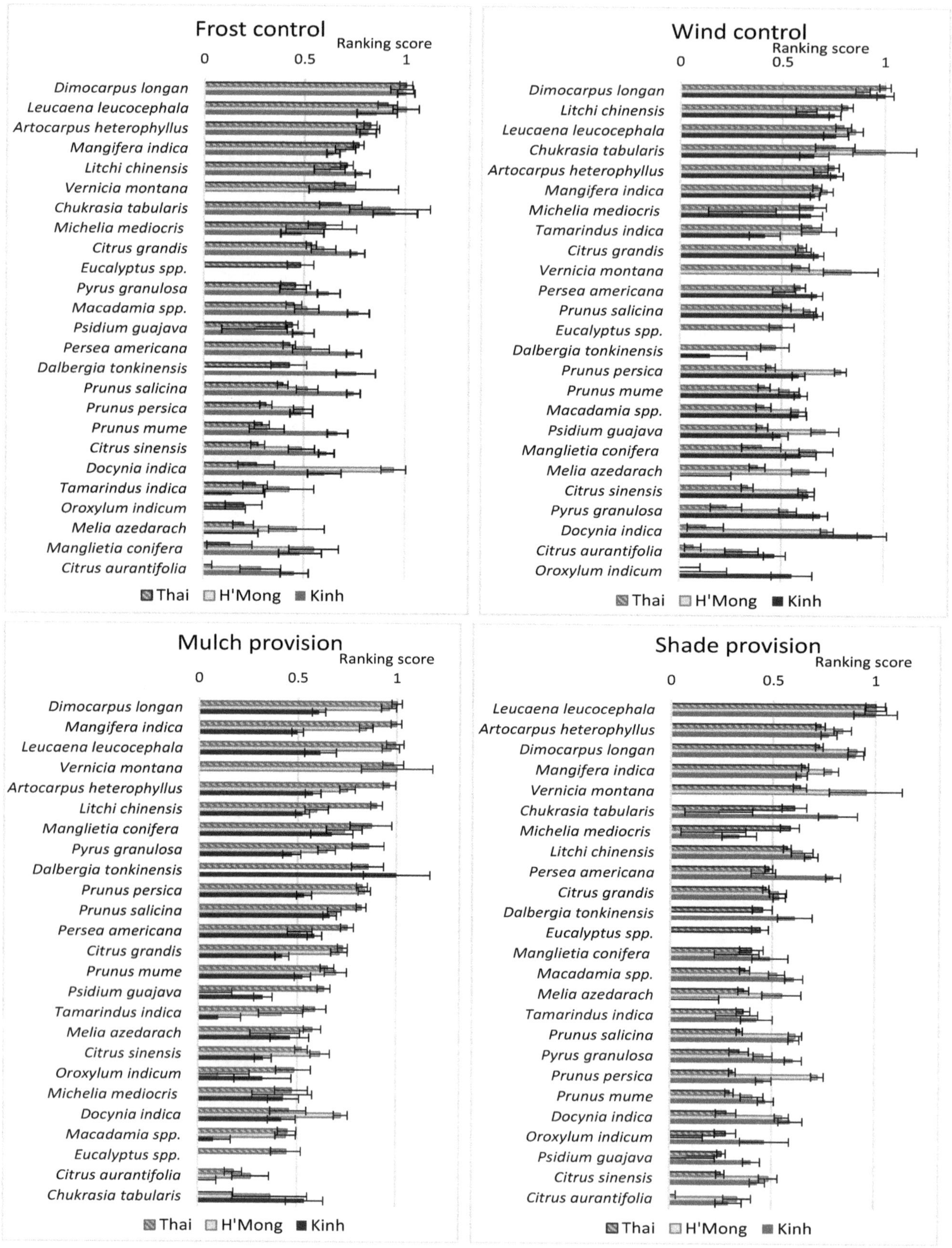

Figure A4. *Cont.*

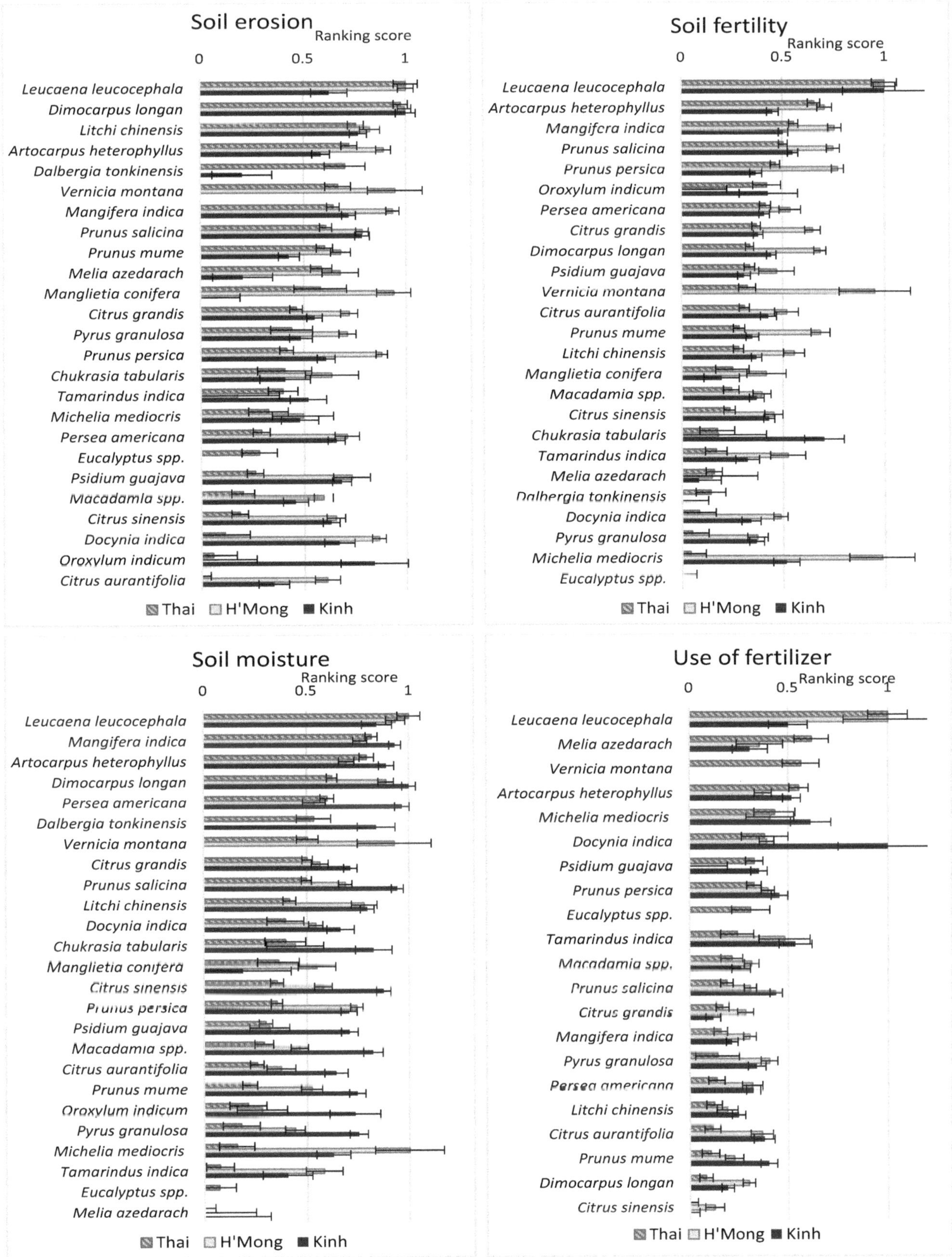

Figure A4. *Cont.*

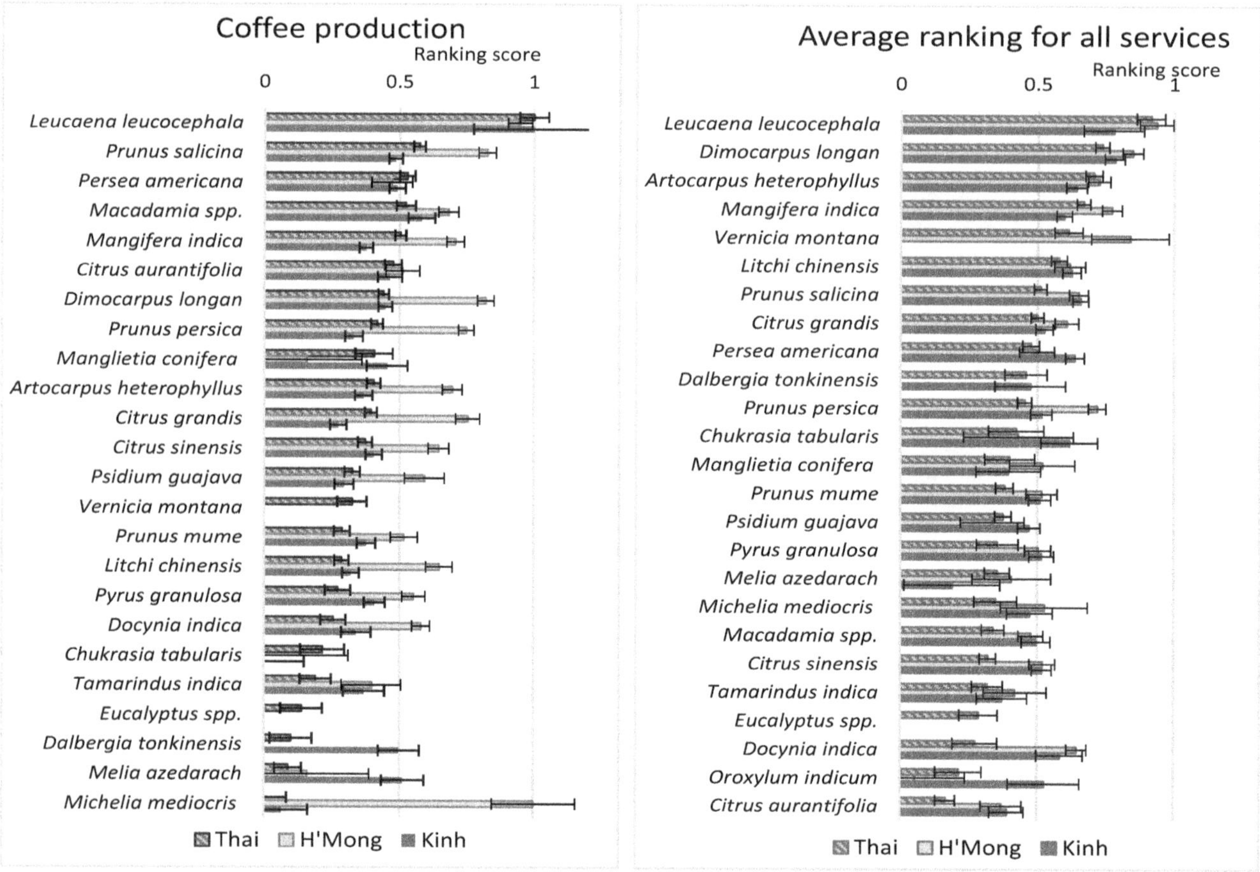

Figure A4. Pairwise ranking of tree species in coffee agroforestry systems by tree service and by ethnic group (Thai, H'Mong and Kinh).

References

1. Tucker, C.M. *Coffee Culture: Local Experiences, Global Connections*, 2nd ed.; Taylor & Francis: Abingdon, UK, 2017; ISBN 9781317392255.
2. Food and Agriculture Organization of the United Nations. *FAOSTAT Statistical Database 2018*; Food and Agriculture Organization of the United Nations: Rome, Italy, 2018.
3. International Coffee Organization. *ICO Annual Review 2010/11 2011*; International Coffee Organization: London, UK, 2011.
4. Amarasinghe, U.A.; Hoanh, C.T.; D'haeze, D.; Hung, T.Q. Toward sustainable coffee production in Vietnam: More coffee with less water. *Agric. Syst.* **2015**, *136*, 96–105. [CrossRef]
5. Vietnam Ministry of Agriculture and Rural Development. *Report on Crop Production in Vietnam*; Crop Production Department: Ha Noi, Vietnam, 2017.
6. Vietnam Customs. *Vietnam International Merchandise Trade Statistics 2018*; Vietnam Customs: Ha Noi, Vietnam, 2018.
7. International Coffee Organization. *Country Coffee Profile: Vietnam*; International Coffee Organization: London, UK, 2019.
8. General Statistics Office of Dien Bien Province. *Agriculture, Fishery and Forestry Census 2018*; General Statistics Office of Dien Bien Province: Vietnam, 2018.
9. General Statistics Office of Son La Province. *Agriculture, Fishery and Forestry Census 2018*; General Statistics Office of Son La Province: Vietnam, 2018.
10. Nghiem, T.; Kono, Y.; Leisz, S.J. The Case of Coffee Production in the Northern Mountain Region of Vietnam. *Land* **2020**, *9*, 56. [CrossRef]
11. Perfecto, I.; Rice, R.A.; Greenberg, R.; Van Der Voort, M.E. Shade coffee: A disappearing refuge for biodiversity: Shade coffee plantations can contain as much biodiversity as forest habitats. *Bioscience* **1996**, *46*, 598–608. [CrossRef]

12. Haggar, J.; Barrios, M.; Bolaños, M.; Merlo, M.; Moraga, P.; Munguia, R.; Ponce, A.; Romero, S.; Soto, G.; Staver, C.; et al. Coffee agroecosystem performance under full sun, shade, conventional and organic management regimes in Central America. *Agrofor. Syst.* **2011**, *82*, 285–301. [CrossRef]
13. Philpott, S.M.; Arendt, W.J.; Armbrecht, I.; Bichier, P.; Diestch, T.V.; Gordon, C.; Greenberg, R.; Perfecto, I.; Reynoso-Santos, R.; Soto-Pinto, L.; et al. Biodiversity loss in Latin American coffee landscapes: Review of the evidence on ants, birds, and trees. *Conserv. Biol.* **2008**, *22*, 1093–1105. [CrossRef]
14. Vaast, P.; Harmand, J.-M.; Rapidel, B.; Jagoret, P.; Deheuvels, O. Coffee and Cocoa Production in Agroforestry—A Climate-Smart Agriculture Model. In *Climate Change and Agriculture Worldwide*; Torquebiau, E., Ed.; Springer: Dordrecht, The Netherlands, 2016; pp. 209–224.
15. Sinclair, F.L. A general classification of agroforestry practice. *Agrofor. Syst.* **1999**, *46*, 161–180. [CrossRef]
16. Boreux, V.; Kushalappa, C.G.; Vaast, P.; Ghazoul, J. Interactive effects among ecosystem services and management practices on crop production: Pollination in coffee agroforestry systems. *Proc. Natl. Acad. Sci. USA* **2013**, *110*, 8387–8392. [CrossRef]
17. Somarriba, E.; Harvey, C.A.; Samper, M.; Anthony, F.; González, J.; Staver, C.; Rice, R.A. Biodiversity conservation in neotropical coffee (Coffea arabica) plantations. In *Agroforestry and Biodiversity Conservation in Tropical Landscapes*; Island Press: Washington, DC, USA, 2004; pp. 198–226.
18. Jha, S.; Bacon, C.M.; Philpott, S.M.; Rice, R.A.; Méndez, V.E.; Läderach, P. A Review of Ecosystem Services, Farmer Livelihoods, and Value Chains in Shade Coffee Agroecosystems. In *Integrating Agriculture, Conservation and Ecotourism: Examples from the Field*; Campell, W.B., Ortiz, S.L., Eds.; Springer: Dordrecht, The Netherlands, 2011; pp. 141–208.
19. Guillemot, J.; le Maire, G.; Munishamappa, M.; Charbonnier, F.; Vaast, P. Native coffee agroforestry in the Western Ghats of India maintains higher carbon storage and tree diversity compared to exotic agroforestry. *Agric. Ecosyst. Environ.* **2018**, *265*, 461–469. [CrossRef]
20. Rahn, E.; Läderach, P.; Baca, M.; Cressy, C.; Schroth, G.; Malin, D.; van Rikxoort, H.; Shriver, J. Climate change adaptation, mitigation and livelihood benefits in coffee production: Where are the synergies? *Mitig. Adapt. Strateg. Glob. Chang.* **2014**, *19*, 1119–1137. [CrossRef]
21. Davis, A.P.; Gole, T.W.; Baena, S.; Moat, J. The Impact of Climate Change on Indigenous Arabica Coffee (Coffea arabica): Predicting Future Trends and Identifying Priorities. *PLoS ONE* **2012**, *7*. [CrossRef]
22. Schroth, G.; Laderach, P.; Dempewolf, J.; Philpott, S.; Haggar, J.; Eakin, H.; Castillejos, T.; Moreno, J.G.; Pinto, L.S.; Hernandez, R.; et al. Towards a climate change adaptation strategy for coffee communities and ecosystems in the Sierra Madre de Chiapas, Mexico. *Mitig. Adapt. Strateg. Glob. Chang.* **2009**, *14*, 605–625. [CrossRef]
23. Vaast, P.; Van Kanten, R.; Siles, P.; Dzib, B.; Franck, N.; Harmand, J.M.; Genard, M. Shade: A Key Factor for Coffee Sustainability and Quality. In Proceedings of the 20th International Conference on Coffee Science, Bangalore, India, 11–15 October 2004.
24. Mancuso, M.A.C.; Soratto, R.P.; Perdoná, M.J. Shaded coffee production. *Colloq. Agrar.* **2013**, *9*, 31–44. [CrossRef]
25. Meylan, L.; Gary, C.; Allinne, C.; Ortiz, J.; Jackson, L.; Rapidel, B. Evaluating the effect of shade trees on provision of ecosystem services in intensively managed coffee plantations. *Agric. Ecosyst. Environ.* **2017**, *245*, 32–42. [CrossRef]
26. Roupsard, O.; Van Den Meersche, K.; Allinne, C.; Vaast, P.; Rapidel, B.; Avelino, J.; Jourdan, C.; Le Maire, G.; Bonnefond, J.-M.; Harmand, J.-M.; et al. Eight years studying ecosystem services in a coffee agroforestry observatory. Practical applications for the stakeholders. In Proceedings of the World Coffee Summit, San Salvador, Salvador, 31 May–3 June 2017.
27. Cerdán, C.R.; Rebolledo, M.C.; Soto, G.; Rapidel, B.; Sinclair, F.L. Local knowledge of impacts of tree cover on ecosystem services in smallholder coffee production systems. *Agric. Syst.* **2012**, *110*, 119–130. [CrossRef]
28. Lamond, G.; Sandbrook, L., Gassner, A.; Sinclair, F.L. Local knowldge of tree attreibutes underpins sepcies selection on coffee farms. *Exp. Agric.* **2019**, *55*, 35–49. [CrossRef]
29. Valencia, V.; West, P.; Sterling, E.J.; García-Barrios, L.; Naeem, S. The use of farmers' knowledge in coffee agroforestry management: Implications for the conservation of tree biodiversity. *Ecosphere* **2015**, *6*, 1–17. [CrossRef]
30. Dumont, E.S.; Bonhomme, S.; Pagella, T.F.; Sinclair, F.L. Structured stakeholder engagement leads to development of more diverse and inclusive agroforestry options. *Exp. Agric.* **2019**, *55*, 252–274. [CrossRef]
31. Van Der Wolf, J.; Jassogne, L.; Gram, G.I.L.; Vaast, P. Turning local knowledge on agroforestry into an online decision-support tool for tree selection smallholger's farms. *Exp. Agric.* **2019**, *55*, 50–66. [CrossRef]
32. Rigal, C.; Vaast, P.; Xu, J. Using farmers' local knowledge of tree provision of ecosystem services to strengthen the emergence of coffee-agroforestry landscapes in southwest China. *PLoS ONE* **2018**, *13*. [CrossRef]

33. Smith Dumont, E.; Gassner, A.; Agaba, G.; Nansamba, R.; Sinclair, F. The utility of farmer ranking of tree attributes for selecting companion trees in coffee production systems. *Agrofor. Syst.* **2019**, *93*, 1469–1483. [CrossRef]
34. Kindt, R.; Coe, R. *Tree Diversity Analysis: A Manual and Software for Common Statistical Methods for Ecological and Biodiversity Studies*; World Agroforestry Centre: Nairobi, Kenya, 2005.
35. Turner, H.; Firth, D. Bradley-terry models in R: The bradleyterry2 package. *J. Stat. Softw.* **2012**, *48*. [CrossRef]
36. Carsan, S.; Stroebel, A.; Dawson, I.; Kindt, R.; Swanepoel, F.; Jamnadass, R. Implications of shifts in coffee production on tree species richness, composition and structure on small farms around Mount Kenya. *Biodivers. Conserv.* **2013**, *22*, 2919–2936. [CrossRef]
37. Nyaga, J.; Barrios, E.; Muthuri, C.W.; Öborn, I.; Matiru, V.; Sinclair, F.L. Evaluating factors influencing heterogeneity in agroforestry adoption and practices within smallholder farms in Rift Valley, Kenya. *Agric. Ecosyst. Environ.* **2015**, *212*, 106–118. [CrossRef]
38. Soto-Pinto, L.; Villalvazo-López, V.; Jiménez-Ferrer, G.; Ramírez-Marcial, N.; Montoya, G.; Sinclair, F.L. The role of local knowledge in determining shade composition of multistrata coffee systems in Chiapas, Mexico. *Biodivers. Conserv.* **2007**, *16*, 419–436. [CrossRef]
39. Meyfroidt, P.; Lambin, E.F.; Erb, K.H.; Hertel, T.W. Globalization of land use: Distant drivers of land change and geographic displacement of land use. *Curr. Opin. Environ. Sustain.* **2013**, *5*, 438–444. [CrossRef]
40. Beer, J. Litter production and nutrient cycling in coffee (Coffea arabica) or cacao (Theobroma cacao) plantations with shade trees. *Agrofor. Syst.* **1988**, *7*, 103–114. [CrossRef]
41. Vaast, P.; van Kanten, R.; Siles, P.; Angrand, J.; Aguilar, A. Biophysical Interactions Between Timber Trees and Arabica Coffee in Suboptimal Conditions of Central America. In *Toward Agroforestry Design*; Jose, S., Gordon, A.M., Eds.; Springer: Dordrecht, The Netherlands, 2008; pp. 133–146.

5

Effects of Agroforestry and Other Sustainable Practices in the Kenya Agricultural Carbon Project (KACP)

Ylva Nyberg [1,*], Caroline Musee [2], Emmanuel Wachiye [2], Mattias Jonsson [3], Johanna Wetterlind [4] and Ingrid Öborn [1,5]

1 Department of Crop Production Ecology, Swedish University of Agricultural Sciences (SLU), P.O. Box 7043, SE-75007 Uppsala, Sweden
2 Vi Agroforestry, P.O. Box 2006, Kitale 30200, Kenya; caroline.musee@viagroforestry.org (C.M.); emmanuel_wachiye@wvi.org (E.W.)
3 Department of Ecology, SLU, P.O. Box 7044, SE-75007 Uppsala, Sweden; mattias.jonsson@slu.se
4 Department of Soil and Environment, SLU, P.O. Box 234, SE-53223 Skara, Sweden; johanna.wetterlind@slu.se
5 World Agroforestry (ICRAF), UN Avenue, P.O. Box 30677-00100 Nairobi, Kenya; ingrid.oborn@slu.se
* Correspondence: ylva.nyberg@slu.se

Abstract: With growing global demand for food, unsustainable farming practices and large greenhouse gas emissions, farming systems need to sequester more carbon than they emit, while also increasing productivity and food production. The Kenya Agricultural Carbon Project (KACP) recruited farmer groups committed to more Sustainable Agricultural Land Management (SALM) practices and provided these groups with initial advisory services on SALM, farm enterprise development and village savings and loan associations. Recommended SALM practices included agroforestry, cover crops, mulching, composting manure, terracing, reduced tillage and water harvesting. The effects of the KACP on the uptake of SALM practices, maize yield, perceived food self-sufficiency and savings during the initial four years were assessed comparing control and project farmers using interviews, field visits and measurements. Farmers participating in the KACP seemed to have increased uptake of most SALM practices and decreased the use of practices to be avoided under the KACP recommendations. Agroforestry and terraces showed positive effects on maize yield. During all four years, the KACP farms had higher maize yield than control farms, but yield differences were similar in 2009 and 2012 and there was no overall significant effect of the KACP. In 2012, the KACP farms had higher food self-sufficiency and tended to have higher monetary savings than control farms.

Keywords: adaptation; carbon sequestration; Kisumu; Bungoma; payment for ecosystem services; village savings and loan associations

1. Introduction

Managing the trade-off between short-term provision of human needs in terms of water, food, shelter and fibre for a growing population and maintaining the capacity of the planet to provide these services in the future is a growing global challenge [1–3]. World agriculture is currently both contributing to global environmental change [3,4] and being severely affected by it [5]. It is, therefore, necessary and urgent to transform farming systems from primarily focusing on productivity to instead having social, economic and environmental sustainability at the core of their development [6]. Most food production world-wide is currently unsustainable and needs modification [7–10]. For example, nutrients need to be recycled, external inputs reduced and farming systems diversified [7,10–12]. Simultaneously, anthropogenic CO_2 emissions are changing the global climate and will continue to do so throughout the 21st century [13]. Crop production and carbon sequestration are ecosystem services with great potential for generating synergies in agricultural landscapes worldwide. Crop production can maintain higher levels of soil carbon through, for example, more perennial crops or returning more crop residues and organic material to fields, and this higher soil carbon level can increase yield [14,15].

The increasing concentration of atmospheric CO_2 is largely attributable to emissions from fossil fuel combustion or land-use change, in the latter case mainly through soil organic carbon (SOC) depletion [13]. Conversion of natural land to agriculture and subsequent soil degradation from, for example, erosion have resulted in average losses of 50–70% of the original SOC pool in agricultural soils [16]. In some areas, this has led to declining yield. There is thus a need for sustainable intensification, where production can increase while the environmental footprint of agriculture decreases [17,18]. This is particularly relevant in developing countries, where most emissions are land-based and where carbon sources can be turned directly into sinks [19]. Sub-Saharan Africa is an area with considerable opportunities for sustainable intensification [17]. Kenya, as a fast-developing country in East Africa, has the potential to take the lead in reaching the Paris Agreement goals on climate change mitigation [20]. In Kenya, 70% of the rural population relies on agriculture for their employment and more than a quarter of the gross domestic product is derived from agriculture [21]. The Kenyan National Climate Change Action Plan 2018–2022 sets goals for agriculture where the main mitigation actions proposed for climate change are limited burning in croplands and more use of conservation tillage and agroforestry [21]. Kenya has the goal of converting 281,000 ha of existing arable and grazing land into agroforestry by 2030 as a climate change mitigation action, and of making climate change-related information and advice an integral part of the Kenyan agricultural advice system [22]. Kenyan smallholders are aware of agroforestry and other sustainable management practices [23] that could increase the soil carbon pool [15,24] and help increase production and income [22,25–28]. The Clean Development Mechanism, which allows countries with carbon emission reduction commitments to implement these reductions in developing countries, can create incentives for smallholders in developing countries to sell carbon sequestered in, for example, agroforestry systems to industrialised countries [29]. Similar incentives can be created within developing countries, through the nationally determined contributions [30]. Engel and Muller [31] identified climate-smart agriculture (e.g., agroforestry) as the most promising practice to be promoted by Payments for Ecosystem Services (PES) among smallholders with limited income. However, the use of carbon finance to incentivise this type of bio-carbon storage is still very low, due to the absence of institutional frameworks, reliable sources of carbon finance and involvement of public and private sector actors [19,32,33]. The number of smallholders needed to achieve an area of land large enough to compensate for project transaction costs also makes carbon finance projects practically unattainable at the current market price for carbon [34,35]. Low carbon prices mean that the incentive for farmers is not the carbon payment, but the benefits arising from emission-reducing farm management. Mbow et al. [36] question whether smallholder farmers can benefit from carbon payments at all and how advisory services can succeed in effectively promoting climate-smart agriculture. More research is needed to identify suitable types of climate-smart agrosystems for different user groups and their results and impacts [36].

The Kenya Agricultural Carbon Project (KACP), a soil and tree carbon project implemented by the non-government organisation (NGO) Vi Agroforestry within the Lake Victoria basin, is targeting small-holder farmers (with <2.5 ha). A majority of the farmers are aiming to be self-subsistent from their farming, largely depending on maize, beans and dairy production, while some are combining the agriculture with off-farm employment or casual jobs. The aim of the KACP is carbon sequestration through the uptake of Sustainable Agricultural Land Management (SALM) practices, enabling smallholder farmers to access the carbon market, as well as increasing yield and productivity and enhancing resilience to climate variability and change [37]. SALM practices include, for example, cover crops and agroforestry to increase biomass production, use of biomass for mulching and composting instead of burning, and avoiding soil erosion through, for example, terracing, reduced tillage and water harvesting [38] (Table A1). The project provided a dedicated advisory package to promote the use of SALM practices. It also included training in farm enterprise development and village savings and loan associations (VSLA), a microfinance intervention for the accumulation of regular savings within a group and rotating credit opportunities, following definitions by, for example, Bouman [39]. The VSLA system is based on trust among people who know each other and is run by the people themselves in one-year cycles with no outside support except for the initial training. A number of studies have reported on the KACP as a novel type of intervention [40–46]. However, no previous study has examined the effects of different SALM practices on maize yield or the actual maize yield response on farms taking part in the project.

The overall aim of this study was to assess the effects of the KACP on farm productivity and livelihoods during the initial four years (2009–2012) using the uptake of SALM practices, maize yield, food self-sufficiency and savings as indicators. Specific objectives were to (i) assess the level and type of SALM implementation among farmers participating in the KACP over time and compare it with that of neighbouring non-participating farms (control farms); (ii) determine the relationship between SALM practices and maize productivity, and compare the productivity over time and between the KACP farms and control farms; and (iii) quantify the level of food self-sufficiency and savings of the KACP farmers and control farmers. Mechanisms for the spread of knowledge and practices between neighbouring farmers were also discussed.

2. Materials and Methods

2.1. Characterisation of the Study Areas and Background to the KACP

The Kenya Agricultural Carbon Project started in 2009 and was implemented in two areas of Kenya, both with around 22,500 ha of potential project area (Figure 1). These were Kisumu (southern part including Kisumu and Siaya counties) and Bungoma (northern part including Bungoma county), with 28 administrative locations (Table A2). The region has a bimodal rainfall pattern with two cropping seasons, normally March–July and August–December (Figure A1). The KACP methodology was developed together with the BioCarbon Fund of the World Bank [37]. In the KACP, smallholder farmers receive carbon credits for soil carbon, which makes it unique since most other community and smallholder carbon projects only support carbon sequestration through tree planting [46]. However, the participating farmers were not given any financial support from the KACP during the study period (2009–2012) and the benefits of using SALM practices were instead emphasised, while the carbon revenue was small and presented to farmers as a co-benefit. Carbon credits were generated and claimed using the approved Verified Carbon Standard methodology, based on background data on soils and climate together with data on the management practices used by the farmers, instead of analysing carbon content in soil samples [47,48]. Carbon revenues were post-paid to farmers in 2014 for the years 2010–2014. Validation and verification were conducted periodically by external teams on emission reductions reported within the KACP [47].

From 2009 to 2012, the advisory system had a fixed number of 28 Vi Agroforestry field advisors, with one advisor in every administrative location, covering approximately 70 km^2. These field advisors offered advisory services to farmers and facilitated monitoring. They identified registered farmer groups or facilitated group formation, and recruited, informed, trained and contracted smallholder farmers to implement a free choice of SALM practices on their farms. Regular interaction and monitoring of project activities by advisors proved important to avoid misunderstandings and identify risks and challenges in agricultural production as early as possible. No new farmer groups were recruited after 2013, so the number of advisors was reduced to four by 2017, to maintain the information flow. The monitoring data were consolidated in maps and tables at the farmer group level for internal monitoring by farmers. They were also collated at the end of each cropping season and entered in databases and ArcGIS for monitoring and evaluation by Vi Agroforestry.

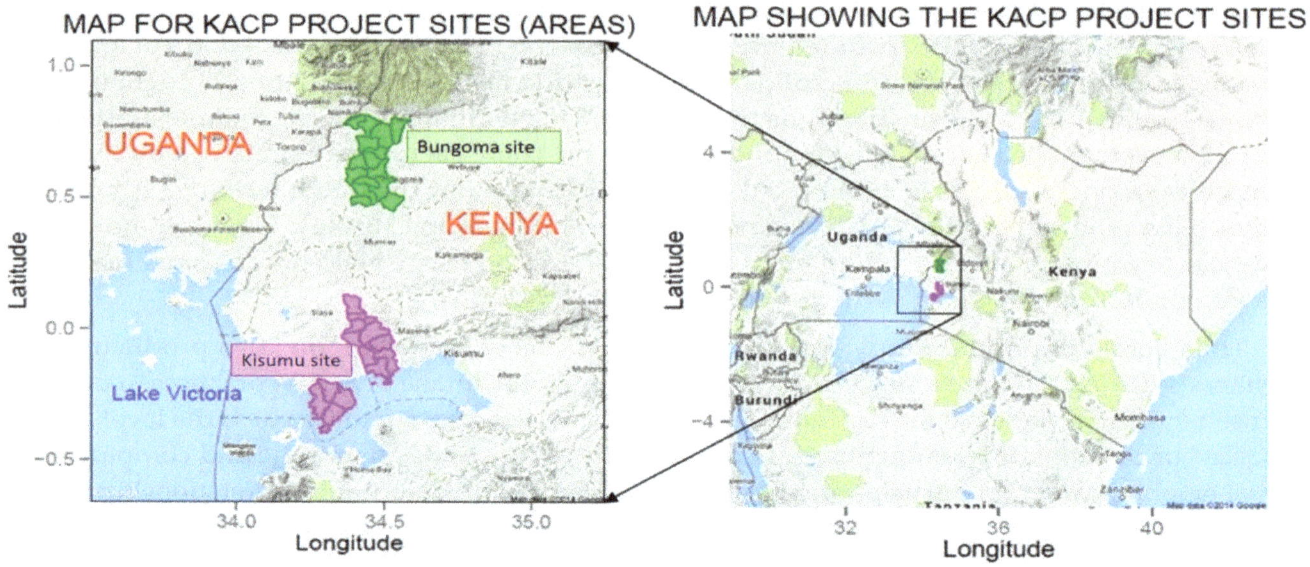

Figure 1. Left: Kenya Agricultural Carbon Project (KACP) areas; the Bungoma sites in Bungoma County (green) and the Kisumu sites in Kisumu and Siaya Counties (purple). Right: The KACP areas within East Africa.

2.2. Farm Sampling and Data Collection

The project started with 660 farmer groups involving 10,873 voluntary farmers [49]. By the end of 2012, the KACP had recruited 1555 verified farmer groups with 26,535 farmers, who implemented SALM on 16,490 ha of eligible cropland or grazing land. Among the initial 10,873 project farms, 200 were selected and monitored more closely by the KACP field advisors, in permanent farm monitoring (PFM). To select the 200 PFM farms (100 in each of the project areas), the areas were stratified into agro-ecological zones (AEZ) and the number of farms in each zone was decided in relation to the zone size. A systematic grid of 1.5×1.5 km^2 for unaligned systematic sampling was applied and the area was divided into clusters formed by the areas between four intersection points of the grid. PFM farms were randomly selected within the clusters according to the agro-ecological stratification [48].

In this study, the PFM (hereafter called "project") data collected by Vi Agroforestry between 2009 and 2012 were used to determine the uptake rate and correlations of SALM practices to maize yield (Figure 2). In addition, in 2012 a set of 160 control farms (80 in each project area) was selected from the 28 locations of the KACP for the purposes of the present study. When selecting control farms, the project farms were used as reference points and the second farm to the north of the project farm was selected if the owners consented. If that farmer belonged to any group that had worked with Vi Agroforestry, it was skipped and the next farm to the north was asked instead, and so on. Thus, only farms that had not previously worked with Vi Agroforestry were selected as control farms. Data collected from all 360 farms (200 project and 160 control farms) included field size, yield and SALM methods used for all maize fields on the farms during two seasons and four years (2009–2012). However, for the control farms, SALM data were only collected for 2012 and maize yield data for 2009–2011 were obtained retrospectively through farmer recall interviews in 2012 (Figure 2). Data on species, numbers and sizes of farm trees were collected, and months of food self-sufficiency and amount and frequency of savings were ranked by the farmers themselves for 2012 only. All parameters were compared between project and control farms except the effects of individual SALM practices on maize yield, which was done across all farms. The project farmers were monitored for two more years (2013–2014) after this study and the project was then converted to a solely self-monitoring system by all farmers.

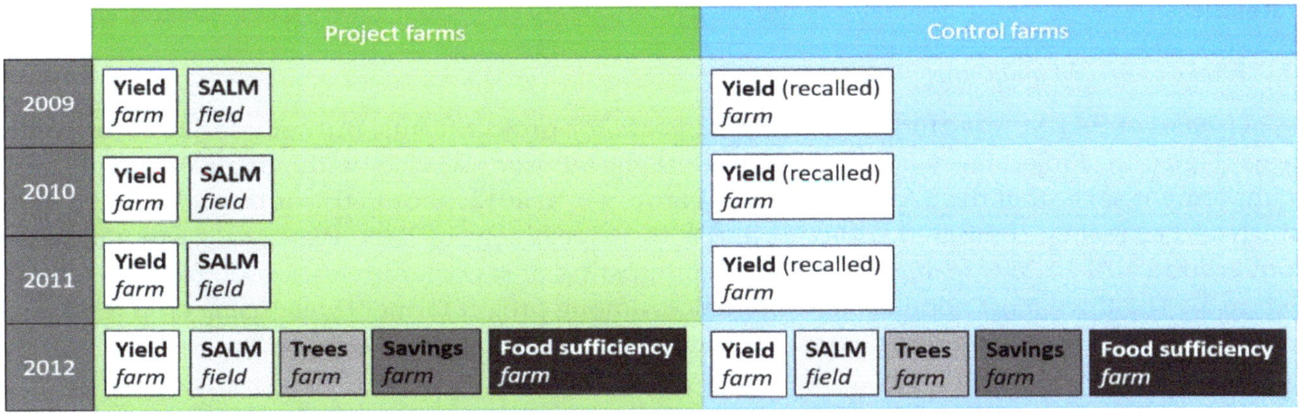

Figure 2. Available data for project and control farms over the years analysed and the scale of the analysis. All data were compared between project and control farms when possible, except for the effects of SALM and trees on maize yield.

2.3. Data Analysis

All statistical analyses were conducted in R 3.4.2 [50]. Correlations between SALM practices and yield were analysed on the field level and for all available years (2009–2012 for project farms, 2012 for control farms). The effects of each SALM practice on maize yield were examined at the field level. For all nine management practices, a linear mixed effect model (the lme function in the nlme package in R) was created, with maize yield as the response variable, usage (or not) of the nine practices, area, year and season as fixed factors, and farm and pair of farms as random factors. The model included recommended SALM practices, such as (1) no tillage, (2) crop residues for direct mulch, (3) raw manure composting, (4) cover crops, (5) terracing and (6) water harvesting structures, and practices to be avoided due to higher carbon emissions, such as (7) removal of crop residues from fields, (8) applying raw manure to fields and (9) burning of residues. Interactions between different practices were not considered, due to large variations in the degree of implementation of each practice. A model simplification procedure was then used to compare and select the model that best explained the variation in the data. The model comparison was carried out using a step-wise Akaike information criterion (AIC) (allowing both forward and backward comparisons). The best-fit model was run using the lmer function in the lme4 package in R to identify potential significant effects of SALM practices.

The effects of agroforestry SALM practices, represented by the number and function/s of farm trees, on maize yield were analysed in a similar way as described above for the SALM practices. The only difference was that the maize yield data used were farm averages from 2012 for each of the two rainy seasons and study areas. For trees, the fixed factors included were area, season and the total number of trees, fodder trees, timber trees and fruit trees. Farm was included as a random factor.

For comparisons of mean seasonal maize yield per farm over time, a linear mixed effect model (lmer function in lme4 package) was used, with four fixed factors (treatment, area, year, season), which were tested as direct effects and with all two-way interactions included in the model. Only project and control farms forming pairs were included in that analysis, which resulted in 78 pairs of farms in Kisumu and 79 in Bungoma. Both individual farms and pairs of farms (project and control) were set as random factors in the model. The paired farms were analysed for four years (2009–2012), and two seasons per year. Contrasts were used to compare the differences in 2009 and 2012 between project and control farms.

The data on food self-sufficiency were categorised into four levels: <6 months food self-sufficiency, 6–7 months, 8–9 months and 10–12 months. A Chi-2 test was used to identify dependencies between food self-sufficiency and KACP participation. The significance level for all analyses was set to $p < 0.05$.

3. Results

3.1. *Uptake of SALM and Other Practices*

Uptake of SALM was studied for all four years for project farms, but only for 2012 for control farms (Figure 2). Project farmers responded well to the advisory services within the KACP and started to implement several of the SALM practices (Figure 3). In 2012, about 60% of the fields were under mulch and terracing, compared with 25% and 40%, respectively, in 2009. Water harvesting increased from around 10% in 2009 to 40% in 2012 and composting of raw manure increased from 50% to 65% (Figure 3). The three most popular SALM practices among project farmers were using crop residues as mulch, composting raw manure before application and terracing fields (Figure 3).

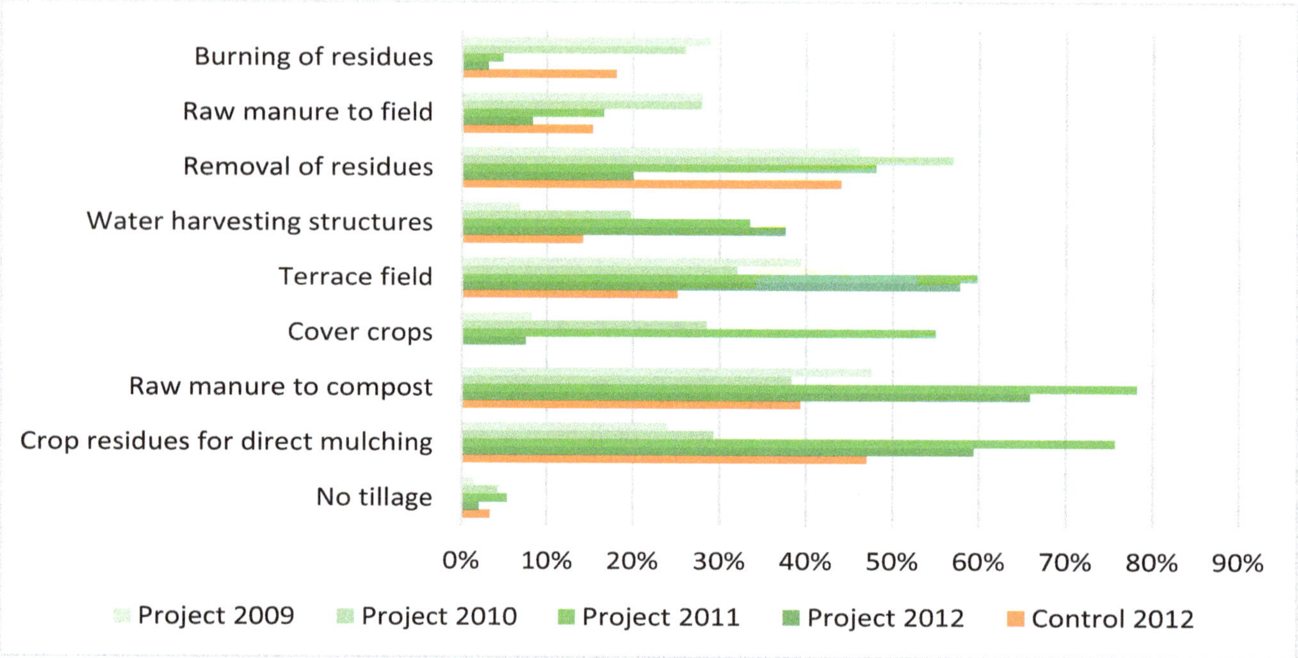

Figure 3. Diagram showing SALM uptake and implementation in fields in the two areas of Kisumu and Bungoma by project farmers and control farmers between 2009 (project start) and 2012. Data for control groups only available for 2012. Data for project farms in Kisumu in 2009 were lost due to a computer malfunction. N = 183 for control 2012, n = 208 for project 2009, n = 481 for project 2010, n = 591 for project 2011 and n = 495 for project 2012.

Some measures never became commonly used, such as no tillage and cover crops. Cover crops increased from 10% in 2009 to 55% in 2011 but decreased again to 10% in 2012. Information on the use of SALM practices on control farms was only available for 2012, at which time crop residues were used for direct mulching on around 50% of control farmers' fields, compared with 25% on project fields in 2009 and 60% in 2012. Raw manure composting was also quite common on control farms, while terracing, use of cover crops and creation of water harvesting structures were rare. The measures to be avoided according to the KACP recommendations were reduced on project farms. Removal of residues decreased from 50% to 20%, raw manure application from 30% to 10% and burning of residues from 30% to close to zero. On the control farms, removing residues from fields and burning of residues were still used to a relatively high degree. In general, by 2012, the project farms had on average more of the promoted practices and fewer practices to be avoided according to the KACP recommendations compared with both the start of the project in 2009 and with the control farms.

Tree planting, especially agroforestry, was another promoted practice taken up by farmers. The majority of trees (counted only in 2012) on the farms were young and planted within the project period. On average, farms in Bungoma had a higher number of trees (98 for project farms, 57 for control farms) than farms in Kisumu (74 for project farms, 40 for control farms) (Figure 4). Most trees on the farms were timber trees (64% on project farms, 75% on control farms). Apart from having on average more trees (86) per farm compared with control farms (48), project farms also had a larger average proportion (19%) of fodder trees than control farms (4%) (Figure 4). In terms of species, on average control farms had more Eucalyptus spp. and project farms had more N-fixing species like Sesbania sesban, Acacia spp. and Calliandra spp. The most common species overall were Grevillea robusta, Markhamia spp. and Albizia spp.

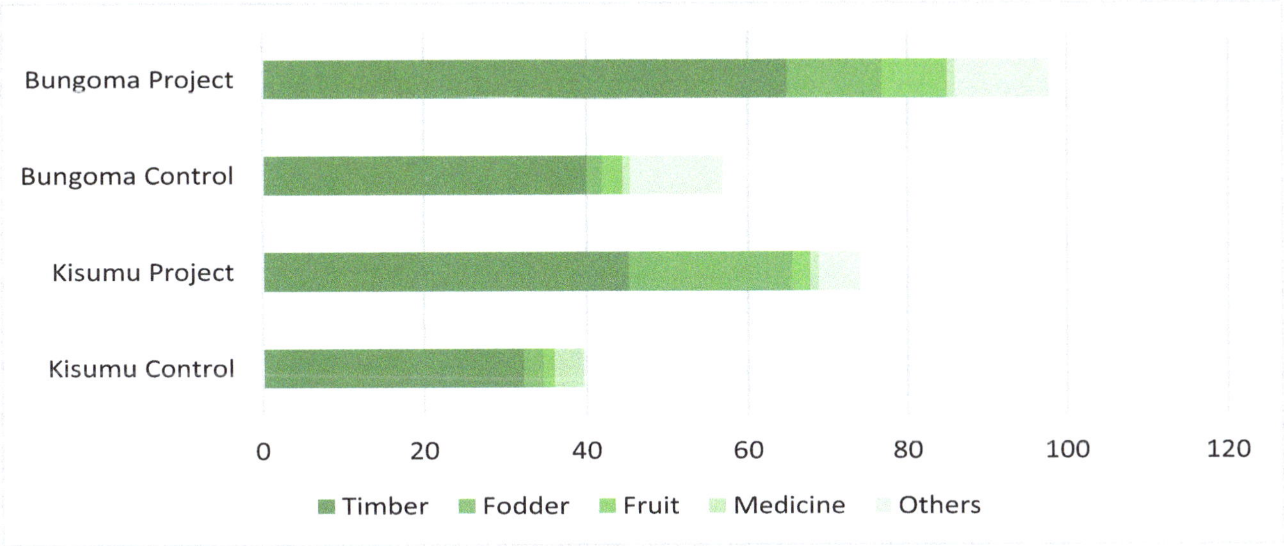

Figure 4. The average number of trees per farm on project and control farms in 2012, divided between fodder, fruit, medicine, timber and other tree types. In total, 65% of all trees were 10 cm or less in diameter at breast height, meaning that they were probably planted within the project period (since 2009). Eucalyptus comprised 4.6% of trees on control farms and 2.7% on project farms. Data from a tree survey in Kisumu (n = 37 control farms, n = 90 project farms) and Bungoma (n = 38 control farms, n = 60 project farms).

3.2. Effects of SALM Practices on Maize Yield

Among the recommended SALM practices, only terracing had a significant ($p = 0.0004$) positive effect on maize yield. However, the management effects were small compared with the highly significant differences ($p < 0.0001$) between years, seasons and regions. Terracing was the only practice that was part of the best explanatory model, together with area, season and year (Yield ~ Area + Year + Season + Terracing + (1|Place/Farm)). When analysing effects from trees, first-season maize yield was not affected by any factor other than region, with Bungoma having a higher yield. Second-season maize yield had the best-fit model that included region and the total number of trees per farm (Yield 2 ~ Region + Total trees + (1|Farm)). Second-season maize yield increased with the increasing total number of trees ($p = 0.02$).

3.3. Maize Productivity

Maize yield varied widely between the areas and was on average 1572–2675 kg ha^{-1} in Bungoma, compared with 725–1661 kg ha^{-1} in Kisumu, among project farms in the first season of the four years (Table 1). There was also a wide variation in second-season yield, which was 518–1054 kg ha^{-1} and 152–678 kg ha^{-1} for Bungoma and Kisumu control farms, respectively. Maize productivity analyses showed higher yield ($p < 0.0001$) for project farms than control farms, higher first-season than second-season yield ($p < 0.0001$) and higher yield ($p < 0.0001$) in Bungoma than Kisumu (Table 1, Figure 5). The four study years also differed ($p < 0.0001$) in terms of yield, with mostly increasing trends (Figure 5). Apart from the main effects, there were three significant interactions (Figure 6): (1) yield differences between project and control farms were larger ($p < 0.0001$) in 2010–2011 than in 2009 or 2012 (Figure 6a); (2) first-season yield increased more ($p = 0.004$) than second-season yield in 2012 (Figure 6b); and (3) there was a larger difference ($p = 0.005$) in the first-season yield compared with the second-season yield between the regions (Figure 6c). Project farms increased their yield mostly in 2010 when the first-season yield declined in control farms. The second-largest increase for project farms was in 2011, while control farms had a one-year lag with their largest increase in 2011 and second-largest in 2012 (although the yield was still lower than on project farms). However, in total, the difference in yield between project and control farms was similar in 2009 and 2012 ($p = 0.15$), and therefore the yield gap between project and control farms did not change significantly during the four initial years of the KACP.

Table 1. Maize yield (kg ha^{-1}) for all project and control farms (measured on field level) in Bungoma and Kisumu. Values are mean ± standard deviation for all four years (2009–2012) and both seasons (1 and 2); n = number of farms included in the analysis.

Year	Season	PROJECT Bungoma N = 96	CONTROL Bungoma N = 55	PROJECT Kisumu N = 85	CONTROL Kisumu N = 44	ALL FARMS N = 280
2009	1	1572 ± 1211	1169 ± 741	725 ± 572	494 ± 397	1069 ± 981
2010	1	1925 ± 1356	1141 ± 934	1055 ± 1037	266 ± 166	1239 ± 1203
2011	1	2647 ± 1756	1222 ± 994	1371 ± 907	702 ± 390	1724 ± 1479
2012	1	2675 ± 1384	2015 ± 1279	1661 ± 1406	1075 ± 840	1939 ± 1397
2009	2	1094 ± 1206	518 ± 411	565 ± 532	321 ± 249	638 ± 781
2010	2	1416 ± 1431	706 ± 469	902 ± 1023	152 ± 101	900 ± 1105
2011	2	1646 ± 1052	783 ± 526	957 ± 757	354 ± 297	945 ± 862
2012	2	1563 ± 941	1054 ± 714	937 ± 744	678 ± 567	982 ± 784

3.4. KACP Effects on Savings and Food Sufficiency

Project farmers added to their savings more often on average than control farmers (Figure 7a) and also added larger amounts per occasion (Figure 7b). More than 45% of project farmers added to savings two or more times per month, while the corresponding figure for control farmers was 21%. Project farmers saved to a larger extent (72%) than control farmers (52%) and Kisumu farmers saved on average more than Bungoma farmers. In the VSLAs within the KACP, farmers were able to borrow up to three times their savings to use for investments.

Among Bungoma farmers, 39% had farm inputs as their main expenditure, compared with just 14% for Kisumu farmers, 60% of whom had food as their main cost. In Bungoma, there was a difference between project and control farmers in that 51% of project farmers spent most on education for their children, while 25% of control farmers still had to spend most on food and thereby only 27% had education as their main expenditure. The majority of farmers in both Kisumu and Bungoma had their main source of income from agricultural products.

Only 4% of control farmers in Kisumu (31% in Bungoma) had enough food for 10–12 months, compared with 16% of project farmers in Kisumu (51% in Bungoma). Moreover, 46% of control farmers in Kisumu and 25% in Bungoma had enough food for less than six months, while the corresponding values for project farmers were 19% and 7%, respectively. In general, farmers in Bungoma had more months of food sufficiency than farmers in Kisumu. Chi-2 tests revealed that food sufficiency was significantly higher ($p < 0.001$) overall for project farmers than control farmers.

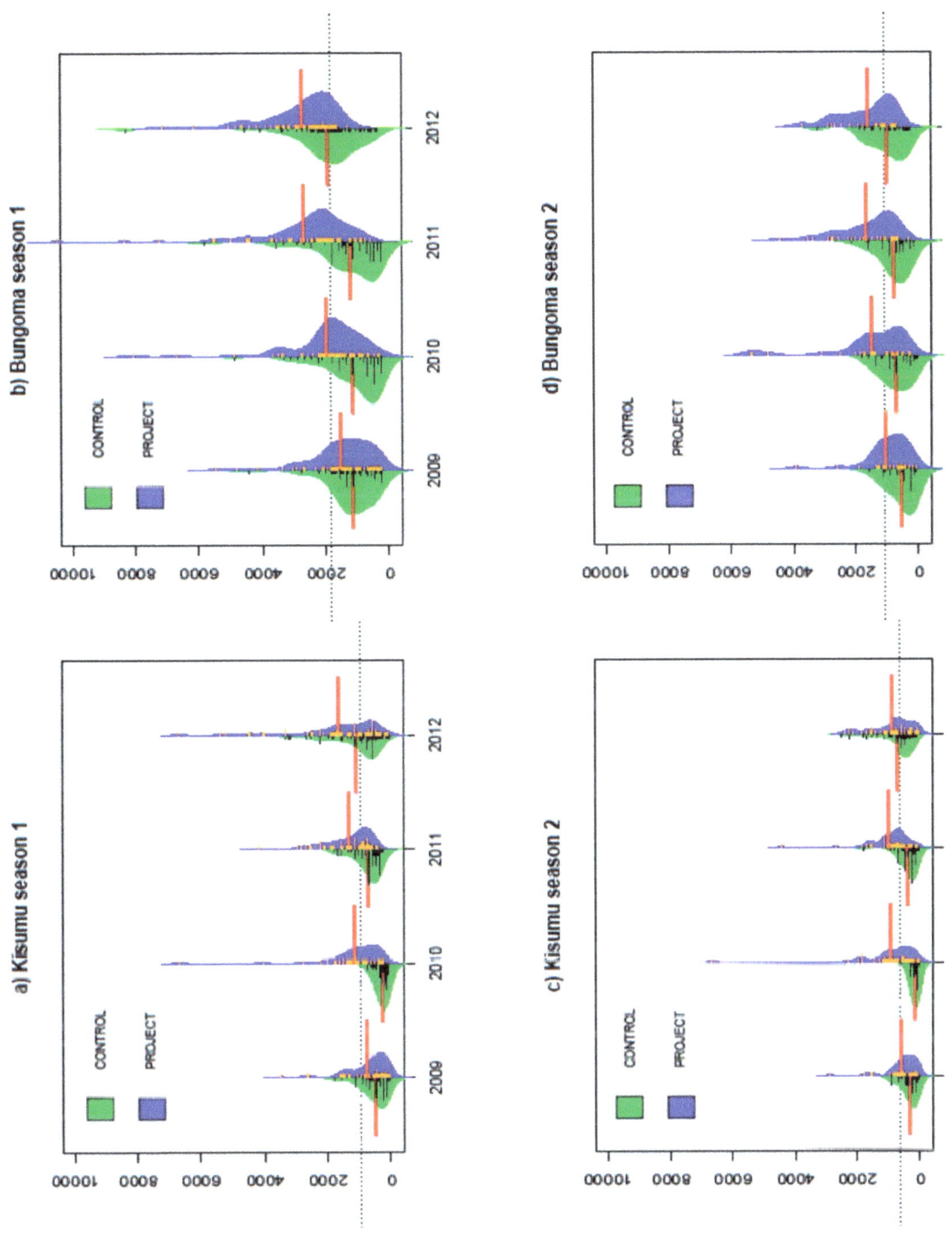

Figure 5. First-season (**a**,**b**) and second-season (**c**,**d**) maize yield kg ha^{-1} on project (blue) and control (green) farms in Kisumu (**a**,**c**) and Bungoma (**b**,**d**), 2009–2012. The lines indicate mean values for each distribution and the dotted line shows the mean for each sub-plot.

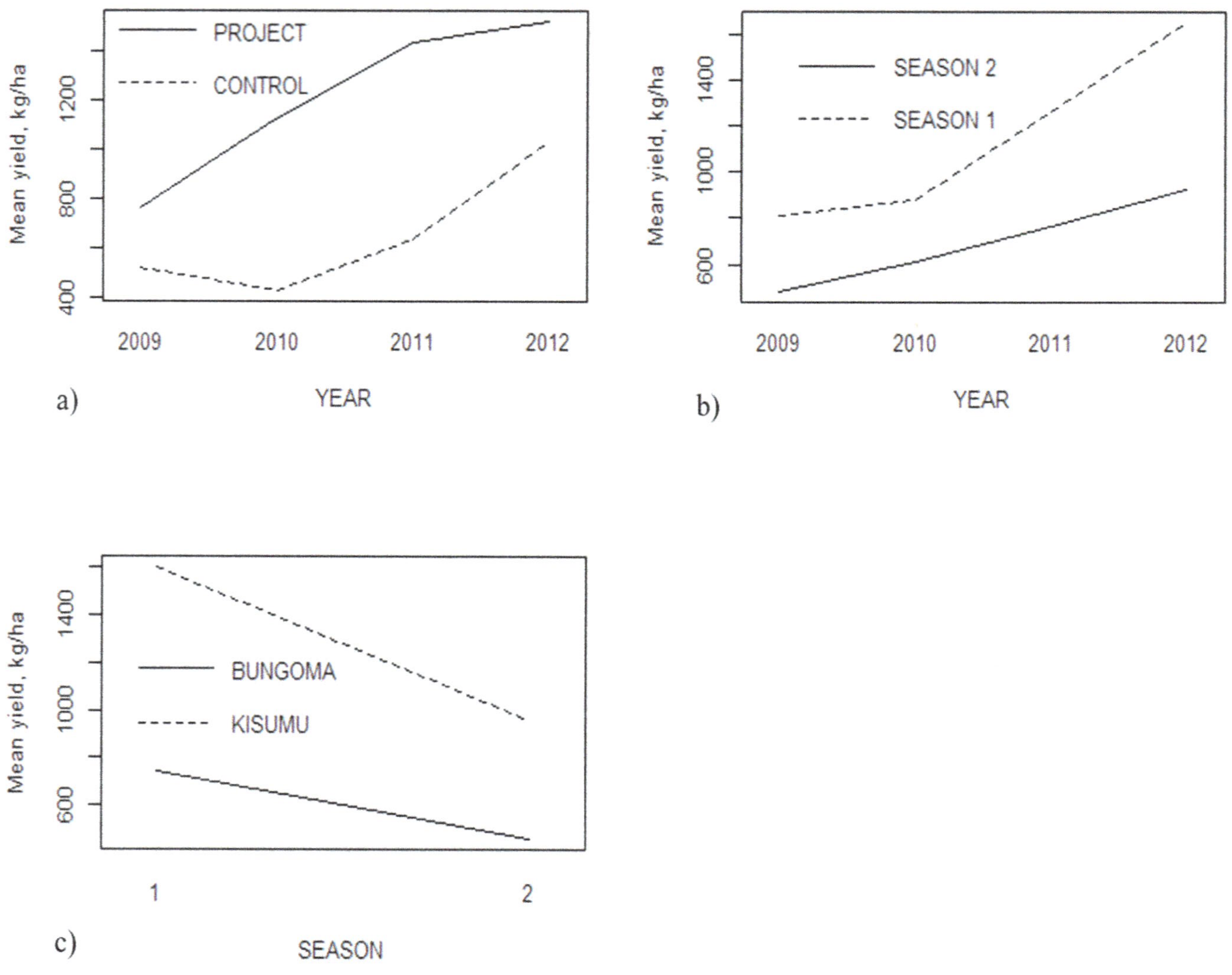

Figure 6. Diagrams showing the three significant interactions identified, between: (**a**) year and treatment ($p < 0.0001$), (**b**) year and season ($p = 0.004$) and (**c**) region and season ($p = 0.005$) on maize yield in kg ha^{-1} for project and control farms.

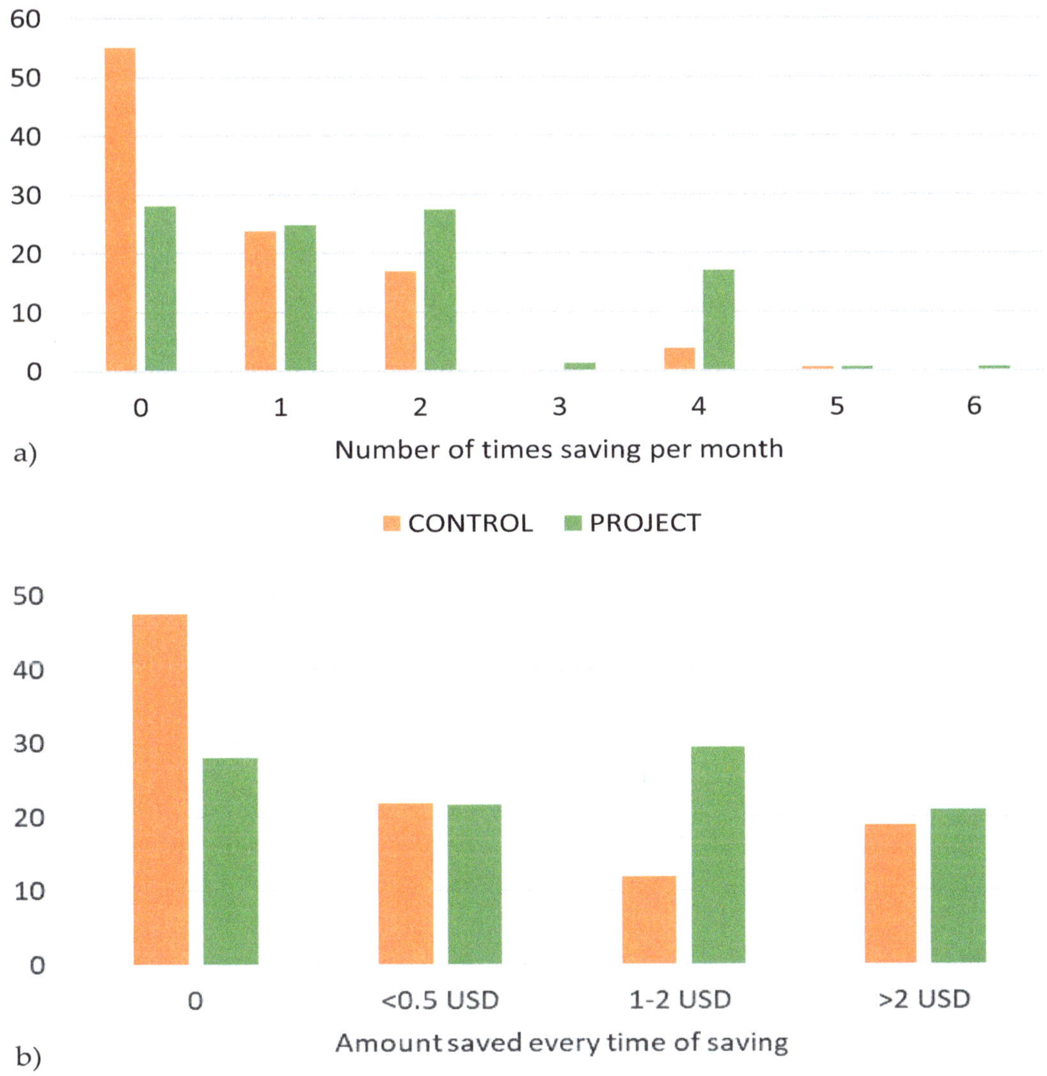

Figure 7. (**a**) Percentage (y-axis) of project and control farm households saving money on different numbers of occasions per month in 2012. (**b**) Percentage of project and control farm households saving different amounts on every saving occasion in 2012.

4. Discussion

4.1. Mulching, Terracing and Trees More Popular Than No Tillage

By 2012, there was a large difference in the use of terraces between project (60%) and control fields (25%) (Figure 3). Terraces can greatly enhance soil moisture recharge [51] but require major initial labour inputs [52]. Cover crop uptake among project farmers increased until 2011, but declined to the initial level in 2012, possibly due to crop rotation patterns or to dissatisfaction with the practice itself (Figure 3). Cover crops/green manure grown for six weeks (planted two weeks earlier than maize), and then incorporated into soil can significantly improve soil fertility by adding nitrogen, increasing maize yield in the same season [53]. However, convincing farmers to grow cover crops mainly for soil fertility reasons was apparently difficult in the KACP (Figure 3), as in other projects [54]. Different SALM practices had different degrees of uptake, for example, mulching and terracing became popular, while no tillage did not. Minimum tillage practices have varying and sometimes negative or contradictory outcomes for smallholders in the region [55,56] and the willingness to use minimum tillage is lower than, for example, mulching [57]. It is also debatable whether minimum tillage options can act as a carbon sink [58,59].

Advice on SALM practices to be used, and practices to be avoided, seemed to be adhered to more by project farmers in 2012 compared with the initial use in 2009, and also compared with control farmers and with the uptake of practices in a similar study [60]. Use of soil fertility management practices is generally low among smallholders in Western Kenya [60] but is influenced positively by, for example, plot size, market and labour access, off-farm earnings and knowledge [60,61]. This implies that the poorest farmers may have less possibility to use such practices.

Trees in the agricultural landscape, that is agroforestry, can have both positive and negative effects on production and other ecosystem services, but most studies show a net positive effect [62]. Fodder trees were promoted by the KACP advisors, which can explain why project farms had a larger share of leguminous fodder trees than control farms (Figure 4). A lack of fodder trees on control farms could also explain the lower use of crop residues for mulch, as more crop residues might be needed for livestock feed. The larger number of trees per farm in Bungoma than Kisumu may be because the farms were larger in Bungoma. It is also easier for tree seedlings to survive in Bungoma since most farmers control their grazing animals and prevent them from browsing on tree seedlings.

4.2. Terraces and Fodder Trees Increased Maize Productivity

Of the six SALM practices and three non-SALM practices analysed, only terraces had a significant positive effect on maize yield. The differences in SALM practices between 2009 and 2010 (when the yield increase was largest for project farms) were mainly the increased use of cover crops and water harvesting structures. However, neither practice showed any significant correlation with yield. The four-year study period can be expected to be sufficient to reveal the effects of most management practices [63]. While the main aim of SALM practices in the KACP is to reduce greenhouse gas emissions and sequester more carbon, they can also improve water availability for crops and soil fertility, both of which improve yield [17,64]. However, few SALM practices showed any significant effects on maize yield in this study. One drawback related to terraces is the relatively high labour demand needed upon establishment, which may restrict farmers from using the measure.

The intensity of the different SALM practices was not studied (since data was not available), but differences in intensity could have masked relationships between management and yield. Most recommended practices are dependent on the type, quantity and quality of, for example, mulch, compost, water harvesting or cover crops. Project farmers, for example, reduced burning and removal of residues (for fodder or fuel) in order to use more as mulch, but 40% of residues were still used for purposes other than mulching in 2012 (Figure 3), suggesting that much is used as fodder and could be returned as manure. Mulching is known to increase and stabilise maize yield [55,57]. However, practices related to crop residues (mulching, composting, green manure, etc.) require trade-offs between several uses, and the end-use often reflects more immediate concerns such as feeding livestock rather than improving soil fertility in the longer term [65,66]. With more fodder trees on the farm, feed and fuel could come from sources other than crop residues.

The use of agroforestry SALM practices was popular in project farms, which had on average around double the number of trees as control farms. A possible drawback with tree–maize intercropping is competition for light, water and nutrients unless the trees are adequately pruned above and below ground [67]. However, the number and type of trees did not have any effect on the first-season maize yield. Timber trees, which were the most common type used, are often planted in woodlots and therefore rarely interfere with crops, while fodder trees are more often intercropped within fields. In this study, second-season maize yield was positively affected by the total number of trees, possibly because farms with many trees had a larger share of fodder trees that are leguminous and can fix the nitrogen from the atmosphere, which has been found earlier [67,68]. Apart from increasing second-season maize yield, trees could provide other services (animal feed, fruits, timber and firewood) which were not considered in this study. Further, a combination of trees and grass is effective in holding and maintaining the soil on hillslopes and terraces [69]. Agroforestry is known to be both knowledge and labour intensive which can limit the uptake of the practice [70].

4.3. Maize Productivity Increased but SALM Practices Only Part of the Explanation

Average maize yield on small, subsistence-focused farms in western Kenya was 0.9 tonnes ha^{-1} during the study period [71]. Nation-wide, mean annual maize yield in Kenya in the four years 2009–2012 was 1294, 1725, 1584 and 1737 kg ha^{-1} [72], compared with 1707, 2139, 2669 and 2921 kg ha^{-1} for both seasons and all farmers in this study. The first cropping season usually has more rain days and larger amounts of rainfall ("long rains") than the second season ("short rains") in both study areas, which largely explains the higher productivity in first-season maize (Figures 5 and A1). Farmers who composted their manure applied it to fields once per year at the beginning of the first season, which may be another factor in the yield differences between seasons. Overall, Bungoma had higher amounts of rainfall and a cooler climate (Table A2 and Figure A1), leaving more water available for crops instead of being lost through evapotranspiration. Therefore, productivity was significantly higher in Bungoma and the pattern was the same for all years and seasons (Figure 5).

Water is not the only limiting factor for maize growth in Western Kenya, as nitrogen and phosphorus are also main limiting factors [64]. Small-scale farms with subsistence-focused production tend to have resource flows showing net nutrient export and low values of carbon and nitrogen especially [71]. Smallholders in the KACP were encouraged to practise integrated soil fertility management, with careful use of inorganic fertiliser and more focus on organic fertilisers. This advice is debatable since many researchers view inorganic fertilisers as important in increasing yield and boosting biomass production for higher carbon cycling [73,74]. The use of inorganic fertilisers may partly explain the different yield patterns for project and control farms (Figure 6a). However, the use of inorganic fertilisers was not included in this study and therefore needs further studies for a better understanding of its effects with or without SALM practices.

The reason why the first-season yield increased more than second-season yield in 2012 (Figure 6b) was likely that yield actually decreased in 2010 for control farms, and also that SALM use was higher for first-season crops and high for both project and control farmers in 2012. The greater yield increase in Bungoma than Kisumu for first-season crops (Figure 6c) could be because Bungoma farmers concentrate on agriculture, while Kisumu farmers were traditionally fishermen and also have more off-farm job opportunities close by, making them less dependent on farming [66]. Farmers who are more focused on farming are likely to be more eager to invest in their farm and use new practices. The Bungoma area (known as "Kenya's breadbasket") also has more responsive soils and a more favourable climate than the Kisumu area.

4.4. Role of Control Farms in Interpreting Results

Immediate effects from SALM practices related to water-holding capacity [64,75] could explain the initial higher yield increases for project farms. Possible reasons why the control farms increased their yield more than project farms in 2011–2012 (Figure 6a) could be reductions in inorganic fertiliser use in project farms, control farms copying practices from project farms or project farmers paying more attention to their crops during the first years of the project, but then relaxing. Learning from neighbours and friends (farmer-to-farmer) is often the most common [76] and most effective [77] way of learning new practices in this region, which could explain both the relatively high SALM uptake in 2012 and delayed yield increases on control farms compared with project farms. Farmer-to-farmer uptake within the KACP areas was reported by Hughes et al. [42], who found significant uptake of certain practices within the project area (not only among the targeted farmer groups) than in neighbouring areas outside the project. However, this does not mean that advisory services were not worth the effort. Farmers generally want better access to advisory services [78], especially in the early stages of adoption [77].

The project and control farmers were all within the KACP areas and lived in the same villages. For control farms, the use of the different SALM practices was only studied in 2012 and by then their use of practices recommended by the KACP was on average higher than the initial (2009) use on project farms. Maize productivity at the start of the project (2009) was higher on project farms than control farms and, while maize yield increased over the years, the difference between project and control farms was similar in 2012. Thus, comparisons of maize yield initially and at the end of the monitoring period showed no overall difference between project and control farms. The value of having control farms in addition to the baseline production in 2009 was to ensure that the yield differences were not normal annual fluctuations due to weather patterns. Control farms were two or three farms away from their paired project farm, so soils and weather conditions were probably similar. However, there might be other differences between project and control farms.

Participation in the KACP was voluntary and participants were generally both asset-poor and income-poor. However, earlier research has found higher adoption of SALM practices among the less poor in the project [41]. This may indicate that it is not the poorest of the poor that risk venturing into such projects, as has been found elsewhere [79]. However, they can perhaps not be expected to participate unless they are specifically targeted with up-front financial support to overcome their risk aversion. In earlier studies, the main factors limiting uptake of SALM practices within the KACP were found to be lack of labour to implement SALM practices, knowledge on how to implement them and land availability [41,44]. In addition, women had problems finding the time to attend training [43,44].

4.5. KACP Farms Had Higher Savings and Food Sufficiency

Kisumu farmers were able to save on average more than Bungoma farmers. This was most likely partly due to more off-farm income opportunities in Kisumu and partly because the yield was lower in Kisumu, so farmers needed cash savings to buy food. On average, project farmers saved to a larger extent, more often and in larger amounts than control farmers, probably because the VSLA concept in the KACP provided opportunities to save (Figure 7a,b). The VSLA methodology has earlier shown positive results for several household indicators, such as the development of income-generating enterprises and access to education and food [80]. Preliminary data from a study in 2016 also indicated higher income from tree products, better fuelwood access and higher milk yield among the KACP farmers compared with neighbouring control areas [42]. All of these can increase the scope of cash savings by households.

Project farmers at both sites had significantly higher food sufficiency than control farmers. This can be directly related to differences in maize yield, but maize is just one of many food crops. The improved food sufficiency may also be attributable to other farm commodities such as sorghum, beans and cassava, yields of which probably also benefit from terracing and other SALM practices. Improved yields and better food sufficiency among the KACP farmers have been reported in an earlier study of women and men farmers [44]. The higher level of food sufficiency in Bungoma most likely reflects the greater farm size and higher average yields in this area than in Kisumu. Agroforestry practices could have added to food sufficiency through, for example, fruits. Earlier studies report maintained yields in combination with such additional benefits from agroforestry [81].

4.6. Limitations of the Study

It is a strength that control farms were included in the study, rather than just monitoring changes over time from the baseline on the project farms. However, the study has some limitations partly related to the control farms. A possibility of farmers participating in the KACP being different, perhaps better-off, than those who did not participate emerged early in the project, so an attempt was made to avoid potential self-selection bias by adding a control group of farmers in 2012. However, this meant that no control farms were monitored from the beginning of the project and there is uncertainty in the data on maize yield based on control farmer recall for the first three years (2009–2011). Moreover, uptake of SALM practices on control farms was only assessed in 2012 and thus data were lacking for 2009–2011. The identified control farms were also perhaps too close to project farms, which may have given a non-representative control group. Another limitation of this study was the "weak definitions" of uptake of practices. The uptake data were binary (practiced or not practiced) and the quantities of mulch, manure, etc. applied were not recorded, which created large variation in the effects of farmers' uptake of practices.

4.7. Implications and Recommendations

Overall, the KACP had a tendency for positive effects regarding uptake and effect of certain SALM practices, development of maize yield, food sufficiency and savings. However, to obtain stronger evidence that the results are due to the intervention, large-scale projects like this would need to include control farms in the design from the beginning, together with clearly defined quantitative indicators of uptake of practices. This could be done jointly by development agencies and research institutions, in order to achieve an optimal design and reduce the costs of monitoring and evaluation. Some of the practices promoted in the KACP were also questionable (e.g., no tillage and careful use of inorganic fertiliser), especially considering the relatively poor target group. A primary aim of the advisory services in such projects should be yield and productivity increases and building resilience, while climate change mitigation objectives should be viewed as a secondary goal [41]. The target group should not be expected to take any risks. Of the SALM practices promoted, only terraces and trees showed positive (but relatively small) effects on yield. Thus, other parts of the advisory services, for example, micro-credit access [82], might have played a greater role in the increased maize yield and savings and need to be explored in future research.

5. Conclusions

This is the first research study on the yield effects of promoted practices in an agricultural soil carbon project for smallholders in sub-Saharan Africa. Advisory services provided to farmers participating in the KACP appeared to increase the uptake of some of the soil management and tree planting (especially fodder trees) practices and to lower the use of non-sustainable practices. Indications of some uptake through farmer-to-farmer horizontal learning were observed for control farmers.

Higher inclusion of trees (agroforestry) and the use of terraces were the only practices that correlated positively with higher maize yield. The lack of effect of practices was likely due to vague definitions of the practices within the project. Maize yield increased from the start of the KACP and continued over the four years studied and was higher among project farmers than control farmers during all years. The yield increase was similar for project and control farmers but the timing differed, since project farms achieved their main increase during the first years, while control farms obtained their main increase during later years.

Participants in the KACP had higher food self-sufficiency and the results indicated greater ability to save money, both more frequently and in greater amounts. Thus, apart from carbon revenues, the KACP farmers seemed to use more sustainable management practices, had higher maize yield, had better food self-sufficiency and tended to save more money than control farmers. This improved the preparedness of the smallholder farmers participating in the KACP and increased their resilience for future challenges. However, it was not possible to determine whether those farmers were already on this path to improvement before joining the KACP.

Author Contributions: Conceptualization, Y.N. and I.Ö.; methodology, I.Ö., Y.N., C.M. and E.W.; software, Y.N., C.M. and E.W.; validation, Y.N., C.M., E.W. and I.Ö.; formal analysis, Y.N.; investigation, C.M. and E.W.; resources, I.Ö.; data curation, Y.N.; writing—original draft preparation, Y.N.; writing—review and editing, Y.N., C.M., E.W., J.W., M.J. and I.Ö.; visualization, Y.N., E.W.; supervision, J.W., M.J. and I.Ö.; project administration, Y.N.; funding acquisition, I.Ö. All authors have read and agreed to the published version of the manuscript.

Acknowledgments: Farmers and advisors around Kisumu and Bungoma are gratefully acknowledged for their comprehensive efforts. Special thanks to Bo Lager, Fred Marani, Wangu Mutua, Amos Wekesa, Peter Wachira and the rest of the Vi Agroforestry team. Thanks also to Johannes Forkman, who assisted in statistical analyses.

Appendix A

Table A1. Sustainable Agricultural Land Management (SALM) practices promoted and implemented in the Kenya Agricultural Carbon Project (KACP) (after [31]).

SALM Categories	SALM Practices
Nutrient management	Mulching, composting, cover crops, nitrogen-fixing crops, manure, restricted chemical fertilisers and chemical management
Soil and water conservation	Terraces, contour bunds, broad beds and furrows, semi-circular bunds, trash lines, diversion ditches and cut-off drains, retention ditches, pitting, trenches, tied ridges, grass strips, irrigation, roof catchment, ground surfaces and rocks, irregular surfaces, tanks, birkas, pans, ponds, dams, wells and boreholes, ecological sanitation, kitchen water
Agronomic practices	Crop rotation, intercropping, green manure, contour strip cropping, relay cropping, use of improved crop varieties
Agroforestry	Plant trees amongst crops, trees and livestock, trees, crops and livestock, trees and insects, trees and water animals, woodlots, boundary planting, dispersed interplanting, fruit orchards
Tillage and residue management	No-tillage/zero-tillage, reduced tillage, pitting systems, stubble and residue mulch tillage, dibble stick planting, strip and spot tillage, ripping, ridge and furrow tillage, residue management
Land restoration and rehabilitation	Natural regeneration, assisted natural regeneration, enrichment planting, fire management, agroforestry
Integrated livestock management	Improved feeding and watering, housing, stall management systems, improved waste management, pest and disease control, improved breeding practices
Sustainable energy	Biomass, biogas, farm residues, energy-efficient stoves, sustainable charcoal production
Integrated pest management	Biological pest control, use of crop-resistant varieties, alternative agricultural practices (spraying, use of fertilisers, pruning), mechanical pest control, pesticides, cultural methods, pest management plan

Table A2. Characteristics of the Kenya Agricultural Carbon Project (KACP) sites at Kisumu and Bungoma.

Parameter	Kisumu Site	Bungoma Site	Data Source
Counties	Siaya, Kisumu	Bungoma	[83]
Divisions	Wagai, Kombewa, Madiany	Bumula, Malakisi, Sirisia	[83]
Location	South Gem, West Gem, North West Gem, North East Gem, South West Gem, South Central Seme, North Central Seme, South West Seme, West Seme, West Uyoma, Central Uyoma, East Uyoma, South Uyoma	Bumula, Kabula, Mabusi, South Bukusu, Khasoko, Siboti, Mukwa Kimaeti, Napara, Kibuke, Malakisi, Namubila, Lwandanyi, Sirisia, Namwela	[83]
Agro-ecological zones [1]	LM1, LM2, LM3, LM4	UM1, UM2, LM1, LM2, LM3	[84]
Soils	Clay content about 39%	Clay content about 20%	[85]
Altitude (m above sea level)	1200–1500 m	1200–1850 m	[86]
Major crops	Maize & sorghum	Maize & sugarcane	Data from project farms n = 200
Average farm size	0.7 ha	1.1 ha	Data from project farms n = 200
Average household size	3.6 adults and 3.2 children	3.7 adults and 4.4 children	Data from project farms n = 200
Mean temperature range and mean annual precipitation	17.4 °C–29.8 °C; 1326 mm	14 °C–27.6 °C; 1884 mm	[83]
Project locations	0°7′45.53″ N, 34°23′38.56″ E; 0°23′34.29″ S; 34°17′58.55″ E	0°27′0.12″ N, 34°31′14.87″ E; 0°48′18.13″ N, 34°24′54.61″ E	[83]
Population density (persons km^{-2})	333 and 465 in Siaya and Kisumu County, respectively	454 in Bungoma County	[87]

[1] LM1 = Lower Midland 1; LM2 = Lower Midland 2; LM3 = Lower Midland 3; LM4 = Lower Midland 4; UM1 = Upper midland 1; UM2 = Upper midland 2.

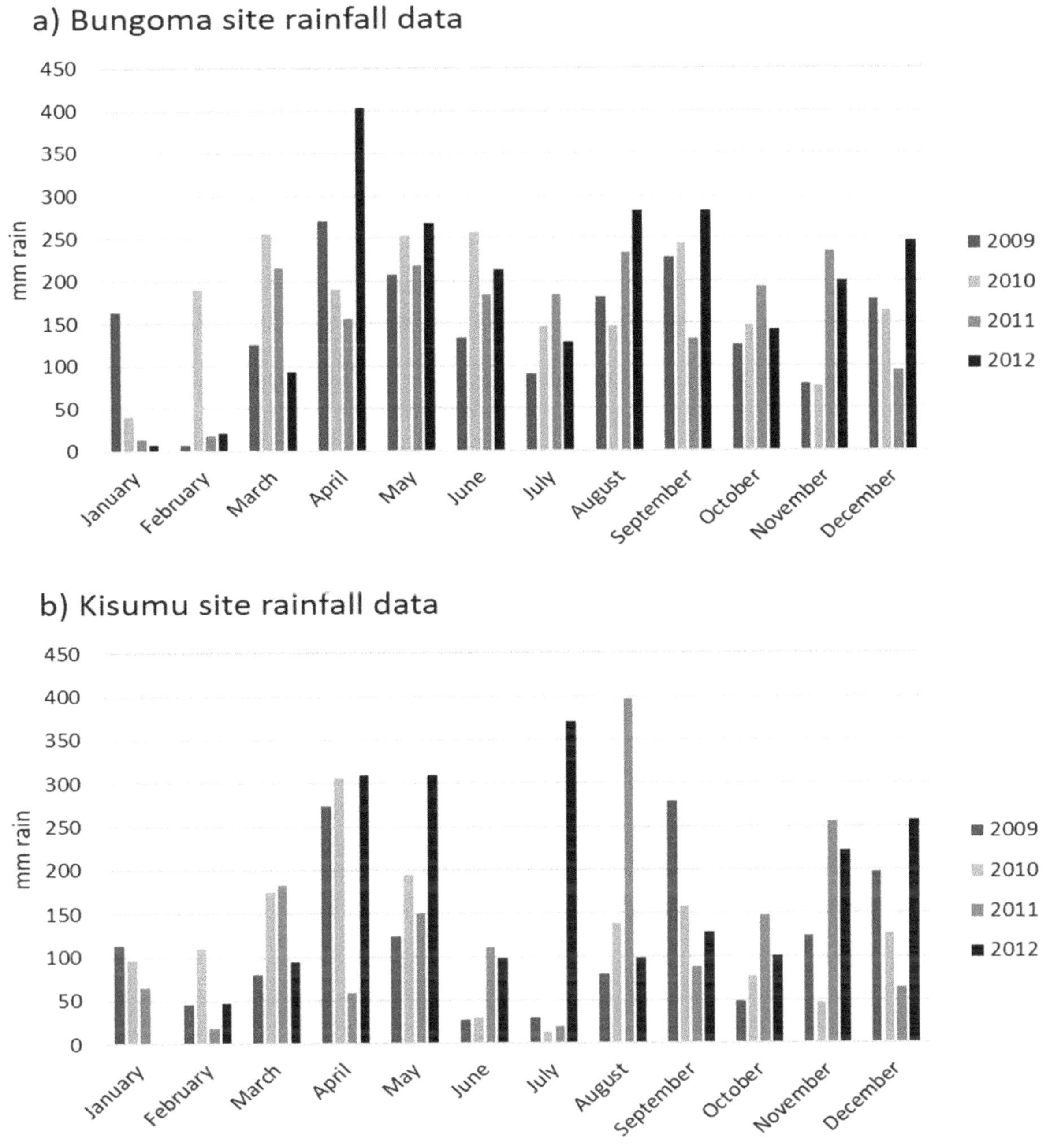

Figure A1. Monthly rainfall data in the study areas, 2009–2012; (**a**) Bungoma site (Kakamega meteorological station; total annual rainfall 2009–2012 = 1789, 2113, 1876 and 2287 mm, respectively). (**b**) Kisumu site (Kisumu meteorological station; total annual rainfall 2009–2012 = 1426, 1477, 1563, and 2036 mm, respectively).

References

1. Elliott, J.; Deryng, D.; Müller, C.; Frieler, K.; Konzmann, M.; Gerten, D.; Glotter, M.; Flörke, M.; Wada, Y.; Best, N.; et al. MConstraints and Potentials of Future Irrigation Water Availability on Agricultural Production under Climate Change. *Proc. Natl. Acad. Sci. USA* **2014**, *111*, 3239–3244. [CrossRef] [PubMed]
2. Foley, B.J.; Melton, F.; Middleton, E.; Hain, C.; Anderson, M.; Allen, R.; McCabe, M.F.; Hook, S.; Baldocchi, D.; Townsend, P.A.; et al. The Future of Evapotranspiration: Global Requirements for Ecosystem Functioning, Carbon and Climate Feedbacks, Agricultural Management, and Water Resources. *Water Resour. Res.* **2017**, *53*, 2618–2626.

3. Foley, J.A.; Ruth, D.; Gregory, P.; Asner, C.B.; Gordon, B.; Stephen, R.; Carpenter, F.; Stuart, C.; Michael, T.; Coe, G.; et al. Global Consequences of Land Use. *Science* **2005**, *309*, 570–574. [CrossRef]
4. Godfray, H.; Charles, J.; Tara, G. Food Security and Sustainable Intensification. *Philos. Trans. R. Soc. B* **2014**, *369*, 20120273. [CrossRef]
5. IPCC. Summary for Policymakers. In *Climate Change 2014: Impacts, Adaptation, and Vulnerability. Part A: Global and Sectoral Aspects. Contribution of Working Group II to the Fifth Assessment Report of the Intergovernmental Panel on Climate Change*; Field, C.B., Barros, V.R., Dokken, D.J., Mach, K.J., Mastrandrea, M.D., Bilir, T.E., Chatterjee, M., Ebi, K.L., Estrada, Y.O., Genova, R.C., et al., Eds.; Cambridge University Press: Cambridge, UK; New York, NY, USA, 2014; pp. 1–32.
6. Johan, R.; Williams, J.; Daily, G.; Noble, A.; Matthews, N.; Gordon, L.; Wetterstrand, H.; DeClerck, F.; Shah, M.; Steduto, P. Sustainable Intensification of Agriculture for Human Prosperity and Global Sustainability. *Ambio* **2017**, *46*, 4–17.
7. Bergström, L.; Bowman, B.T.; Sims, J.T. Definition of Sustainable and Unsustainable Issues in Nutrient Management of Modern Agriculture. *Soil Use Manag.* **2005**, *21*, 76–81. [CrossRef]
8. Horrigan, L.; Robert, S.; Walker, P. How Sustainable Agriculture Can Address the Environmental and Human Health Harms of Industrial Agriculture. *Environ. Health Persp.* **2002**, *110*, 445–456. [CrossRef]
9. Will, S.; Richardson, K.; Rockström, J.; Sarah, E.; Fetzer, C.I.; Bennett, M.E.; Biggs, R.; Carpenter, S.R.; de Vries, W.; de Cynthia, W. Planetary Boundaries: Guiding Human Development on a Changing Planet. *Science* **2015**, *347*. [CrossRef]
10. Wivstad, M.; Dahlin, A.S.; Grant, C. Perspectives on Nutrient Management in Arable Farming Systems. *Soil Use Manag.* **2005**, *21*, 113–121. [CrossRef]
11. Altieri, M.; Clara, A.; Nicholls, I.; Rene, M. Technological Approaches to Sustainable Agriculture at a Crossroads: An Agroecological Perspective. *Sustainability* **2017**, *9*, 349. [CrossRef]
12. Buttel, F.H. Sustaining the Unsustainable. In *Handbook of Rural Studies*; Paul, C., Ed.; Terry Marsden and Patrick Mooney; SAGE Publications Ltd.: London, UK, 2006.
13. IPCC. Summary for Policymakers. In *Global Warming of 1.5 °C. An IPCC Special Report on the Impacts of Global Warming of 1.5 °C above Pre-Industrial Levels and Related Global Greenhouse Gas Emission Pathways, in the Context of Strengthening the Global Response to the Threat of Climate Change, Sustainable Development, and Efforts to Eradicate Poverty*; Masson-Delmotte, P., Zhai, H.O., Pörtner, D., Roberts, J., Skea, P.R., Shukla, A., Pirani, W., Moufouma-Okia, C., Péan, R., Pidcock, S., et al., Eds.; World Meteorological Organization: Geneva, Switzerland, 2018.
14. Abraha, M.; Hamilton, S.K.; Chen, J.; Robertson, G.P. Ecosystem Carbon Exchange on Conversion of Conservation Reserve Program Grasslands to Annual and Perennial Cropping Systems. *Agric. For. Meteorol.* **2018**, *253*, 151–160. [CrossRef]
15. Lal, R. Soil Carbon Sequestration Impacts on Global Climate Change and Food Security. *Science* **2004**, *304*, 1623–1627. [CrossRef]
16. Lal, R. Global Potential of Soil Carbon Sequestration to Mitigate the Greenhouse Effect. *Crit. Rev. Plant. Sci.* **2003**, *22*, 151–184. [CrossRef]
17. Mueller, N.; James, D.; Gerber, S.; Matt, J.; Deepak, K.R.; Navin, R.; Jonathan, A.F. Closing Yield Gaps through Nutrient and Water Management. *Nature* **2012**, *490*, 254–257. [CrossRef] [PubMed]
18. Öborn, I.; Vanlauwe, B.; Phillips, M.; Thomas, R.; Brooijmans, W.; Atta-Krah, K. *Sustainable Intensification in Smallholder Agriculture: An. Integrated Systems Research Approach*; Öborn, I., Vanlauwe, B., Phillips, M., Thomas, R., Brooijmans, W., Atta-Krah, K., Eds.; Earthscan Food and Agriculture Series: London, UK; Routledge Taylor and Francis Group: New York, NY, USA, 2017.
19. Swallow, M.B.; Goddard, T.W. Value Chains for Bio-Carbon Sequestration Services: Lessons from Contrasting Cases in Canada, Kenya and Mozambique. *Land Use Policy* **2013**, *31*, 81–89. [CrossRef]
20. Paris Agreement. *Paper Presented at the Conference of the Parties to the United Nations Framework Convention on Climate Change*; 21st Session; FCCC: Paris, France, 2015.
21. Government, K. *National Climate Change Action Plan 2018-2022*; Ministry of Environment and Forestry: Nairobi, Kenya, 2018.
22. Government, Kenya. *National Climate Change Action Plan 2013–2017. 258*; Ministry of Environment and Mineral Resources: Nairobi, Kenya, 2013.

23. Ylva, N.; Jonsson, M.; Ambjörnsson, E.L.; Wetterlind, J.; Öborn, I. Smallholders' Awareness of Adaptation and Coping Measures to Deal with Rainfall Variability in Western Kenya. *Agroecol. Sustain. Food* **2020**, *44*, 1–29.
24. Sikstus, G.; Sumeni, S.; Sabodin, R.; Muqfi, I.H.; Nur, M.; Hairiah, K.; Useng, D.; van Noordwijk, M. Soil Organic Matter, Mitigation of and Adaptation to Climate Change in Cocoa–Based Agroforestry Systems. *Land* **2020**, *9*, 323.
25. Elizabeth, B.; Ringler, C.; Okoba, B.; Roncoli, C.; Silvestri, S.; Herrero, M. Adapting Agriculture to Climate Change in Kenya: Household Strategies and Determinants. *J. Environ. Manag.* **2013**, *114*, 26–35.
26. FAO. *Adapting to Climate Change through Land and Water Management in Eastern Africa—Results of Pilot Projects in Ethiopia, Kenya and Tanzania*; FAO, Ed.; FAO: Rome, Italy, 2014.
27. Amy, Q.; Neufeldt, H.; McCabe, J.T. Building Livelihood Resilience: What Role Does Agroforestry Play? *Clim. Dev.* **2019**, *11*, 485–500.
28. Vermeulen, S.J.; Aggarwal, P.K.; Ainslie, A.; Angelone, C.; Campbell, B.M.; Challinor, A.J.; Hansen, J.W.; Ingram, J.S.I.; Jarvis, A.; Kristjanson, P. Options for Support to Agriculture and Food Security under Climate Change. *Environ. Sci. Policy* **2012**, *15*, 136–144. [CrossRef]
29. Nair, P.K.R.; Vimala, D.; Nair, M.; Kumar, B.; Solomon, G.H. Soil Carbon Sequestration in Tropical Agroforestry Systems: A Feasible Appraisal. *Environ. Sci. Policy* **2009**, *12*, 1099–1111. [CrossRef]
30. UNFCCC. Nationally Determined Contributions (Ndcs). United Nations Framework Convention on Climate Change. Available online: https://unfccc.int/process-and-meetings/the-paris-agreement/nationally-determined-contributions-ndcs#eq-2 (accessed on 30 August 2019).
31. Stefanie, E.; Muller, A. Payments for Environmental Services to Promote Climate-Smart Agriculture? Potential and Challenges. *Agric. Econ.* **2016**, *47*, 173–184.
32. Kelley, H.; Goldstein, A. *Ahead of the Curve—State of the Voluntary Carbon Markets 2015*; Peters-Stanley, M., Gonzalez, G., Eds.; Ecosystem Marketplace—A Forest trends initiative: Washington, DC, USA, 2015.
33. Stringer, L.C.; Dougill, A.J.; Thomas, A.D.; Spracklen, D.V.; Chesterman, S.; Speranza, C.I.; Rueff, H.; Riddell, M.; Williams, M.; Beedy, T.; et al. Challenges and Opportunities in Linking Carbon Sequestration, Livelihoods and Ecosystem Service Provision in Drylands. *Environ. Sci. Policy* **2012**, *19–20*, 121–135. [CrossRef]
34. Kelley, H.; Goldstein, A. *Raising Ambition: State of the Voluntary Carbon Markets 2016*; Ecosystem Marketplace: Washington, DC, USA, 2016.
35. Henry, M.H.; Tittonell, P.; Manlay, R.J.; Bernoux, M.; Albrecht, A.; Vanlauwe, B. Biodiversity, Carbon Stocks and Sequestration Potential in Aboveground Biomass in Smallholder Farming Systems of Western Kenya. *Agric. Ecosyst. Environ.* **2009**, *129*, 238–252. [CrossRef]
36. Cheikh, M.; van Noordwijk, M.; Luedeling, E.; Neufeldt, H.; Minang, P.A.; Kowero, G. Agroforestry Solutions to Address Food Security and Climate Change Challenges in Africa. *Cur. Opin. Environ. Sustain.* **2014**, *6*, 61–67.
37. Johannes, W. *Project Information Document (Pid) Kenya Agricultural Carbon Project (Kacp)*; World Bank: Washington, DC, USA, 2010.
38. Amos, W.; Jönsson, M. Sustainable Agriculture Land Management. A Training Material. Vi Agroforestry. Available online: http://www.viagroforestry.org/who-we-are/resources/publications/ (accessed on 8 January 2015).
39. Bouman, F.J.A. Rotating and Accumulating Savings and Credit Associations: A Development Perspective. *World Dev.* **1995**, *23*, 371–384. [CrossRef]
40. Atela, J.O. *The Politics of Agricultural Carbon Finance: The Case of the Kenya Agricultural Carbon Project, STEPS Working Paper 49*; STEPS Centre: Brighton, UK, 2012; 44p. Available online: www.steps-centre.org (accessed on 5 June 2017).
41. Cavanagh, C.J.; Chemarum, A.K.; Vedeld, P.O.; Petursson, J.G. Old Wine, New Bottles? Investigating the Differential Adoption of 'Climate-Smart' Agricultural Practices in Western Kenya. *J. Rural Stud.* **2017**, *56*, 114–123. [CrossRef]
42. Karl, H.; Morgan, S.; Baylis, K.; Oduol, J.; Smith-Dumont, E.; Vagen, T.; Mutemi, M.; LePage, C.; Kegode, H. Assessing the Downstream Socioeconomic Impacts of Agroforestry in Kenya. In *ICRAF Working Paper No 291*; World Agroforestry: Nairobi, Kenya, 2018.
43. Lee, J. Farmer Participation in a Climate-Smart Future: Evidence from the Kenya Agricultural Carbon Market Project. *Land Use Policy* **2017**, *68*, 72–79. [CrossRef]

44. Lee, J.L.; Martin, A.; Kristjanson, P.; Wollenberg, E. Implications on Equity in Agricultural Carbon Market Projects: A Gendered Analysis of Access, Decision Making, and Outcomes. *Environ. Plan. A* **2015**, *47*, 2080–2096. [CrossRef]
45. Seth, S.; Heiner, K.; Kapukha, M.; Kiguli, L.; Masiga, M.; Kalunda, P.N.; Ssempala, A.; Recha, J.; Wekesa, A. Building Local Institutional Capacity to Implement Agricultural Carbon Projects: Participatory Action Research with Vi Agroforestry in Kenya and Ecotrust in Uganda. *Agric. Food Sec.* **2016**, *5*, 13.
46. Öborn, I.; Wekesa, A.; Natongo, P.; Kiguli, L.; Wachiye, E.; Musee, C.; Kuyah, S.; Neves, B. Who Enjoys Smallholder Generated Carbon Benefits? In *Co-Investment in Ecosystem Services: Global Lessons from Payment and Incentive Schemes*; Namirembe, S., Leimona, B., van Noordwijk, M., Minang, P., Eds.; World Agroforestry Centre: Nairobi, Kenya, 2017; pp. 1–10.
47. Mutua, W. Kenya Agricultural Carbon Project Monitoring Report. Vi Agroforestry. Available online: https://www.vcsprojectdatabase.org/#/project_details/12252012 (accessed on 12 December 2018).
48. Matthias, S.; Tennigkeit, T.; Zanchi, G.; Bird, N. *Technical Guidelines—Activity Baseline and Monitoring Survey Guideline for Sustainable Agricultural Land Management Practices (Salm)*; UNIQUE forestry consultants: Freiburg, Germany, 2010.
49. Lager, B. *Vcs Project Description Kenya Agricultural Carbon Project. Kisumu*; Vi Agroforestry: Kisumu, Kenya, 2012.
50. R: A Language and Environment for Statistical Computing. R Foundation for Statistical Computing, Vienna, Austria. Available online: http://www.r-project.org/index.html (accessed on 5 May 2019).
51. Wei, W.; Feng, X.; Yang, L.; Chen, L.; Feng, T.; Chen, D. The Effects of Terracing and Vegetation on Soil Moisture Retention in a Dry Hilly Catchment in China. *Sci. Total Environ.* **2019**, *647*, 1323–1332. [CrossRef]
52. Tejendra, C.; Raizada, M.N. Agronomic Challenges and Opportunities for Smallholder Terrace Agriculture in Developing Countries. *Front. Plant. Sci.* **2017**, *8*. [CrossRef]
53. Ngome, A.F.E.; Becker, M.; Mtei, K.M. Leguminous Cover Crops Differentially Affect Maize Yields in Three Contrasting Soil Types of Kakamega, Western Kenya. *J. Agric. Rural Dev. Trop.* **2011**, *112*, 1–10.
54. Sommer, R.; Mukalama, J.; Kihara, J.; Koala, S.; Winowiecki, L.; Bossio, D. Nitrogen Dynamics and Nitrous Oxide Emissions in a Long-Term Trial on Integrated Soil Fertility Management in Western Kenya. *Nutr. Cycl. Agroecosyst.* **2016**, *105*, 229–248. [CrossRef]
55. Biamah, E.K.; Gichuki, F.N.; Kaumbutho, P.G. Tillage Method and Soil and Water Conservation in East Africa. *Soil Till Res.* **1993**, *27*, 105–123. [CrossRef]
56. Ndoli, A.; Baudron, F.; Sida, T.S.; Schut, A.G.T.; van Heerwaarden, J.; Giller, K.E. Conservation Agriculture with Trees Amplifies Negative Effects of Reduced Tillage on Maize Performance in East Africa. *Field Crop. Res.* **2018**, *221*, 238–244. [CrossRef]
57. Kiboi, M.N.; Ngetich, K.F.; Diels, J.; Mucheru-Muna, M.; Mugwe, J.; Mugendi, D.N. Minimum Tillage, Tied Ridging and Mulching for Better Maize Yield and Yield Stability in the Central Highlands of Kenya. *Soil Till. Res.* **2017**, *170*, 157–166. [CrossRef]
58. Potma, G.; Carlos, D.R.J.; Mishra, U.; Furlan, F.J.F.; Ferreira, L.A.; Inagaki, T.M.; Romaniw, J.; Ferreira, A.; Briedis, C. Soil Carbon Inventory to Quantify the Impact of Land Use Change to Mitigate Greenhouse Gas Emissions and Ecosystem Services. *Environ. Pollut.* **2018**, *243*, 940–952. [CrossRef]
59. Sommer, R.; Paul, B.K.; Mukalama, J.; Kihara, J. Reducing Losses but Failing to Sequester Carbon in Soils—the Case of Conservation Agriculture and Integrated Soil Fertility Management in the Humid Tropical Agro-Ecosystem of Western Kenya. *Agric. Ecosyst. Environ.* **2018**, *254*, 82–91. [CrossRef]
60. Kamau, M.; Smale, M.; Mutua, M. Farmer Demand for Soil Fertility Management Practices in Kenya's Grain Basket. *Food Sec.* **2014**, *6*, 793–806. [CrossRef]
61. Phuong, L.T.H.; Biesbroek, G.R.; Sen, L.T.H.; Wals, A.E.J. Understanding Smallholder Farmers' Capacity to Respond to Climate Change in a Coastal Community in Central Vietnam. *Clim. Dev.* **2018**, *10*, 701–716. [CrossRef]
62. Shem, K.; Öborn, I.; Jonsson, M.; Dahlin, A.S.; Barrios, E.; Muthuri, C.; Malmer, A.; Nyaga, J.; Magaju, C.; Namirembe, S.; et al. Trees in Agricultural Landscapes Enhance Provision of Ecosystem Services in Sub-Saharan Africa. *IJBESM* **2016**, *12*, 255–273.
63. Esteban, L.M.; Zema, D.A.; Carrà, B.G.; Cerdà, A.; Plaza-Alvarez, P.A.; Cózar, J.S.; Gonzalez-Romero, J.; Moya, D.; de las Heras, J. Short-Term Changes in Infiltration between Straw Mulched and Non-Mulched Soils after Wildfire in Mediterranean Forest Ecosystems. *Ecol. Eng.* **2018**, *122*, 27–31.

64. Kihara, J.; Bationo, A.; Waswa, B.; Kimetu, J.M.; Vanlauwe, B.; Okeyo, J.; Mukalama, J.; Martius, C. Effect of Reduced Tillage and Mineral Fertilizer Application on Maize and Soybean Productivity. *Exp. Agric.* **2011**, *48*, 159–175. [CrossRef]
65. Castellanos-Navarrete, A.; Tittonell, P.; Rufino, M.C.; Giller, K.E. Feeding, Crop Residue and Manure Management for Integrated Soil Fertility Management—A Case Study from Kenya. *Agric. Syst.* **2015**, *134*, 24–35. [CrossRef]
66. Lagerkvist, C.J.; Shikuku, K.; Okello, J.; Karanja, N.; Ackello-Ogutu, C. A Conceptual Approach for Measuring Farmers' Attitudes to Integrated Soil Fertility Management in Kenya. *NJAS Wagen. J. Life Sci.* **2015**, *74–75*, 17–26. [CrossRef]
67. Ndoli, A.; Baudron, F.; Antonius, G.; Schut, T.; Athanase, M.; Giller, K.E. Disentangling the Positive and Negative Effects of Trees on Maize Performance in Smallholdings of Northern Rwanda. *Field Crop. Res.* **2017**, *213*, 1–11. [CrossRef]
68. Abubeker, H.; Talore, D.G.; Tesfamariam, E.H.; Friend, M.A.; Mpanza, T.D.E. Potential Use of Forage-Legume Intercropping Technologies to Adapt to Climate-Change Impacts on Mixed Crop-Livestock Systems in Africa: A Review. *Reg. Environ. Chang.* **2017**, *17*, 1713–1724.
69. Kurniatun, H.; Widianto, W.; Suprayogo, D.; van Noordwijk, M. Tree Roots Anchoring and Binding Soil: Reducing Landslide Risk in Indonesian Agroforestry. *Land* **2020**, *9*, 256.
70. Ylva, N.; Wetterlind, J.; Jonsson, M.; Öborn, I. The Role of Trees and Livestock in Ecosystem Service Provision and Farm Priorities on Smallholder Farms in the Rift Valley, Kenya. *Agric. Syst.* **2020**, *181*, 102815.
71. Diwani, T.N.; Asch, F.; Becker, M.; Mussgnug, F. Characterizing Farming Systems around Kakamega Forest, Western Kenya, for Targeting Soil Fertility-Enhancing Technologies. *J. Plant. Nutr. Soil Sci.* **2013**, *176*, 585–594. [CrossRef]
72. FAO. Faostat—Maize Yields Kenya 2009–2012. In *Food and Agriculture Data*; FAOSTAT: Food and Agriculture Organization of the United Nations: Rome, Italy, 2013. Available online: http://www.fao.org/faostat/en/#data/QC (accessed on 17 May 2019).
73. Yongshan, C.; Camps-Arbestain, M.; Shen, Q.; Singh, B.; Cayuela, M.L. The Long-Term Role of Organic Amendments in Building Soil Nutrient Fertility: A Meta-Analysis and Review. *Nutr. Cycl. Agroecosyst.* **2018**, *111*, 103–125.
74. Vanlauwe, B.; Barrios, E.; Robinson, T.; van Asten, P.; Zingore, S.; Gérard, B. System Productivity and Natural Resource Integrity in Smallholder Farming: Friends or Foes? In *Sustainable Intensification in Smallholder Agriculture: An. Integrated Systems Research Approach*; Öborn, I., Vanlauwe, B., Phillips, M., Thomas, R., Brooijmans, W., Atta-Krah, K., Eds.; Routledge: London, UK, 2017; pp. 159–176.
75. Erenstein, O. Smallholder Conservation Farming in the Tropics and Sub-Tropics: A Guide to the Development and Dissemination of Mulching with Crop Residues and Cover Crops. *Agric. Ecosys. Environ.* **2003**, *100*, 17–37. [CrossRef]
76. Kalungu, J.W.; Filho, W.L. Adoption of Appropriate Technologies among Smallholder Farmers in Kenya. *Clim. Dev.* **2018**, *10*, 84–96. [CrossRef]
77. Pramila, K.; Patnam, M. Neighbors and Extension Agents in Ethiopia: Who Matters More for Technology Adoption? *Am. J. Agric. Econ.* **2014**, *96*, 308–327.
78. Stefanovic, J.O.; Yang, H.; Zhou, Y.; Kamali, B.; Ogalleh, S.A. Adaption to Climate Change: A Case Study of Two Agricultural Systems from Kenya. *Clim. Dev.* **2017**, *11*, 319–337. [CrossRef]
79. Helth, L.J.; Rasmussen, O.D. Can Microfinance Reach the Poorest: Evidence from a Community-Managed Microfinance Intervention. *World Dev.* **2014**, *64*, 460–472.
80. Brannen, C.F. An Impact Study of the Village Savings and Loan Association (Vsla) Program in Zanzibar, Tanzania. Bachelor's Degree, Wesleyan University, Middletown, CT, USA, 2010.
81. Kearney, S.P.; Fonte, S.J.; García, E.; Siles, P.; Chan, K.M.A.; Smukler, S.M. Evaluating Ecosystem Service Trade-Offs and Synergies from Slash-and-Mulch Agroforestry Systems in El Salvador. *Ecol. Indic.* **2017**, *105*, 264–278. [CrossRef]
82. Moushumi, C.; Ajayi, O.C.; Hellin, J.; Neufeldt, H. Climate Change Adaptation and Social Protection in Agroforestry Systems: Enhancing Adaptive Capacity and Minimizing Risk of Drought in Zambia and Honduras. In *ICRAF Working Papers*; World Agroforestry Centre: Nairobi, Kenya, 2011.
83. Andersson, A. Kenya Agricultural Carbon Project Monitoring Report. Vi Agroforestry. 2016. Available online: https://www.vcsprojectdatabase.org/#/project_details/1225 (accessed on 1 September 2018).

84. FAO. *Agro-Ecological Zoning Guidelines*; Food and Agriculture Organisation of the United Nations: Rome, Italy, 1996. Available online: http://www.fao.org/docrep/W2962E/W2962E00.htm (accessed on 13 September 2014).
85. FAO/IIASA/ISRIC/ISSCAS/JRC. Harmonized World Soil Database (Version 1.1). FAO, Rome, Italy and IIASA, Luxenburg, Austria. Available online: http://www.iiasa.ac.at/Research/LUC/External-World-soil-database/HTML/ (accessed on 15 March 2018).
86. USGS. Shuttle Radar Topography Mission (Srtm) 1 Arc-Second Global. Elevation Data. US Geological Survey. Available online: https://lta.cr.usgs.gov/SRTM1Arc (accessed on 3 February 2018).
87. KNBS. Population Distribution by Sex, Number of Households, Area and Density by County and District. In *2009 Population and Housing Census*; Kenya National Bureau of Statistics 5, Ed.; KNBS: Nairobi, Kenya. Available online: https://www.knbs.or.ke/download/population-distribution-by-sex-number-of-households-area-and-density-by-county-and-district/ (accessed on 14 August 2017).

6

Agroforestry as Policy Option for Forest-Zone Oil Palm Production in Indonesia

Edi Purwanto [1,*], Hery Santoso [2], Idsert Jelsma [1], Atiek Widayati [1], Hunggul Y. S. H. Nugroho [3] and Meine van Noordwijk [4,5]

1 Tropenbos Indonesia, Bogor 16163, Indonesia; ijelsma@gmail.com (I.J.); atiekwidayati@tropenbos-indonesia.org (A.W.)
2 Java Learning Centre (JAVLEC), Klidon, Yogyakarta 55581, Indonesia; herys@javlec.org
3 Ministry of Environment and Forestry, Makassar 90243, Indonesia; hunggul@balithutmakassar.org
4 World Agroforestry (ICRAF), Bogor 16155, Indonesia; m.vannoordwijk@cgiar.org
5 Plant Production Systems, Wageningen University and Research, 6708 PB Wageningen, The Netherlands
* Correspondence: edipurwanto@tropenbos-indonesia.org

Abstract: With 15–20% of Indonesian oil palms located, without a legal basis and permits, within the forest zone ('Kawasan hutan'), international concerns regarding deforestation affect the totality of Indonesian palm oil export. 'Forest zone oil palm' (FZ-OP) is a substantive issue that requires analysis and policy change. While spatial details of FZ-OP remain contested, we review literature on (1) the legal basis of the forest zone and its conversion, (2) social stratification in oil palm production (large-scale, plasma and independent growers), and (3) environmental consequences of forest conversion to FZ-OP, before discussing policy options in a range of social and ecological contexts. Policy options range from full regularization (as FZ-OP stands could meet international forest definitions), to conditional acceptance of diversified smallholder plantings in 'agroforestry concessions', to gradually phasing out FZ-OP and eviction/destruction. A nuanced and differentiated approach to FZ-OP is needed, as certification of legality along supply chains is vulnerable to illegal levies and corruption. Corporate actors trading internationally can avoid use of uncertified raw materials, effectively shifting blame and depressing farmgate prices for domestic-market palm oil, but this will not return forest conditions or stop further forest conversion. We discuss an agenda for follow-up policy research.

Keywords: certification; deforestation; palm oil; forest classification; Jambi; legality; independent smallholders; agroforestry concessions; Sumatra; West Kalimantan

---

1. Introduction

A reader cannot safely assume to understand what the word 'forest' means in any new context, as the ecological (vegetation-based) and social (institutional, rule-based) meanings of the word 'forest' only partially overlap [1,2]. There can be 'trees outside a forest' [3] and 'forests without trees' [4]. Conversion of natural forest to plantation forestry is not considered to be deforestation, an issue that sparked debate when the UNFCCC climate change convention considered forest-specific policies [5]. What is excluded from the 'tree' concept, at the heart of common forest definitions, also matters: rubber (*Hevea brasiliensis*) can be considered to be a forest tree when grown for timber, or an agricultural commodity when tapped for its latex [6]. A number of global studies, however, have included the conversion of rubber monoculture (regardless of its primary purpose) to an oil palm monoculture under global 'deforestation' statistics [7]. The debate whether or not palms are included under the 'tree' concept has direct implications for whether or not the conversion of natural forest to an oil palm monoculture can be called 'deforestation' [8] and for how much a 'deforestation-free' commitment of the palm oil industry means [9]. National definitions of the tree and forest concepts can differ from the global ones [10,11]. At the social-ecological system scale of a landscape, the concept of 'agroforest' describes tree-based vegetation managed by farmers, who often see labeling this as a forest as a threat [12,13]. Institutionally such land use may be legalized under 'community-based forest management' rules [14], but these arrangements may maintain forest authorities in the 'landlord' role, expecting a share in any harvestable goods or sellable services that the land may generate [15]. The absence, at least until recently, of formal recognition for agroforestry as a valid form of land use intermediate to 'forest' and 'agriculture' has not prevented the existence and spread of such land uses that defy the rules [16] (Figure 1).

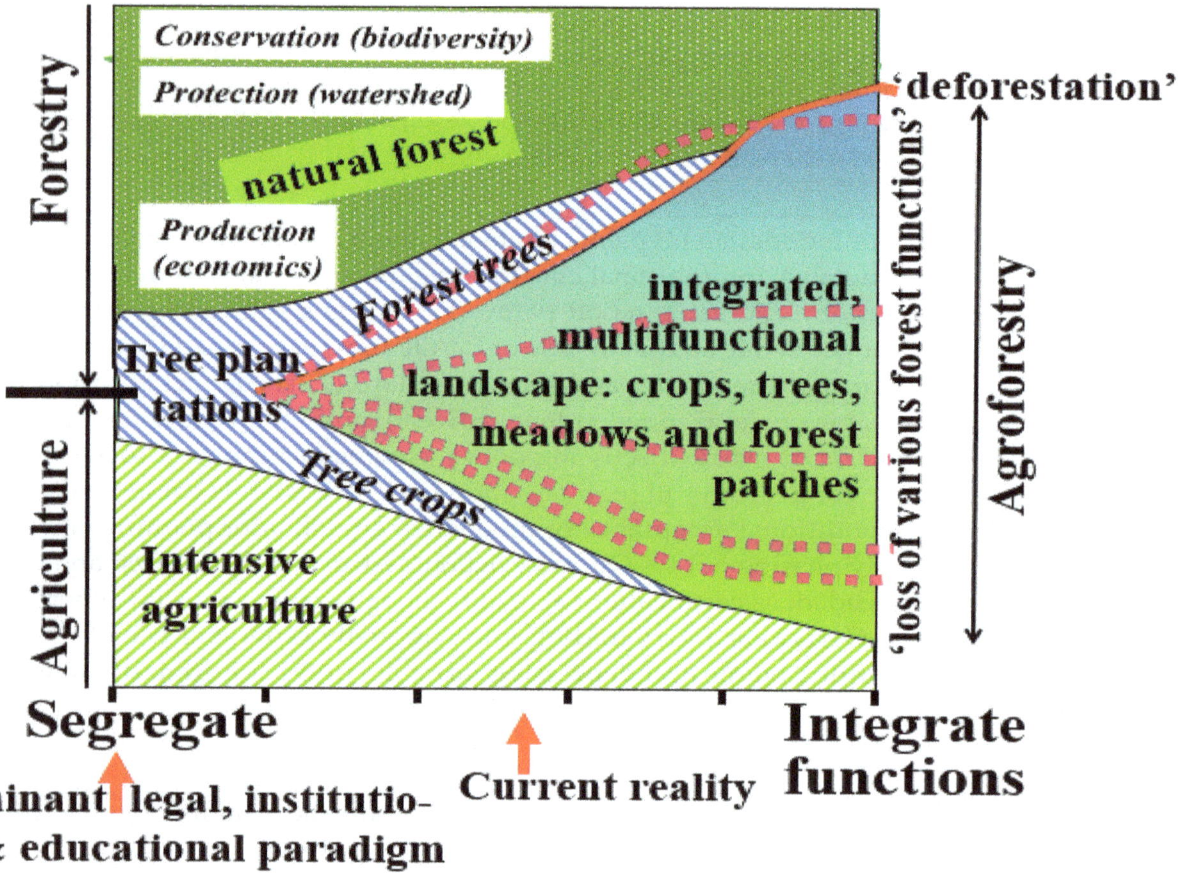

Figure 1. Forest-Agriculture interface with the current segregated policy interpretation distinct from the more fluid 'integrated' reality in the landscape, with consequences for the meaning of 'deforestation' and 'loss of forest functions' [17].

In the international debate over consequences of tropical commodity production, 'forest' is primarily seen as a vegetation type or ecosystem, observable through remote sensing [18], with a rather arbitrary tree cover threshold used in a dichotomous classification. Global data sets based on a continuum interpretation with tree cover as metric are available and have been used to monitor changes in tree cover in agricultural lands [19]. Concerns about deforestation, and the concomitant loss of biodiversity and net C emissions, have led to institutional support in the climate change convention for government action to support 'reforestation' [20] and 'reducing emission from deforestation and forest degradation' [21,22]. In parallel, citizens in importing countries expressing concern about the footprint of their consumption have induced corporate actors to dissociate themselves from deforestation-linked supply chains [23]. Palm oil has been a primary target for the emergence of voluntary private standards [24,25]. The Roundtable on Sustainable Palm Oil (RSPO) and related institutions have triggered responses in Malaysia and Indonesia (jointly around 85% of global palm oil production) that saw their national sovereignty at risk in international markets [26]. A study of RSPO impacts in Kalimantan suggested that it increased forest conversion in agricultural lands but decreased such within the forest estate [27].

Recent estimates, discussed below, show that, based on remote sensing, 15–20% of Indonesian oil palms are located within the 'forest zone' (Kawasan hutan), a primarily institutional indication of land status. We indicate these as Forest Zone Oil Palm (FZ-OP). The FZ-OP percentage can be interpreted in multiple ways. To an international audience that has come to believe that most, if not all, of Indonesian palm oil is linked to deforestation, the number may seem to be small. To the Indonesian palm oil sector, it may give the impression of some bad apples that need to be dealt with, as they spoil the good name of the vast majority of those falsely accused of deforestation. To the forest authorities, the numbers are a clear sign of illegality, as current rules do not allow oil palm to be planted within the forest zone as FZ-OP. Actually, oil palms are the most readily recognizable of the four major tree crops (coffee, cocoa, rubber and oil palm) that occur both outside and inside the forest zone [28], but part of the smallholder oil palm may be part of a finer-grained mosaic and not as easily noticeable. Key questions regarding FZ-OP are: (1) What came first: local land use or designation of the land as part of the forest zone?, (2) How does the land use with tree crops relate to the primary designated forest function in (A) production forest (income generation), (B) protection forest (watershed protection) and (C) Conservation areas (Biodiversity), with recent additional interests in restricting drainage of peat soils in the production forest zone, (3) How do the various stakeholders along the palm-oil value chain respond to the lack of legality of forest based oil palm production?

Issues related to oil palm (and its product palm oil) require an understanding of the whole value chain, as concerns of end-users, often based on limited or biased information they have access to, inform the product manufacturers, who want to be able to claim a 'clean' product, and their suppliers, the partly refined products derived from Crude Palm Oil (CPO), processed from Fresh Fruit Bunches (FFB) transported from farmgate to mill, and grown by smallholders or large-scale plantations. Current leverage on 'illegal' palm oil may primarily come from making mills responsible for their source areas. However, it is not clear whether these efforts mostly lead to 'shifting blame' [29,30] or whether they actually contribute to changing land-use and effectively protecting forests.

While uncertainty over the exact numbers and locations on FZ-OP continues, we set out to formulate lessons from earlier interactions between forest authorities and the other tree crops (coffee, cocoa, rubber) and to contextualize and analyze policy options and how these might be applied to the various contexts that contribute to the overall issue. We will review the available data and literature on:

- The history of forest legality and oil palm expansion in Indonesia as context for current forest-related issues of local land use versus designation of the land as part of the forest zone;
- Spatial analysis of FZ OP at the intersection of forests and tree crops for Indonesia as a whole and zoomed in on two provinces with higher-resolution data;
- Social and economic concerns on oil palm expansion and the role of smallholders in FZ-OP;
- Environmental concerns on tree crops in relation to the primary designated forest function in (A) production forest (income generation), (B) protection forest (watershed protection) and (C) Conservation areas (Biodiversity);
- Policy options linked to forest-related contexts, informed by the responses of various stakeholders along the palm-oil value chain to FZ-OP.

We conclude by formulating more specific policy research questions as a follow-up.

2. Methods

2.1. Literature Review

Starting from several recent reviews of oil palm in Indonesia, we used a 'snowball' method to identify more recent papers citing these sources and followed up on the citation network thus established to document a synthetic view on the various sub-topics.

2.2. Maps

For a more detailed analysis of the different forest designations within the forest zone and for the time frames of expansion within the forest zone, we compiled data for Jambi, a province in Sumatra around the Equator, and West-Kalimantan. The two provinces were selected to represent the two islands where oil palm is major (Sumatra and Kalimantan) while they had different historical settings on traditional land use of agroforest and its dynamics. Availability of the data also determined the choice of the two provinces. We traced the development of oil palm and agroforestry areas in the 2000–2020 period in each of the Forest-zone categories and sought the trajectories of changes and observed interlinks between oil palm, forest and agroforestry. Where the data were available, we incorporated the other major agroforestry tree-crops; otherwise, we utilized the common class of 'mixed agricultural tree-crops' mapped in most datasets. The following dataset was utilized for the analyses (Table 1).

Table 1. Maps utilized for case studies.

No	Map Title	Year (s)	Extracted Class (es)	Source
1	Peta Penutupan Lahan Indonesia (Indonesia Land Cover Map)	2000, 2009, 2018	Agroforest	[31]
2	Ecological Vegetation Map of West Kalimantan	2019	Agroforest Rubber plantations Oil palm plantations	[32]
3	Global map of smallholder and industrial closed canopy oil palm plantations	2019	Oil palm plantations Independent smallholder oil palm	[33]
4	National Main Commodity Maps in Indonesia	2019	Oil palm Rubber	[28]
5	Kawasan hutan Provinsi Kalimantan Barat (Forest-zone Lands of West Kalimantan Province)	2014		[34]
6	Kawasan hutan Provinsi Jambi (Forest-zone Lands of Jambi Province)	2014		[35]

2.3. Public Consultation

In a webinar in September 2020, attended by 374 participants, the issue of overlap between Indonesian forest estate and oil palm was highlighted and a number of the basic data presented; the questions and discussions formed the start of this manuscript. Specific aspects discussed are listed in Appendix A at the end of this manuscript.

3. Indonesian Context

3.1. Forest Legality

While efforts to agree on a forestry law during the Colonial period stranded [36], the post-independence (1945) constitution clarified that all the country's resources were for the benefit of the Indonesian people—with subsequent debate on the degree to which the Government of Indonesia has to respect prior claims of customary communities ('Masyarakat Hukum Adat') [37]. The Basic Agrarian Law of 1960 that recognizes colonial period documents as a basis for valid rights to land and the Forestry Law of 1967 was considered to be complementary but had gaps between them [38]. The 1999 revision of the

Forestry Law specified that state ownership claims within the designated 'forest zone' ('Kawasan Hutan') required completion of a gazettement procedure (verifying that there are no valid pre-existing claims), which has been very slow to implement [39], leaving doubts on the legality of government forestry policies that treat the Forest-zone as government-owned.

The total area of forest-zone lands in Indonesia (63% of total land area) is subdivided by primary function, as 'Production Forest' (68.8 M ha; 36.1% of the country), 'Protection Forest' (29.7 M ha; 15.6% of the country) and 'Conservation Areas' (22.1 M ha; 1.6% of the country) [40]. Logging is only allowed in the production forests, with further restrictions in part of the zone, and conversion to monocultural plantations for the pulp and paper industry in other parts. Although logging-based economic interests prevail in this zone, the production forests still have a recognized role in biodiversity conservation, carbon storage and maintenance of watershed functions. Most of the Protection Forest (the Indonesian term could also indicate '(watershed) protecting' forests) are on mineral soils and have been defined through a scoring system emphasizing slope and similar criteria for soil and water protection. For peatland areas, there are separate criteria for determining a Protection Function ('Fungsi Lindung') indication, based on peat depth and the peatland moratorium policy [41]. The total area of Protection Function in Indonesia's peatlands is 12.3 M ha [42].

Building on the specific example of the damar agroforests in Krui (West Lampung, Sumatra) where government ownership claims to forest-like vegetation could not be substantiated [43,44], it was realized that 'agroforests', of various specific histories and intensities of management, were included in the Forest-zone, either with Production or Protection Forest designations, but also claimed by local communities as existing before the Forestry law was formalized [45]. Since then, a government commitment to 'community-based forest management' and 'village forests' has allowed some conflict zones to be resolved, but much remains to be done [46,47]. Meanwhile, calls to redefine the Indonesian forest estate and get the boundaries of its Forestry Regulatory Framework and Agrarian rules right [12,48–51] were largely unheeded. Global interest in and country-level expectations of, new financing mechanisms through Reducing Emissions from Deforestation and (forest) Degradation (REDD+) created a new dynamic where issues of 'forest legality' obtained new relevance [52,53]. Indonesia's challenge to balance the new global environmental agenda on forest conservation, with the existing development deficits is part of a wider global pattern [54].

3.2. *Complexity of Frontier Situation*

In Southeast Asia's long engagement with global markets, cocoa, coffee, fast-growing trees, oil palm, rubber and shrimp have all had periods of rapid expansion or 'booms' when the region proved to be a low-cost producer in which land, labor, know-how, and market access could be rapidly mobilized [55]. The 'land grab' literature tries to answer key questions on who seeks to exercise control over land to grow export-oriented crops under boom conditions, how would-be producers navigate regulatory powers and market forces to gain control over land, and how booms differentially affect areas with secure and insecure land control relations.

Since the 19th century, commodities such as rubber [56], coffee [57] and cocoa [58] have boomed in specific parts of Indonesia, creating new 'forest frontier' conditions, involving local people, attracting investors to come in, but also stimulating people from other location to move in, spontaneously and/or in government-sponsored transmigration [59]. The relationships between spontaneous migrants, large-scale plantation companies, local communities and various branches of government have become complex in many locations [60–63]. Due to its location, boom crop production is closely associated with the issue of deforestation and environmental degradation [64].

Indigenous people and local communities (jointly described here as IP/LC) in forest frontier areas have had a long tradition of a 'dual economy' [65] where food security continued to depend on swiddening, while cash-crop agroforestry (rubber, coffee, resin trees and now oil palm) provided a financial basis for their livelihoods [66]. Where the terms of trade were sufficiently favorable, staple food production could be 'outsourced' [67]. Accordingly, most of the oil palm grown by IP/LC, just not to say smallholders, has not been a direct cause of deforestation as they developed in a village environment of orchards, old rubber groves and swidden fallows. Smallholders also tend to use fertilizer sparingly, as they are unable to access subsidized fertilizer supplies that are monopolized by larger plantations [68].

While global policies and markets are often held responsible for accelerated deforestation in the tropics [69], local knowledge is generally assumed to lead to overall positive outcomes in matters of conservation [70,71]. Even in matters of primary forest conversion, local communities are usually presented as the best managers to reconcile conservation and development [69]. Increased attention for forest protection in the context of REDD+ may have increased pressure on converting agroforests outside the forest zone to become oil palm monocultures [72].

The Indonesian forest frontier is home to approximately 26 thousand villages (often defined as 'forest villages') and more than 37 M peoples, with some 18% still struggling to escape from poverty based on the national standard [73]. Most of the forest villages are also not formally registered yet, as nationally only 30% of villages are already registered, mostly in Java. They have no clear and defined boundary and are dominated by state forest land, with formal restrictions to use for crops to generate food or income. In such landscapes, boom crop production, such as oil palm, has taken place. Expansion of agriculture into the forest area, often following onto a logging phase that brought people and road access (however poor) into the forest zone, is considered because of these complexities, that for decades have remained unresolved.

3.3. *Total Oil Palm Area*

According to the most recent official figure, the oil palm area in Indonesia was 14.7 M ha in 2019 [74]. Several organizations have also mapped oil palm areas for different objectives and with a range of methods (Table 2). Bahktiar et al. [75] claimed a substantially larger area (16.8 M ha), by including areas cleared for oil palm but without stands recognizable by remote sensing yet. The lowest recent estimate was 11,530,000 ha [33], in a study that only included fully-developed stands. Time-series data of oil palm for Borneo were analyzed for forest loss or deforestation [76] and biodiversity impacts [77]. The breakdowns of oil palm areas varied, with one study differentiating 'industrial' and 'smallholder' oil palm [33], while another study [75] focused on oil palm areas within forest zone lands. Condro et al. [28] mapped major commodities including coffee and cacao, and obtained area estimates close to the data from the Ministry of Agriculture. Across Indonesia, the largest oil palm areas are in Sumatra and the second largest in Kalimantan, covering respectively 6.4 M ha and 4.9 M ha in 2018 [74], while another study [75] mentioned 10.5 M ha and 5.7 M ha, respectively, for these two islands. Oil palm is expanding in W and SE Sulawesi, and in Papua, but current areas remain relatively small.

Table 2. Published estimates of the total area of oil palm in Indonesia.

No	Oil Palm Hectarage	Year	Approach and Methods	Notes	Reference
1	16,800,000	2015–2017	Multi-data analyses	Including unplanted area for industrial oil palm within forest zone	[75]
2	14,896,964	2019	Automatic classification using multi-data in GEE	RS methods; Oil palm and other major commodities	[28]
3	14,724,420	2019	Using trade statistics and yield data as a basis	Official Oil Palm statistics; industrial (private and state) and smallholders	[74]
4	11,530,000	2019	Automatic classification of sentinel imageries with NN	RS methods; Differentiation of industrial and smallholders	[33]
5	11,100,000	2015–2017	Visual interpretation of Landsat imageries	Oil palm and deforestation	[78]

The distribution of oil palm in Indonesia is uneven and strongly related to climatic conditions. In North Sumatra (Figure 2A), the province with the most even rainfall distribution and virtually no dry season oil palm can reach up to 50% of the area at district (Kabupaten) level; elsewhere in lowland Sumatra and the southern half of Kalimantan it can reach up to 30% of the area, but in most of Indonesia, oil palm forms less than 10% of the landscape or is virtually absent. For Indonesia as a whole oil palm covers about 8% of the total land area. The fraction of oil palm under smallholder ownership (see below for the distinctions within this category) is approaching 50% for Sumatra, but half of that in most of Kalimantan (Figure 2B). The recent expansion of oil palm, indicated by relatively young stands, is evident outside of the areas that already have a large oil palm fraction (Figure 2C).

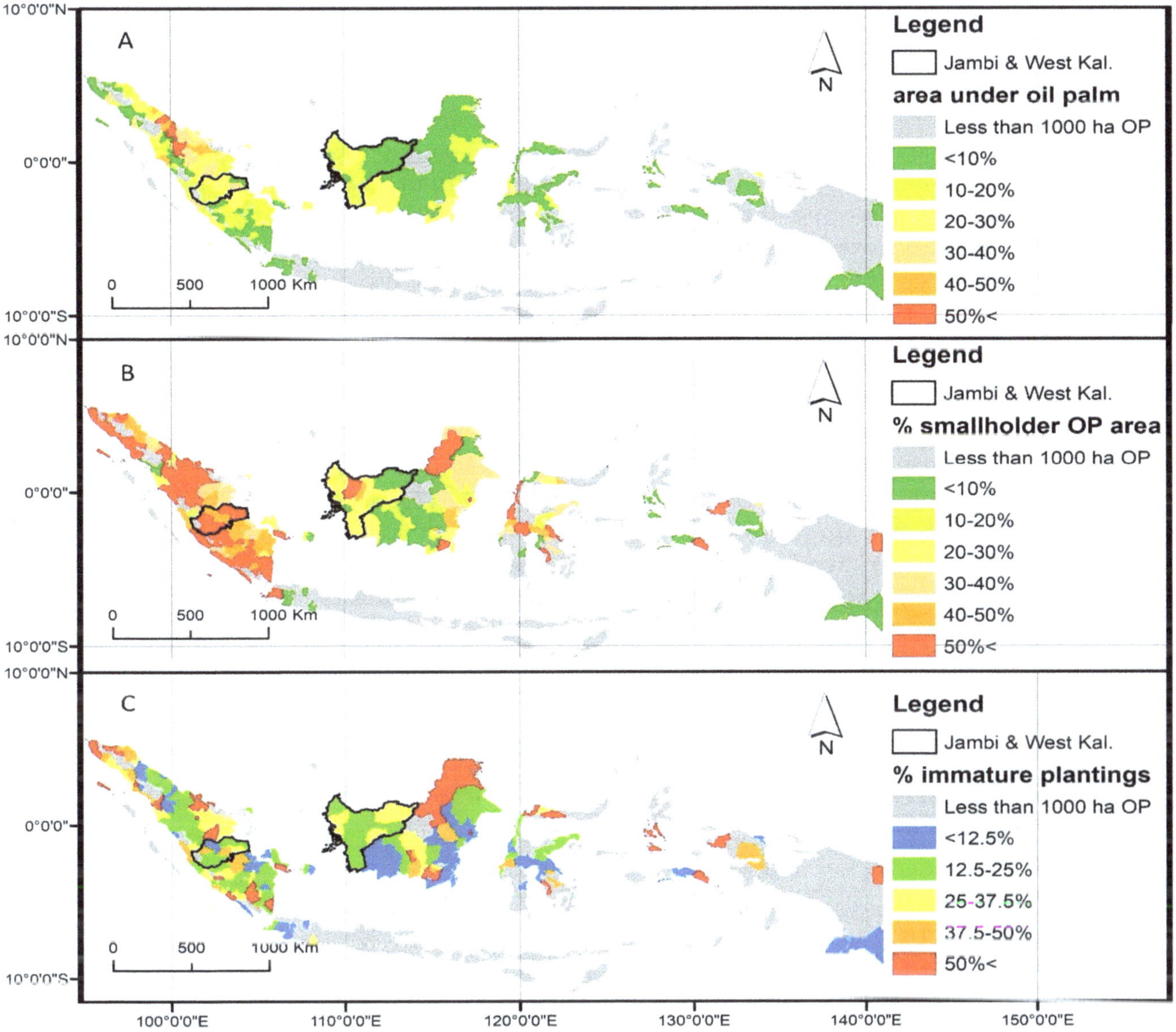

Figure 2. Maps on 2017 oil palm distribution at the district level in Indonesia, highlighting Jambi and West Kalimantan provinces; (**A**) Area fraction under oil palm; (**B**) Share of oil palm cultivated by smallholders; (**C**) Share of oil palm that does not yield yet as it is too young (usually less than three years old); Sources: Own maps, oil palm data from [79]; district boundary and area data from [80].

Several authors have quantified oil palm development within the forest zone in Indonesia [76,78]. Estimates vary from 2.5 [76] to 3.4 M ha [75]. This implies that 15–20% of Indonesian oil palms grow within the designated forest zone. The dataset of [75] allows some further exploration of the geographical patterns of Indonesia's Forest-zone oil palm (FZ-OP). The relative share in Indonesia's FZ-OP at the province level does not match shares in total oil palm production. With 31.5% of

Indonesian oil palm, the two provinces of Riau + C Kalimantan have 66.8% of its FZ-OP (Riau 20.7% OP, 42.1% of FZ-OP; C. Kalimantan 10.9% of OP and 24.7% of FZ-OP). With 41.2% of Indonesian OP, the rest of Sumatra has 24.0% of its FZ-OP; with 23.7% of Indonesian OP, the rest of Kalimantan has 7.7% of its FZ-OP and with 3.6% of Indonesian OP, Eastern Indonesia + Java have 1.4% of its FZ-OP. Across Indonesia, 2.96% of FZ-OP is designated as conservation area, 4.71% as (watershed) protection forest, 15.15% as production forest with restrictions, 45.4% as regular production forest, and 34.8% as production forest indicated for conversion. The total area of FZ-OP is 2.88 times the amount of production forest indicated as 'conversion forest' (planned, legal deforestation).

For further exploration of that pattern as well as to showcase the development of oil palm gardens as part of agroforestry pathways, we focused on two provinces, one each in Sumatra (Jambi ranked #7 in oil palm production and #5 in FZ-OP) and Kalimantan (W. Kalimantan, ranked #3 in oil palm production and #6 in FZ-OP). Both are outside the historical oil palm core area in N. Sumatra, but they were part of the expansion since the 1990s; the district-level presence of oil palm within these provinces varies from 0–30%. The two provinces have harmonized their Kawasan Hutan areas with the provincial planning (RTRWP) with ministerial decrees that legalized the status of 'Forest-zone lands.' Therefore, issues of agroforest and oil palm practices in these two provinces are not burdened by different interpretations of forest legality between the central government (MoEF) and the provincial governments.

4. Oil Palm in the Forest Zones of West Kalimantan and Jambi

4.1. Data by Province and Forest Category

The total area of Forest-zone lands, encompassing all Conservation Areas (National Park and forest reserves), Protection Forest and all Production Forest status (HP, HPT, HPK) in West Kalimantan is 8.1 M ha or 56% of the province area, while for Jambi it is 2 M ha or 43% of the province.

For West Kalimantan, agroforest area occupied 38% of all Production Forest status and 16% of Protection Forest areas [31], while according to [32], agroforest covered <10% (Table 3). Some of the agroforests in [32] are mapped by [31] as disturbed forest, and that was the main differentiating factor of forest and agroforest areas in the two maps. Major commodities of oil palm and rubber are negligible in West Kalimantan's Forest-zone lands, the highest being oil palm in Production Forest areas (4%) mapped by [28].

Table 3. Areas of agroforest and major tree crops in West Kalimantan and Jambi in 2018–2019.

Agroforest and Major Tree Crops	Source	Production Forest (ha)	Protection Forest (ha)	Conservation Areas (ha)
West Kalimantan, total area		**4,444,111**	**2,295,424**	**1,423,567**
Agroforest	[32] +#	318,521 (7%)	66,631 (3%)	10,576 (1%)
	[31] +	1,676,606 (38%)	363,714 (16%)	35,053 (2%)
Rubber plantations	[32]	7889 *		14 *
	[28]	8157 *	764 *	264 *
Oil palm plantations	[32]	80,771 (2%)	6663 *	1852 *
	[28]	186,037 (4%)	29,753 (1%)	4610 *
	[33]	29,382 *	3017 *	59 *
Independent smallholder oil palm	[33]	8267 *	906 *	53 *
Jambi, total area		**1,235,639**	**180,170**	**634,431**
Agroforest	[31] +#	231,112 (19%)	12,463 (7%)	35,594 (6%)
Rubber plantations	[28]	735 *		
Oil palm plantations	[28]	99,848 (8%)	2964 (2%)	20,065 (3%)
	[33]	40,399 (3%)	337 *	902 *
Independent smallholder oil palm	[33]	38,164 (3%)	1032 (1%)	1915 *

+ processed/reclassified for this paper, # utilized for further analyses, * percentage lower than 1% of the respective forest-zone land category.

In Jambi, agroforest occupied 19% of the land with Production Forest status, and <10% for Protection Forest and Conservation Areas [31] (Table 3). Rubber and oil palm monocultures were <10% in all Forest-zone land categories in Jambi [28,33]. A breakdown of FZ-OP by elevation and forest category (Figure A1) shows most are below 100 m above sea level.

Taking the total amount of palm oil produced in each of the provinces as point of references, the likely source areas differ between the two provinces, and according to three data sources specified in Table 4, with 6–17% derived from the forest zone. Within the forest zone, the fractions derived from production forest are highest (even after correction for the total area), followed by protection forest and conservation areas.

Table 4. Likely source area of palm oil produced in West Kalimantan and Jambi according to three spatial data sources, expressed as %, assuming homogenous productivity per unit land.

Province	West Kalimantan			Jambi	
Data Source	[75]	[33]	[28]	[33]	[28]
Non-forest land	93.65	96.30	89.93	89.05	83.05
Production forest	5.70	3.30	8.44	10.40	13.78
Protection forest	0.47	0.34	1.35	0.18	0.41
Conservation areas	0.13	0.01	0.20	0.37	2.77

4.2. Data by Time Period—Land Cover Changes Involving Agroforest and Smallholder Oil Palm

Agroforest in West Kalimantan was mostly as rubber gardens and *tembawang* (Dayak traditional land-use system) having been practiced for generations [81,82]. Inside the Forest-zone lands in West Kalimantan, the majority (76%) of 300,000 ha agroforest areas in 2019 was already agroforest in 2000, and only 11% was forest (Figure 3)

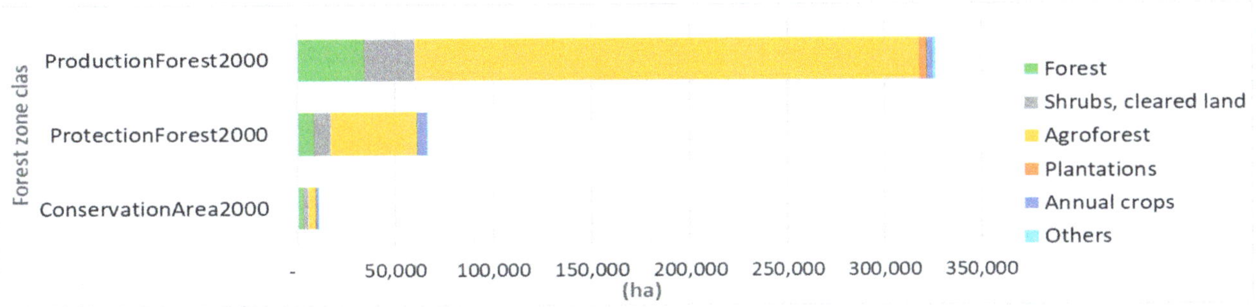

Figure 3. Land cover trajectory (2000) of 2019 agroforest in each Forest-zone class in West Kalimantan (analyzed from [31,32]).

The development of independent smallholder oil palm was closely linked to the conversions of the locally managed agroforest areas due to the lucrative oil palm market in the vicinity of the villages and the decreasing popularity of rubber due to its volatile price [82]. That trend is observed in the land cover trajectory of independent smallholder areas in West Kalimantan, i.e., 57% of which was agroforest in 2000 and 54% in 2009 (Figure 4). Only 13% and 6% of independent oil palm was a forest in 2000 and in 2009, respectively.

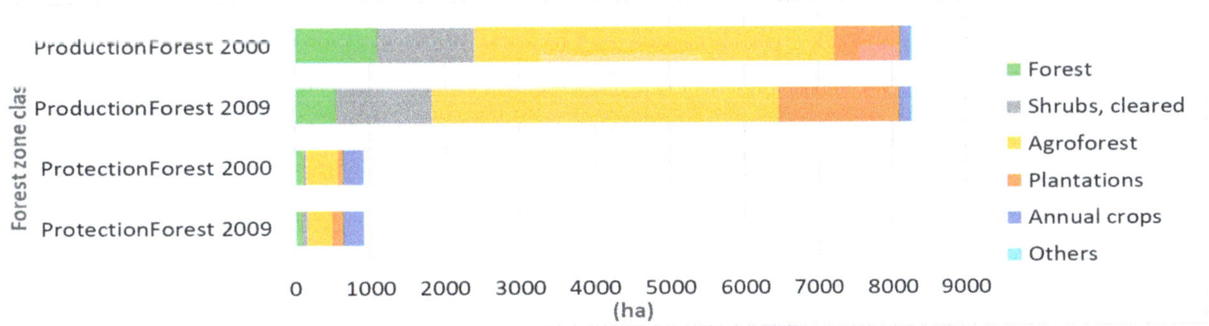

Figure 4. Land cover trajectories (2000 and 2009) of 2019 independent smallholder oil palm in each Forest-zone class in West Kalimantan (based on [31,33]).

For Jambi, agroforest was originally developed by local communities with rubber as the major crop introduced in the beginning of the 1900s during the Dutch period [56,82,83]. It dominated private lands (APL), but its growth reached the areas designated as Forest-zone lands. The 2018 agroforest in Jambi's Forest-zone lands was equally agroforest (44%) and forest (44%) in 2000 (Figure 5), proving the historical presence of this locally managed land use in state's lands.

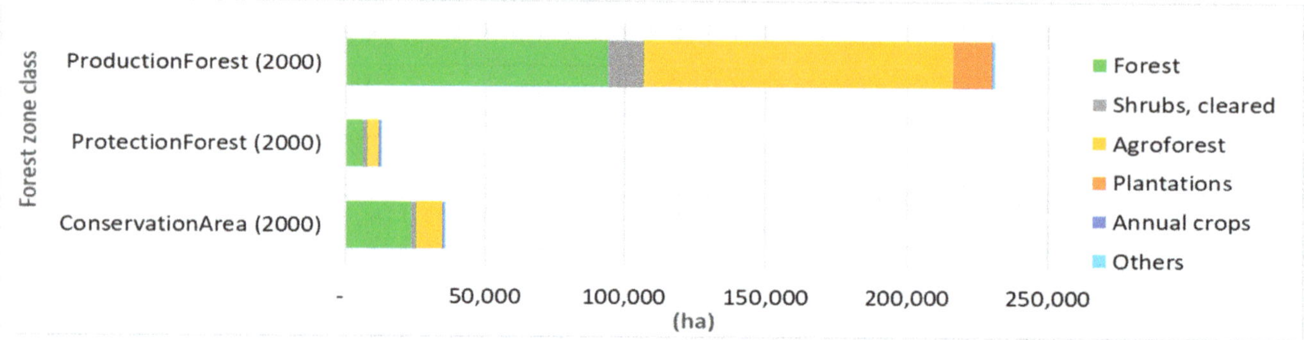

Figure 5. Past land cover types of agroforest in Forest-zone lands in Jambi (analyzed from [31]).

Expectations of high profitability of conversion to oil palm attracted many agroforest rubber farmers in Jambi, also signifying the changes from food sufficiency to cash-cropping [84], in addition to the volatile price mentioned earlier. Independent smallholder oil palm in Jambi's Forest-zone lands was 39% agroforest and 15% plantations in 2009, while in 2000 it was 11% agroforest and 21% plantations (Figure 6). These demonstrate the development of independent smallholder oil palm in Forest zone lands involving agroforest and plantations, mostly rubber. Growth of oil palm also reached degraded areas, shown by 15% shrubs in 2000 and 34% in 2009 (Figure 6), which included peatlands in the lowland areas [85,86].

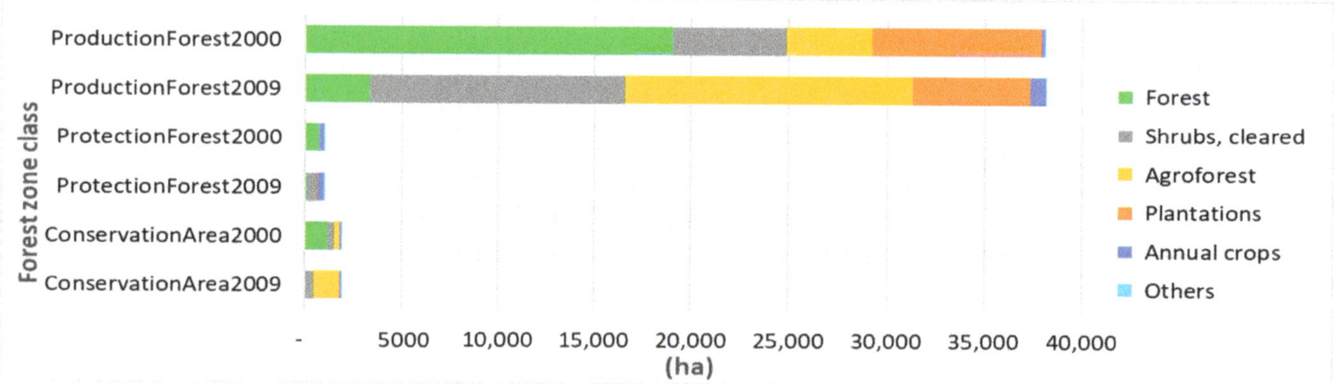

Figure 6. Land cover trajectories (2000 and 2009) of 2019 independent smallholder oil palm in each Forest-zone class in Jambi (based on [31,33]).

Agroforest that has long been managed by local communities has evolved in Forest-zone lands, with the highest proportions and areas in Production Forest status (Figures 4 and 6). With the long history, this land use had faced unresolved forest zone issues since before the 1999 Forestry Law (see Section 3.1) until now, under the *Kawasan Hutan* officiated in 2014 for the two provinces. Since the past decade, oil palm has emerged to become part of the dynamics of this local farming, including in Forest-zone lands. The data show that these dynamics mostly occupied the Production Forest status (Table 3), the class allocated to support various production functions. Looking at the 'predecessor' cash-crop of rubber as part of the ever-evolving mixed-crop agroforest land use [56,81,82], independent oil palm might grow similarly, spatially and or temporally, as part of community-based production functions in land with Production Forest status.

5. Social Dimensions

5.1. Overview Indonesian Oil Palm Sector

The palm oil value chain has a distinct hourglass shape (Figure 7). Upstream there are vast conglomerates that own plantations covering hundreds of thousands of hectares and may be involved in all stages of the value chain. At the other extreme are millions of smallholders that own only a small plot and sell their produce at the farmgate. In between are many shades of grey, ranging from smaller groups, independent large-scale plantations, investors owning a hundred hectares, farmer groups and independent farmers who accumulated multiple small plots over time. Their produce is eventually processed in just over 1000 mills [87], subsequently sold to a few dozen refinery companies, which subsequently supply a few trading companies, that sell their products to a vast number of stakeholders as consumer goods companies and retailers [88].

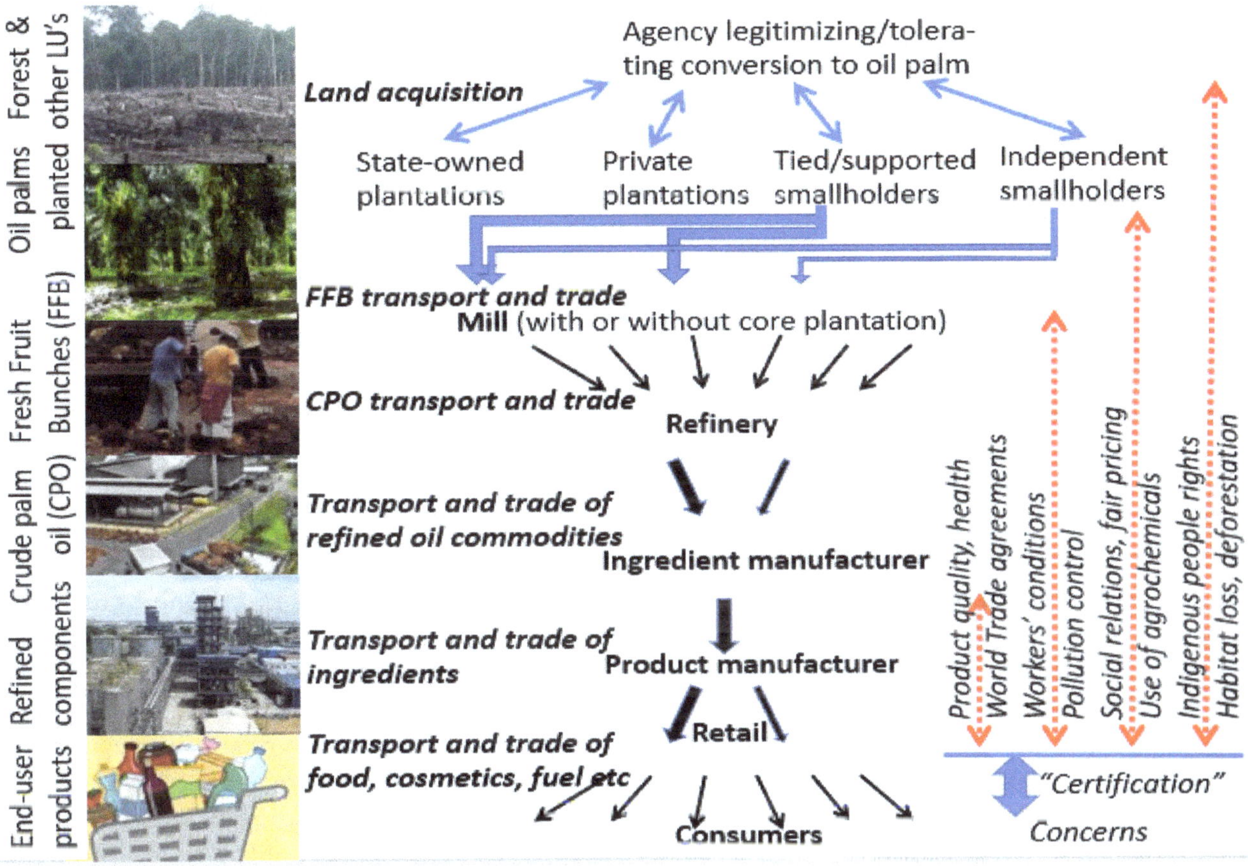

Figure 7. Schematic structure of the oil palm value chain [26].

5.2. Heterogeneity and Expansion amongst Oil Palm Smallholders

Starting on Sumatra's east coast plantation belt, oil palm has been cultivated commercially in Indonesia since 1911 [89]. Smallholders, however, were not involved in oil palm cultivation till the late 1970s, when the New Order regime assigned State-owned plantation companies to develop oil palm plantations that included smallholder farmers. This policy was directed to support the socio-economic development of new settlers and local people in Indonesia's outer islands[1] and develop export commodities that reduce dependency on mineral oil exports. Whereas smallholders first appeared

[1] Outer islands usually refer to islands beyond densely populated Java, Madura and Bali islands.

in oil palm statistics in 1979, covering a mere 3125 ha [90], by 2018 their area covered an estimated 5.8 million hectares, which is equivalent to 40.6% of Indonesia's palm oil area [79]. Whereas oil palm smallholders are often categorized in national statistics as a single entity it is increasingly clear that this vast sub-sector is highly heterogeneous and there is a need for targeted policy measures to improve the environmental and socio-economic performance of the different types of smallholders [91,92]. A key differentiation amongst oil palm smallholders is the scheme vs. independent smallholder dichotomy [93,94].

Scheme smallholder plantations, often also referred to as plasma or '*kemitraan*', are smallholder plantations that have been developed by a plantation company and usually involve company credit for plantation establishment and other agro-inputs, knowledge sharing, and a take-off agreement between smallholders and partner company. Smallholders are usually allocated 2–3 hectares [95]. However, partnership agreements can be highly diverse depending on the regulatory framework present at the time of partnership development, and the negotiations between stakeholders [95,96]. For example, early schemes allocated 60–80% of the plantation to smallholders, but under the influence of World Bank policies and to accommodate private sector investments, this shifted to 40% for smallholders in the late 1980s and 1990s [97,98]. In 2007 the new plantation law stipulated that companies needed to allocate 20% of their concession to smallholders within their concession boundary, and in 2013 this was changed to 20% for smallholders with land possibly outside the company concession [68]. Many companies do not reach this target, however, and this requirement may well be deleted in the new 'omnibus' law that Indonesia is currently preparing [99].

Whilst in some partnerships smallholders are strongly involved in plantation management decisions, participate in plantation labor, and may even outperform their partner company in terms of yields [100], in other arrangements smallholder plantations are fully managed by the company. This last category is often referred to as one-roof management, in which smallholders are effectively mere shareholders [96,101]. Whereas production levels in schemes are generally close to company plantations and thus relatively high for smallholder producers, such schemes frequently suffer transparency issues and especially beneficial to companies [68,102]. Over time government policy regarding oil palm smallholders has increasingly shifted from being a poverty reduction strategy towards one that strengthens industry interests and requires minimal government investments.

Independent oil palm smallholders quickly emerged once the benefits of oil palm cultivation by scheme farmers became obvious. Mills started buying produce outside their plasma and own plantations, and basic infrastructure developed by logging, plantation, and mineral oil companies opened up lands that were previously uneconomic to develop [103,104]. Findings from Jambi indicate that direct economic profits of independent oil palm plantations were significantly higher than forests [105]. Also, oil palm demands less labor compared to competitor cash-crops as rubber, thus allowing for more land to be cultivated or freeing time to engage in other economic activities [106,107]. Independent smallholders generally have no direct links to mills and do not receive extension services from the government or companies. Subsequently, independent smallholders generally have limited knowledge of good agricultural practices, use poor planting material, apply minimal and unbalanced fertilizer regimes, receive low prices for their produce, and suffer yields well below companies and plasma farmers [96,108,109].

The independent oil palm smallholder sector is highly heterogeneous, with many different layers in society engaging in smallholder oil palm development. The independent oil palm smallholders sector includes small local farmers with a few dozen oil palms, migrants attracted by cheap land, oil palm company employees investing in nearby land, local government officials and shop owners looking for investment opportunities and purchasing 5–10 ha, local or urban elites that engage in semi-corporate oil palm plantations and everything in between [93,101]. Multiple typologies on independent oil palm smallholders in Indonesia have been developed [93,101,110], showing presence and ratios between types of smallholders differ in different landscapes. For example, remote peat frontiers often involve relatively large investors whilst transmigration areas or traditionally relatively densely populated areas have a relatively large share of smaller farmers that converted their traditional land uses [93,101].

Whereas land rights and land ownership documentation are generally well arranged in plasma, this is often not the case amongst independent oil palm smallholders [93]. Krishna et al. [111] suggested that especially indigenous populations can obtain forested land and develop oil palm or cash crops in the forestry domain, but sales of such land are limited as it is undervalued due to lack of marketability. Although migrants often do not have nationally recognized land ownership, they require more security in land titles and pay higher prices for land. These authors [111] therefore warn that land titling programs may well lead to the increasing value of land, indigenous people selling their land, claiming new land, and thereby triggering new deforestation [112]. Purnomo et al. [113] provide a detailed analysis of the stakeholders involved in the conversion of land into independent smallholder oil palm plantations in Riau and highlight the considerable profits that accrue to those involved at different stages of the land conversion, including the land mafia. It is increasingly clear that the diversity amongst smallholders and the different landscapes in which they operate needs to be acknowledged for developing adequate policies that foster more sustainable landscapes.

Although reliable current data on plasma vs. independent smallholder are not available, smallholders in plasma schemes were numerically overtaken in 2005 by 'independent' smallholders (including those in various partnership schemes) [114]. Since then some existing plasma plantations have transformed into 'independent' smallholder plantations and oil palm sector growth rates between 2015 and 2018 show that private sector large-scale oil palm plantations grew by 1.91 million ha and that smallholder oil palm area increased by 1.28 million ha [79]. If companies meet their plasma obligations, this would mean a 382,345 ha increase in plasma area and a 901,143 ha increase in independent smallholder area, highlighting that it is especially the independent smallholder sector that is expanding. Whereas companies and associated plasma smallholdings are relatively easy to identify, monitor, and sanction due to the size of their plantations and relative ease of targeting managements, this is not the case with independent smallholders. These independent smallholders are huge in the number of management units that occupy relatively small areas, making it highly complex to manage independent smallholder oil palm expansion. With limited external support and little monitoring, the independent smallholder oil palm has developed into a cheap buffer for the industry; this relates to setting sustainability standards as well as organizing a supply chain.

6. Ecological Dimensions

The economic benefits of oil palm production reviewed above have been discussed in tradeoff with negative impacts on ecosystem services [25,115–117]. However, specific issues of FZ-OP have not been discussed as such in the literature we reviewed. Oil palm cultivation, but especially its continued expansion, has led to severe negative effects on environmental quality such as biodiversity depletion [118,119], loss of hydrological function [118,120–123], increased carbon emissions [85,124], and reduction of water and soil quality [116,118,121,125,126]. The magnitude, and sometimes the sign, of the changes in ecosystem service levels, depend on what land cover is compared to oil palm, and to a smaller extent on the specific way oil palm is grown. Where conversion of natural forest to oil palm represents a drastic change in conditions, comparison of oil palm with the monocultures of fast-growing timber for the pulp- and paper industry, a legal change within production forest, may come out in favor of oil palm in several ecological dimensions. Although the Indonesian rules for logging in the production forest were supposed to secure sustainability across multiple harvest and recovery cycles, there are very few examples where such has been actually achieved in the reality of social-ecological systems.

6.1. Biodiversity

The conversion of natural forests to oil palm plantations threatens biodiversity according to many authors [115,127–129]. Forest fragmentation and large-scale homogenization landscape, whether for the development of plantation forestry or oil palm, pose a serious threat to tropical biodiversity [118,130,131], reducing species richness [128,132,133] and leading to a strongly declined functional diversity [134,135]. However, the response of biodiversity to land-cover change depends upon the extent to which natural habitat features are replicated while the sensitivity of species to change varies [130].

Several global studies have highlighted that past and predicted expansion of oil palm plantations into forest areas threatens mammal [136] and bird species with extinction [134,137,138]. In Indonesia, both Sumatran and Bornean orangutans (*Pongo abelii, P. pygmaeus*) and proboscis monkeys (*Nasalis larvatus*) population trends are decreasing due to the expansion of large-scale oil palm plantations [139–144]. Herds of elephants, tigers and rhinos are reported to be critically threatened due to this expansion [137,145]. In peat swamp forests, the expansion of oil palm monocultures is likely to have negative effects on, among others, macrofungal biodiversity [146].

A study [131] quantifying the impact on the herpetofauna of the Pacific lowlands of Costa Rica found that total species richness of amphibians and reptiles was reduced to 45–49% compared to forest area due to the almost complete absence of leaf litter, understory vegetation, and woody debris and the more open canopy. However, different research [147] conducted in Sumatra (Indonesia) showed different results. Rare amphibians were much more abundant in riparian forests and common amphibians were more prevalent in oil palm plantations. Indeed, reptile richness and abundance were higher in oil palm plantations than in all other habitats. Surprisingly, a meta-analysis [137] found that average invertebrate species richness did not differ significantly between oil palm and forest sites when published studies were compiled. It is commonly found that different aspects of biodiversity are influenced in opposite ways during land-use change [148].

Protection of any forested habitats and enhancing understory vegetation can help improve opportunities for some species [131]. Forest patches, even when small, fragmented and degraded, i.e., riparian sites, doubtlessly are required to sustain the species in human-transformed landscapes [139,146]. Substantial biodiversity loss can only be avoided if future oil palm expansion is managed by avoiding deforestation and cultivating non-forested and abandoned areas for sustainable oil palm cultivation [78,143,149]. The small-scale oil palm production due to greater vegetation heterogeneity is also likely to provide greater ecosystem services provision and should be considered as one of the strategies for achieving sustainable oil palm production [150]. By reducing deforestation and ecosystem degradation, stopping land clearing techniques by land burning, and stopping the development of oil palm plantations on unsuitable land, sustainable palm oil management can be achieved [120].

6.2. Watershed Functions

Oil Palm plantations are known to alter the hydrology of the subwatersheds in which they develop, by increasing the risk of flooding, increasing soil erosion (especially in the planting phase) and nutrient leaching, polluting ground- and surface water [124]. Once established, oil palm-dominated landscapes are 'water greedy' [122], responsible for decreasing local water tables and water supplies as well as changes in streamflow levels and water quality [118,122,151]. Several studies reported that the conversion of forests to young plantations initially decreased evapotranspiration (ET), but also decreased infiltration rates [122,124]. Researchers in Brazil [152] reported that water shortages have occurred more often since oil palm cultivation has become the dominant land use and large-scale deforestation has taken place. A study of land-use change impact on flooding frequency in Batanghari Watershed, Indonesia, showed that land cover change from forest to rubber and oil palm plantation contribute to the higher flooding frequency [153].

Though water availability, air, and water quality were perceived to be the most heavily impacted ecosystem services by oil palms [151], there is a knowledge gap about the magnitude of these ecohydrological changes and their variations over the oil palms lifetime [118,123]. It is unclear whether young and mature plantations have similar or different ET rates when compared to native forests [118,123]. The impacts of oil palm on the major components of the hydrological cycle (e.g., total water yields, dry season baseflow, streamflow dynamics, evaporative return to the atmosphere) are still at an early stage [118,121,152,154,155].

The constraint is that in general, countries with large oil palm plantations are developing countries with limited resources to assess the impact of plantations throughout the growing period of oil palm on hydrological functions [119]. Moreover, oil palm plantations are generally homogeneous monoculture stands of various age classes, with varying water use characteristics [154]. Most of the hydrological studies in oil palm plantations are carried out at the plot scale, while oil palm plantations can reach thousands of hectares across several watersheds [121]. Indeed, research on the hydrologic impacts of oil palm plantations at the scale of watersheds is rare [153].

Quantitative data from field measurement on the relationship between water availability at each stage and water yield are still limited as well as the actual water use of oil palm at the field level, and the minimum amount of water needed to optimize oil palm productivity [156]. Research on the impact of oil palm cultivation on hydrology is still focused mainly on the study of plot-scale ecohydrological fluxes, such as the impact of tree age on canopy rainfall interception [157,158] and transpiration rates in palms with varying ages and grown on different slopes [152,159].

However, in general, large scale conversions of forests to intensively managed plantations, result in significant changes in the hydrological cycle including periodic water scarcity [122,160]. In natural ecosystems, such as forests, most of the rainfall water is absorbed by the soil and plays a role in increasing plant transpiration and replenishing groundwater. In oil palm monoculture plantations, water does not penetrate the eroded and compacted soil properly [122]. Decreased infiltration reduces water storage [124] and increases surface runoff [121] potentially reducing the access to usable water and increasing the risk of flooding [161].

It is well known that highly productive monoculture stands are dependent on an abundant supply of water over time [150]. Indeed, drainage on oil palm plantations can lead to a substantial reduction in streamflow during dry seasons or droughts, soil subsidence, and potentially increasing future flood risks [162]. Thus, the expansion of oil palm plantations will affect water consumption which, in the long term, could affect local water resources [118]. Rather than to high water use of oil palms per se, local water scarcity seems connected to the modified redistribution of water after precipitation at the landscape scale, reducing effective buffering [122].

There may be a potential trade-off between water use and management intensity of oil palm plantations [154]. Optimizing water use and oil palm production and increasing cultivated varieties can result in more efficient land use and reduced conversion of natural landscapes [153]. The practices of retaining old palms during crop rotation and cultivating ground cover crops can mitigate some of the impacts [118]

Retaining a riparian buffer zone helped reduce the negative impacts of oil palm plantations on streams [163,164]. A good riparian zone with dense native vegetation of complex structure adjacent to the streams that flow through the plantations can help restrict the transport of soil, sediment from the oil palm land, and in preventing chemicals pollutant from reaching waterways. They can also serve as valuable natural habitats for riparian and terrestrial species, as well as carbon storage areas. Appropriate riparian zone management, combined with effective monitoring, is essential to maintain or enhance the ecological function and biodiversity of stream ecosystems [165].

6.3. Greenhouse Gas Emissions

Forest conversion to oil palm is responsible for significant net greenhouse gas (GHG) emissions [116,124,128,139,166]. However, information of net carbon emissions throughout the life of the oil palm plantation, especially on peat soils, is still very limited [167]. Aboveground C stocks in oil palm plantation are, averaged over a plantation life-cycle, around 40 t C ha^{-1}, which is lower than most predecessor vegetation (unless this is 'grassland'), leading to a 'carbon debt' in the first cycle [168,169]. On mineral soils, good-practice management can maintain soil carbon levels over a production cycle (compensating for early losses) [170]. Intensive use of fertilizer to increase yields adds greenhouse gas emissions as well, and the environmental optimum, minimizing the footprint per unit of palm oil, depends on context [171]. A synthesis of peat-oxidation emission values for tropical peatland by the Intergovernmental Panel on Climate Change [172], suggested default values of 51 Mg CO_2 ha^{-1} y^{-1} for smallholder systems, 55 Mg CO_2 ha^{-1} y^{-1} for commercial plantations (oil palm, industrial timber), and 10 Mg CO_2 ha^{-1} y^{-1} for disturbed secondary forest. A recent study in smallholder oil palm on peat soils in Jambi suggested values may be higher [173] than that. While the 'moratorium' has restricted large-scale oil palm development on peat soils, smallholder expansion continues and becomes a more prominent part of the overall problem.

7. Policy Responses and Options

7.1. FZ-OP as a Policy Issue

From the evidence so far, we conclude that FZ-OP is a substantive issue that cannot be denied or ignored at the policy level. Credible spatial data sets indicate that 15–20% of Indonesian oil palms are growing within the forest zone, with FZ-OP identified in all provinces in which oil palm is grown. FZ-OP is, however, unevenly distributed in Indonesia, with the provinces of Riau and Central Kalimantan, where agreement between provincial and national authorities over the boundaries of the forest zone was only reached in 2014, as the main contributor. But even if these two provinces are seen as a 'policy anomaly', the case studies in the neighboring provinces of Jambi and West Kalimantan showed that FZ-OP is real, but of mixed history and characteristics that need to be understood before appropriate policies can be designed. FZ-OP is most common within the 'production forest' category, but also includes 'watershed protection forest' and 'conservation areas'. The probability that a unit of land in the elevation and climate zone suitable for oil palm has been converted to oil palm varies from more than 10% to around 1% in conservation areas (Figure A1). Taking palm oil production as a point of reference for the two case-study provinces, up to 14% may derive from 14% production forests and up to 3% from protection forests and conservation areas. As there is no legal basis for oil palm presence within the forest zone, any FZ-OP indicates a lack of legality. Whether it indicates a loss of forest functions or not depends on the comparator land cover—which in the case of production forest can be an industrial timber plantation, with properties similar to oil palm, but should be naturally established in the protection and conservation forests, with substantially higher diversity. Meanwhile, three types of oil palm owners (large-scale plantations, tied or vertically integrated smallholders, and independent smallholders) represent different social contexts and likely require differentiated policy responses. Such responses can focus on rights and permits in land use and water management, or on the transport and processing stages of the value chain.

7.2. Institutional Responses

Given the importance of palm oil exports for the Indonesian economy (around 7% of total export value), the initial 'denial' phase of environmental issues around its production was promoted by the industry and its supporters in the Ministry of Economic Affairs. Yet, with 15–20% of the production area (and a probably lower share of the harvested produce) potentially spoiling the national reputation of all palm oil, the phase of 'shooting the messengers' ended, and communication efforts shifted to demonstrating active responses, while emphasizing the social dimensions of smallholder interests in escaping from rural poverty [26].

7.2.1. RSPO—Market Segmentation ('Shifting Blame')

The Round Table for Sustainable Palm Oil (RSPO) is a private sector-led initiative set up in 2004 to counter environmental and social concerns surrounding the rapid expansion of oil palm. It is often regarded as the most credible sustainability scheme in the oil palm sector and has among the most stringent and explicit principles and criteria [174,175]. Still, more critical analyses emphasize that RSPO certification merely provides a technical managerial solution to satisfy mainly Northern sustainability concerns but does not touch upon the underlying social relations in the production of the commodity that needs to be improved for truly sustainable palm oil production [176]. Examples of this are land and labor rights issues in Indonesia, unequal distribution of profits, but also the fact that even RSPO-certified oil palm monocultures present a huge loss of biodiversity compared to the forests, whether primary, secondary, or agro-forestry systems [171]. Moreover, the RSPO is voluntary and in 2015 covered an estimated 15% of Indonesia's oil palm area. With plenty of 'brown' supply chains to accommodate plantations that do not join RSPO, it is clear that this initiative cannot change the sector as a whole [88,177]. Santika et al. [178] also note that RSPO certification has not improved the wellbeing of local communities. They associate this with the considerably larger size of RSPO-certified plantation compared to non-certified plantations, which leaves little space for, especially forest-dependent, communities to maintain their livelihoods. Although the RSPO acknowledges that the inclusion of small and medium-sized growers is a key target [179], these small and medium actors are struggling with the costs, knowledge and institutional requirements needed for certification. With RSPO definitions of sustainability, there appears a clear risk that especially large corporate actors are classified as 'sustainable', as these are better able to fulfill RSPO criteria. Smaller actors on the other hand run the risk of being marginalized and blamed for the broader ills as land management and poor rural development strategies, for which oil palm is only one of many accelerants.

7.2.2. ISPO—National Sovereignty

Transnational business initiatives as the RSPO do not include governments. The Government of Indonesia, as well as the Indonesian Palm Oil Association (IPOA/GAPKI), regarded such initiatives as primarily reflecting Northern concerns, hollowing out the role of the State, and threatening the socio-economic development within Indonesia [180,181]. Therefore, in 2011, the Indonesian government launched the Indonesian Sustainable Palm Oil (ISPO) initiative to regain momentum and authority in sustainability discussions and to develop a sustainability framework that bypasses hard-to-comply-with sustainability criteria that are not demanded in key Southern markets such as China, India, and Indonesia itself [182]. However, ISPO requires compliance with existing legislation and whereas at first it is only mandatory for company palm oil plantations, with the latest update in 2020 it will obligatory for smallholders in five years, as well. The large area of smallholder SF-OP cannot be certified by ISPO under current rules and will prevent the target of full ISPO coverage to be achieved—unless the rules change and existing SF-OP can become legal. This will require the Ministry of Environment and Forestry and the Ministry of Agriculture to reach an agreement on how this can be done.

From the start, the ISPO organization suffered a lack of capacity to implement its mandate [183], and its principles and criteria are often deemed insufficient for guaranteeing sustainable development [169,184]. Although ISPO has been revised and supposedly strengthened in 2015 and 2020, the latest update of ISPO still poorly addresses human rights issues, lacks protection for non-primary forests, and lacks clear definitions, procedures and independent monitoring [185]. Again, as with RSPO, current smallholder certification only reaches a fraction of the total number of smallholders and smallholder exclusion looms.

7.2.3. Deregulation and Crisis Responses

Recently the Indonesian parliament adopted an 'omnibus' law that, to facilitate business development and job creation, removes legal obligations for environmental impact assessment, simplifies procedures for obtaining permits, and abolishes the requirement for plantations to support smallholder producers as part of their land concessions. The law has been critiqued as both environmental and social (worker rights) concerns appear to lose existing safeguards. Its consequences for the SF-OP issue are not yet clear.

7.3. Current Policy Options Based on Land and Water Management

Several policy options have been discussed in the public debate so far, summarized in Table 5. Just as 'stop the bleeding' is the first line of medical defense in dealing with injuries, the first efforts made, through a 'moratorium' are efforts to stop a further expansion of tree crops within the forest zone. Once some opportunity has been created for evaluating longer-term options, these can be linked to the diversity of contexts, in a 'policy options by context' table (Table 5), that requires an evaluation from the full range of relevant stakeholders before negotiations can progress and decisions with a chance of implementation success can be made. The options can target the land-base for oil palm production and/or the transport and entry into the mills and further value chains (compare Figure 7).

7.3.1. Legalize by Including Oil Palm in Forest Definition

The simplest solution to any illegality issue, at least from a formal governance perspective, is to legalize it. As the internationally used forest definition is ambiguous as to the 'tree' concept that underpins quantitative criteria about canopy cover and spatial scale, an argument can be made that palms, including *Elaeis guineensis*, are included in the woody perennial category, while quantitative criteria on potential tree height, the density of the canopy cover and longevity are easily within the range of what are commonly understood to be 'forestry trees'. Such an approach would avoid complexity, but not address any of the environmental and social concerns of the status quo. It would, however, put the pulp and paper industry, supported by Forestry authorities as core economic activity, and the oil palm industry on a level playing field. It will make all oil palm production deforestation-free by definition and greatly help in achieving the target that all of the Indonesian palm oil can be ISPO-certified, without major changes on the ground.

However, as simple and attractive as the option may seem in the short term, it would severely undermine Indonesia's international standing as a country balancing development and environmental concerns. It would also open the door for a much further expansion of oil palm within forests, without legal means to control such expansion.

7.3.2. Grandfathering

As part of an overall policy package, a grandfathering approach here refers to defining a cut-off date that tolerates (although not fully legalizes) oil palm cultivation in the forest zone if started before an agreed date. It is a common practice where 'new' regulations emerge, such as the RSPO standards that refer to the time standards were published. It would normally be restricted to the current plantation cycle and not allow replanting of oil palm.

A 'grandfather' approach could be a policy option that shares responsibility for the status quo and its path dependency where government agencies have been involved in the facilitation of illegal activities. Where smallholders from the local or indigenous communities shifted to oil palm from past rubber involvement in their swidden/fallow rotations they may have done so on land that is classified as 'production forest' but has not been formally gazetted. In the two provinces with the highest fraction of forest-zone oil palm, the designation of forest zone has remained contested between provincial and national forest authorities until 2014, creating a grey zone in which law enforcement was a low priority.

A case could also be made for migrants who moved to the frontier area for income generation, not for investment and capital accumulation (as 'white-collar' farmers). Part of these migrants (or their parents) were mobilized by the government through the "transmigration program" and looked for opportunities to increase their land, acquiring land from informal local land markets that disrespect forest zone categorization. Where this happened in the past, a 'generic pardon' or grandfather rule could close the books on it—but at risk of setting precedents that current rules can still be breached.

Table 5. Policy options to deal with oil palm within the forest zone, by applicability domain.

Policy Option	Applicability Domain				Expected Consequences		
	Conservation Areas	Protection/Production Forest on Peat	Protection Forest on Mineral Soil	Production Forest on Mineral Soil	Social	Economic	Environmental
Land-focused:							
1.Legalize	NA	NA	NA	NA	++	++	–
2. Grandfathering	NA	NA	A	A	+	+	+/-
3. Evict	NA	NA	NA	NA	–	-	+/-
4. Charge	A	NA	A	A	+/-	+/-	+/-
5. Agroforestry concessions	NA	A	A	A	+	+/-	+/-
6. Focus on high-value locations	A	A	A	A	+/-	+/-	+
7. Rewet peatlands	NA	A	NA	NA	+/-	+/-	+
Value-chain based:							
8. Mill certification	A	NA	A	A	-	-	+/-
9.Transport permits	A	NA	A	A	-	-	+/-
10.Segment markets	A	NA	A	A	-	+/-	+/-

Remarks: A (applicable), NA (Not applicable) from a forestry institutional perspective.

7.3.3. Evict Farmers, Destroy the Crops

On the other side of the spectrum from a blanket legalization of all forest-zone oil palm, there are voices for strict law enforcement. Evicting farmers and destroying their crops, to prevent their return, has been implemented before, e.g., in the 1990s in coffee encroachment areas in Sumatra. However, this sparked further conflicts, while the uprooting may have aggravated environmental issues in the short term, as did the type of reforestation with fast-growing exotic tree species in the longer term from an environmental perspective, as documented for Sumberjaya in Lampung (Sumatra) [185] and Manggarai (Flores) [186].

During the 'New Order' regime (1965–1998) military enforcement of government-set rules was common, but not free from corruption where economic interests were involved. After the regime change in 1998, reliance on court procedures, prisons and fines has shifted to the role of local governments in illegal procedures. However, attempts to sue provincial authorities in C Kalimantan and Riau for transgressing forest-zone rules backfired when the court found the forest zone in these provinces had not been gazette according to the prevailing Forestry Law [187].

In areas where production forests have been largely transformed into oil palm plantations for many years, such as in parts of Riau [91], it is unlikely that evicting small farmers, local upper-middle-class families, elites and companies with vested interests is feasible. They are not going to accept the destruction of their investment, and therefore this option may well be politically unachievable and risky. Already there have been cases where 'farmer groups' have held captive inspectors from the national authorities [188]. As highlighted elsewhere [111], the land mafia is making large profits with land conversion and in certain frontiers, this goes well beyond poor farmers looking for a few hectares of land [91]. Furthermore, the mere destruction of crops will not necessarily lead to increased environmental performance as oil palm plantations capture considerably more carbon than the degraded lands and *Imperata* (alang-alang) fields that destroyed plantations are likely to turn into. Heavily degraded lands without clear ownership and proper management are prone to fire and a key source of repeatedly occurring mass forest fires in Indonesia, with their detrimental effect on GHG emissions and human health [111,189].

7.3.4. Charge Land-Owner Benefit Shares to Pay for Forest Management Elsewhere

In Indonesia 'share-cropping' has a long tradition of arrangements where a land owner allows others to use farmland, based on a share of the crop yield obtained. Such arrangements are often preferred by both parties over hired labor or land rent contracts; the land owners share of the yield, the payment for and accounting of agricultural inputs tend to vary with local circumstances and the supply–demand balance for contracts. Often patron–client relations are involved with social dimensions beyond economic rationality. Rules can apply to trees, as well as to land as a production factor [190]. In the tradition of the Java Social Forestry program, forest authorities allowed farmers to grow annual crops, perennial fodder grass, or low-stature tree crops such as coffee under similar rules of a yield-share for the forest authorities [15]. Reference to such a system could lead to a financial charge on oil palm in the forest zone, for example within production forest lands where gazettement has been completed and the state is the legal owner.

Currently, forest management, from protected areas to production forests that are in the recovery phase, is severely restricted by lack of funding. Charging a land owner's share of palm oil produced within the production forest zone could provide funding for forest management elsewhere.

Of the 3.4 M ha oil-palm (OP) plantation in state forest areas, 700,000 ha is smallholder OP [75]. Rules may have to be differentiated by scale, for example, distinguishing between oil palm farm sizes of >100, 25–100, 5–25, and 0–5 ha, as only the last category is likely to be a family farm without external labor. Such differentiation could be part of the Social Forestry (SF) and Agrarian Reform (AR)/UUCK schemes. A legal basis for such charges could be found in Law 32/2009 on Environment protection and management.

If ways can be found to implement this, it will be attractive to local forest authorities—possibly too attractive and providing incentives for further oil palm expansion within the forest zone.

7.3.5. Agroforestry Concession

Peru, dealing with similar issues of past undocumented and illegal agricultural encroachment into forest zones, has created a legal category of an 'agroforestry concession' [191]. These schemes allow for agreements that re-establish government authority over forest lands but allow current tree-based land uses to continue, within agreed conditions. They resolve the issue that the current illegality of the land use is a bottleneck for government support of any type in the area, including a lack of extension services that restrain the development of socially and environmentally desirable practices. Similar issues have been noted in Riau, where support for companies and especially oil palm farmers in the forestry domain has been minimal to non-existent as illegal activities cannot formally be supported [91,92].

Agroforestry concessions as a legal instrument would complement current 'community-based forest management' and 'village forest' schemes, without claims that oil palm plantations are considered 'forests'. They would also allow investment in upgrading different types of production systems, systems that are more acceptable than mere oil palm monocultures. Current research interests in testing diversified oil palm agroforestry systems are opening new perspectives that might match farm economies [192–194].

7.3.6. Swaps with High-Value Legal Deforestation Locations

Side-by-side with the 3.6 M of illegal oil palm in the forest zone, there is still a substantial area (1.2 M ha by recent estimates [195]) of old-growth forest that has been legally transferred (PKH/release) to large scale oil palm development (as a land bank), but has not yet been converted into oil-palm plantations. Earlier efforts under a 'land swap' umbrella to exchange land for companies with such concession rights have not had much impact [196], but in combination with a substantial landowners' charge on existing illegal forest-zone oil palm, it may be feasible to recover high-value forest while giving up (from a forestry perspective) production forest land already converted. The challenge is, however, that such swaps usually involve locations in different districts or even provinces. Other high-value locations that deserve priority in resolving existing forest-zone oil palm are riparian zones and ecological corridors.

Specific ideas have been formulated on how such swaps could, in the context of the Oil Palm moratorium, return existing non-converted OP concessions to government control. It could be implemented by buying out rightsholders using funds obtained by charging FZ-OP units of more than 25 ha based on Law 32/2009 on Environmental protection and management, charging land-rents to 'big smallholder OP' (5–25 ha), while 'real smallholder OP' (<5 ha) could be exonerated in a Social Forestry (SF) and Agrarian Reform (AR)/UUCK perspective.

What has yet to be explored, however, is how the records on spatial boundaries of forest zone would be adjusted to 'legalize' the FZ-OP for the longer term. Moreover, it is unclear how such swaps would be understood internationally, and whether the smallholder OP can, after such swaps, enter international markets within a 'deforestation-free' label. One would have to explain that the oil palms were planted on illegally deforested land, but became legalized because elsewhere deforestation had been legalized, did not happen and its rights were withdrawn. This will be quite a mouthful of legalese, that only has a chance if the Government that implements such schemes has a record of transparency beyond what currently exists.

7.3.7. Rewet Peatlands

Specific solutions are needed for peatland forests converted to oil palm. The current situation of illegality makes it hard for government agencies to engage. For example, oil palm plantations located in Riau peatlands that were part of the Forestry Domain suffered severe water table management issues, with larger farmers just digging canals without coordination, leading to drought and flooding, leaving the area prone to peat fires [91]. Such undesirable practices, which are relevant in other landscapes as well [92], may be countered by (partial, temporary or full) legalization of their plantations, charging farmers for their activities in the forestry domain and using these proceeds to finance and coordinate proper water table management. Such an approach may lead to landscape-wide improved water table management and associated reductions in draught, fire, flooding and associated GHG emissions. The government rules about maintaining groundwater tables in peatlands no deeper than 40 cm below the surface are not incompatible with growing oil palm—at least not under smallholder, non-mechanized management.

7.4. Policy Options Based on Transport and Processing

7.4.1. Impose Legality Checks at the Mills

Current sustainability initiatives as RSPO and ISPO heavily depend on legality checks at the mill, restricting certification to mills with adequate traceability of the fresh fruit bunches they obtain. However, supply chain characteristic for especially independent smallholders, who usually sell to middlemen that mix and may resell their produce to other traders, are unable to fully trace all produce that enters the mills. This system appears to benefit especially the larger actors, who have the most integrated value chains, shifting blame and costs to others. As noted elsewhere [29,30] certification may solve problems of guilt for downstream users who do not want to be part of value chains with negative social and environmental consequences, but it may primarily shift blame, rather than transforming other means of production, unless economic signals, accounting for increased transaction costs, are sending a clear message.

7.4.2. Apply a Transport Permit System

Forestry has a long history of trying to control illegal logging by applying transport systems and road checks. As documented for charcoal trade in Africa [197], it can increase transaction costs by legal and illegal levies, with little impact other than reducing farmgate profitability at the sites of production. This approach thus suffers similar drawbacks as legality checks at the mills. It is prone to corruption. In areas where there are a lot of roads, this becomes cumbersome. However, in some contexts where there is less infrastructure, it may work to some extent.

7.4.3. Segment Markets

Legality checks at the mills and chain of custody rules [93], potentially supported by transport permit systems, can support the bifurcations into 'green' and 'brown' supply chains, targeting different market segments (e.g., international and domestic). It will not solve the issue of illegal forest-zone oil palm, but it may affect the profitability of the lower-grade production system if transaction costs of the certification are controlled or subsidized. The new EU anti-deforestation policy and RSPO are heading in this direction, recognizing that importing countries have to respect the sovereignty of producing countries within world trade rules. Keep your own street clean and shift blame for existing problems to others. The EU with its deforestation-free supply chains is picking the low-hanging fruits, with a 'grandfather' cut-off date for deforestation before 2008, which can easily cover EU demands, whilst leaving the tougher issues to other parties.

7.5. Follow-Up Policy Research

Table 5 provides an early assessment of ten policy options for four subcategories of FZ-OP, from three overarching perspectives. The tentative evaluation of options could be the start of a further process of stakeholder consultations, that may modify perspectives on expected effectiveness and acceptability by main stakeholder groups. As some of the options proposed, especially the 'agroforestry concession' idea would be new to the Indonesian forest management framework, deeper analysis of legal opportunities and consequences will be needed before specific recommendations can be made.

8. Conclusions

Up to one-fifth of oil palms in Indonesia are located within the 'forest zone' (Kawasan hutan), especially in lands indicated as production forest, with smaller fractions in protection forest and conservation areas and substantial variation between the provinces and islands. Two (out of 34) provinces are responsible for two-thirds of the forest-zone oil palm, while harboring nearly one-third of Indonesian oil palms. While state-owned and private large-scale plantations dominated in the early expansion phase of oil palm, smallholders have become prominent, especially in Sumatra, where 60% of oil palm capacity is located, especially those without a contract with mills ('independents'). Part of the smallholder oil palm within the forest zone has been derived from earlier agroforests (often rubber-based).

As there is no legal basis for oil palm within the forest zone, one-fifth of Indonesia's oil palms cannot meet current criteria for sustainability certification, whether the voluntary international RSPO standard or the domestic ISPO standard. Solutions for the ensuing policy problem will have to differentiate between the primary forest functions indicated for different parts of the forest zone, with the smallest gap between desired function and current reality in the production forest zone, where most of the oil palms are located, and largest gap for the small fraction (<1%) of oil palms in conservation areas. Policies will have to differentiate between the scale of the production unit, the opportunities to charge a 'land owner's' share by forest authorities, the history of the conversion process, and the expected impact through discouraging further expansion.

Policy options for dealing with forest-zone oil palm can focus on the land where oil palms are grown and/or on the transport and mills that are processing the fresh fruit bunches produced. Where mill-based regulation can be effective in segregating the market and ensuring that part of the supply chain is 'deforestation-free', it is unlikely to close marketing channels for forest-zone oil palm products. Land-based options will have to explore a middle ground as neither blanket legalization of oil palm as a forest commodity, nor evictions and destroying of the crops are realistic in the institutional context. Combinations of financial charges and permits for agroforestry development may be the most feasible basis of a negotiated policy package. Further policy research and stakeholder consultations will be needed before specific policy recommendations can be made, adjusted to the legal environment and responsibilities of national, provincial and district-level government authorities, as well as likely responses of international trade and domestic industry.

Author Contributions: Conceptualization, E.P., H.S., I.J., A.W. and M.v.N.; Investigation, E.P., H.S., I.J., A.W., H.Y.S.H.N. and M.v.N.; Methodology, E.P., A.W. and M.v.N.; Writing—original draft, E.P., H.S., I.J., A.W., H.Y.S.H.N. and M.v.N.; Writing—review & editing, E.P., H.S., I.J., A.W. and M.v.N. All authors have read and agreed to the published version of the manuscript.

Acknowledgments: The authors acknowledge the availability of spatial data sets through Yves Laumonier (CIFOR), Lilik Prasetyo and Aryo Condro. The National Main Commodity Maps in Indonesia (Condro et al. (2020)) were produced under collaboration between the Faculty of Forestry, IPB, and the United Nations Development Programme (UNDP). We acknowledge questions and inputs from speakers and participants of the on-line seminar on 26 September 2020.

Appendix A

Appendix A.1. Details of the Initial Stakeholder Consultation

The webinar "The Future of Agro-Commodities in Forest Areas" on 26 September 2020 had 373 registered participants. They attended the on-line event geographically spread over 33 (out of 34) provinces; institutionally, 30% represented government agencies (local, national; forestry, non-forestry),

3% research organizations, 44% universities, 11% private sector, 13% NGO's. Demographically 35% was female and 27% below 25 years of age.

As first speaker the Director General of Natural Resources and Ecosystem Conservation (KSDAE), Ministry of Environment and Forestry (KLHK), Ir. Wiratno, MSc, marked the shift from a timber focus in the past to a much wider range of economic commodities as potentially derived from and grown in forests, as long as the societal functions of forests can be secured. Speakers from NGO's reviewed data on forest-associated poverty levels (around 21% of poor people in Indonesia are forest villagers) and presence of tree crops, with most specific debate about oil palm.

The discussion focused on the reliability of data of various sources and on opportunities to resolve issues through implementation of Law No.34/2014 on Soil and Water Conservation and Government Regulation No.46/2017 on Environmental Economic Instruments. Panelists emphasized a need for the involvement of government agencies in intensive assistance for large-scale plantations and smallholders in forest areas, and for law enforcement for deliberate occupation in forest areas and resolving tenure issues. The various strands of discussion informed the current manuscript.

Appendix A.2. Analysis of FZ-OP by Elevation Zone and Forest Class

The spatial data on FZ-OP and oil palm on non-forest lands are differentiated by forest class and elevation in Figure A1.

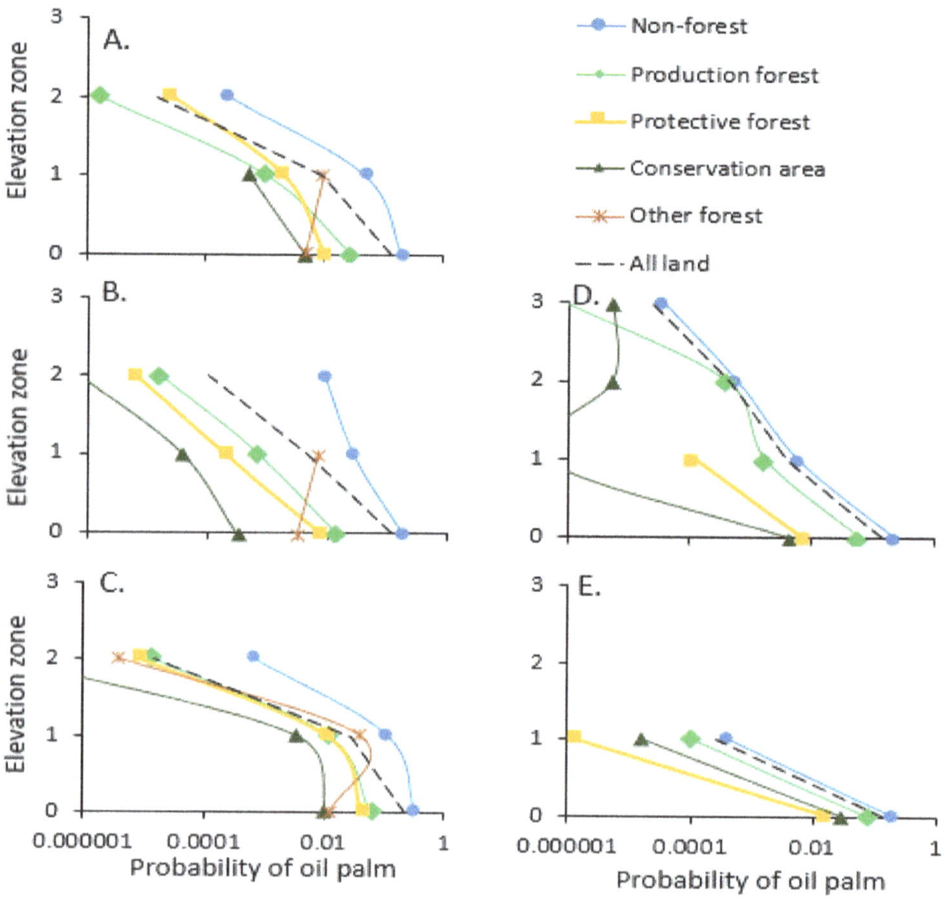

Figure A1. Probability of land having oil palm as land cover at four elevation zones (0 = 0–100; 1 = 100–300; 2 = 300–1000; 3 = >1000 m a.s.l.) in two provinces ((**A–C**), West Kalimantan; (**D,E**) Jambi), for specified land use designations outside and inside the forest-zone, according to three data sources ((**A**) [75]; (**B,D**) [33]; (**C,E**) [28]); (NB log scale on the X-axis).

References

1. Chazdon, R.L.; Brancalion, P.H.; Laestadius, L.; Bennett-Curry, A.; Buckingham, K.; Kumar, C.; Moll-Rocek, J.; Vieira, I.C.G.; Wilson, S.J. When is a forest a forest? Forest concepts and definitions in the era of forest and landscape restoration. *Ambio* **2016**, *45*, 538–550. [CrossRef]
2. van Noordwijk, M.; Minang, P.A. *If We Cannot Define It, We Cannot Save It*; ASB Policy Brief 15; ASB Partnership for the Tropical Forest Margins: Nairobi, Kenya, 2009; p. 4. Available online: http://www.asb.cgiar.org/pdfwebdocs/ASBPB15.pdf (accessed on 24 October 2020).
3. de Foresta, H.; Temu, A.; Boulanger, D.; Feuilly, H.; Gauthier, M. *Towards the Assessment of Trees Outside Forests: A Thematic Report Prepared in the Framework of the Global Forest Resources Assessment 2010*; Food and Agriculture Organization of the United Nations: Rome, Italy, 2013.
4. van Noordwijk, M.; Suyamto, D.A.; Lusiana, B.; Ekadinata, A.; Hairiah, K. Facilitating agroforestation of landscapes for sustainable benefits: Tradeoffs between carbon stocks and local development benefits in Indonesia according to the FALLOW model. *Agric. Ecosyst. Environ.* **2008**, *126*, 98–112. [CrossRef]
5. Romijn, E.; Ainembabazi, J.H.; Wijaya, A.; Herold, M.; Angelsen, A.; Verchot, L.; Murdiyarso, D. Exploring different forest definitions and their impact on developing REDD+ reference emission levels: A case study for Indonesia. *Environ. Sci. Policy* **2013**, *33*, 246–259. [CrossRef]
6. FAO. *Global Forest Resources Assessment 2010*; FAO Forestry Paper No. 163; UN Food and Agriculture Organization: Rome, Italy, 2010.
7. Keenan, R.J.; Reams, G.A.; Achard, F.; de Freitas, J.V.; Grainger, A.; Lindquist, E. Dynamics of global forest area: Results from the FAO Global Forest Resources Assessment 2015. *For. Ecol. Manag.* **2015**, *352*, 9–20. [CrossRef]
8. van Dijk, K.; Savenije, H. *Oil Palm or Forests? More Than a Question of Definition*; Policy Brief; Tropenbos International: Wageningen, The Netherlands, 2010.
9. van Noordwijk, M.; Dewi, S.; Minang, P.; Simons, T. 1.2 Deforestation-free claims: Scams or substance. In *Zero Deforestation: A Commitment to Change*; No. 58; Tropenbos International: Wageningen, The Netherlands, 2017; pp. 11–16.
10. Sasaki, N.; Putz, F.E. Critical need for new definitions of "forest" and "forest degradation" in global climate change agreements. *Conserv. Lett.* **2009**, *2*, 226–232. [CrossRef]
11. Zomer, R.J.; Trabucco, A.; Verchot, L.V.; Muys, B. Land area eligible for afforestation and reforestation within the Clean Development Mechanism: A global analysis of the impact of forest definition. *Mitig. Adapt. Strateg. Glob. Chang.* **2008**, *13*, 219–239. [CrossRef]
12. Fay, C.; Michon, G. Redressing forestry hegemony when a forestry regulatory framework is best replaced by an agrarian one. *For. Trees Livelihoods* **2005**, *15*, 193–209. [CrossRef]
13. Silvianingsih, Y.A.; Hairiah, K.; Suprayogo, D.; van Noordwijk, M. Agroforests, swiddening and livelihoods between restored peat domes and river: Effects of the 2015 fire ban in Central Kalimantan (Indonesia). *Int. For. Rev.* **2020**, *22*, 382–396. [CrossRef]
14. Akiefnawati, R.; Villamor, G.B.; Zulfikar, F.; Budisetiawan, I.; Mulyoutami, E.; Ayat, A.; van Noordwijk, M. Stewardship agreement to reduce emissions from deforestation and degradation (REDD): Case study from Lubuk Beringin's Hutan Desa, Jambi Province, Sumatra, Indonesia. *Int. For. Rev.* **2010**, *12*, 349–360. [CrossRef]
15. Cahyono, E.D.; Fairuzzana, S.; Willianto, D.; Pradesti, E.; McNamara, N.P.; Rowe, R.L.; van Noordwijk, M. Agroforestry Innovation through Planned Farmer Behavior: Trimming in Pine–Coffee Systems. *Land* **2020**, *9*, 363. [CrossRef]
16. Van Noordwijk, M.; Zomer, R.J.; Xu, J.; Bayala, J.; Dewi, S.; Miccolis, A.; Cornelius, J.P.; Robiglio, V.; Nayak, D.; Rizvi, J. Agroforestry options, issues and progress in pantropical contexts. In *Sustainable Development through Trees on Farms: Agroforestry in Its Fifth Decade*; Van Noordwijk, M., Ed.; World Agroforestry (ICRAF): Bogor, Indonesia, 2019.
17. Van Noordwijk, M.; Williams, S.; Verbist, B. Towards integrated natural resource management in forest margins of the humid tropics: Local action and global concerns. In *ASB Lecture Notes*; ICRAF: Bogor, Indonesia, 2001.
18. Hansen, M.C.; Potapov, P.V.; Moore, R.; Hancher, M.; Turubanova, S.A.; Tyukavina, A.; Thau, D.; Stehman, S.V.; Goetz, S.J.; Loveland, T.R.; et al. High-resolution global maps of 21st-century forest cover change. *Science* **2013**, *342*, 850–853. [CrossRef] [PubMed]

19. Zomer, R.J.; Neufeldt, H.; Xu, J.C.; Ahrends, A.; Bossio, D.; Trabucco, A.; van Noordwijk, M.; Wang, M. Global tree cover and biomass carbon on agricultural land: The contribution of agroforestry to global and national carbon budgets. *Sci. Rep.* **2016**, *6*, 29987. [CrossRef] [PubMed]
20. Thomas, S.; Dargusch, P.; Harrison, S.; Herbohn, J. Why are there so few afforestation and reforestation Clean Development Mechanism projects? *Land Use Policy* **2010**, *27*, 880–887. [CrossRef]
21. Corbera, E.; Schroeder, H. Governing and implementing REDD+. *Environ. Sci. Policy* **2011**, *14*, 89–99. [CrossRef]
22. Minang, P.A.; Van Noordwijk, M.; Duguma, L.A.; Alemagi, D.; Do, T.H.; Bernard, F.; Agung, P.; Robiglio, V.; Catacutan, D.; Suyanto, S.; et al. REDD+ Readiness progress across countries: Time for reconsideration. *Clim. Policy* **2014**, *14*, 685–708. [CrossRef]
23. Garrett, R.D.; Levy, S.; Carlson, K.M.; Gardner, T.A.; Godar, J.; Clapp, J.; Dauvergne, P.; Heilmayr, R.; de Waroux, Y.L.P.; Ayre, B.; et al. Criteria for effective zero-deforestation commitments. *Glob. Environ. Chang.* **2019**, *54*, 135–147. [CrossRef]
24. Sheil, D.; Casson, A.; Meijaard, E.; Van Noordwijk, M.; Gaskell, J.; Sunderland-Groves, J.; Wertz, K.; Kanninen, M. *The Impacts and Opportunities of Oil Palm in Southeast Asia: What do We Know and What Do We Need to Know*; Center for International Forestry Research: Bogor, Indonesia, 2009. [CrossRef]
25. Sayer, J.; Ghazoul, J.; Nelson, P.; Boedhihartono, A.K. Oil palm expansion transforms tropical landscapes and livelihoods. *Glob. Food Secur.* **2012**, *1*, 114–119. [CrossRef]
26. van Noordwijk, M.; Pacheco, P.; Slingerland, M.; Dewi, S.; Khasanah, N. *Palm Oil Expansion in Tropical Forest Margins or Sustainability of Production? Focal Issues of Regulations and Private Standards*; World Agroforestry (ICRAF): Bogor, Indonesia, 2017. [CrossRef]
27. Heilmayr, R.; Carlson, K.M.; Benedict, J.J. Deforestation spillovers from oil palm sustainability certification. *Environ. Res. Lett.* **2020**, *15*, 075002. [CrossRef]
28. Condro, A.A.; Setiawan, Y.; Prasetyo, L.B.; Pramulya, R.; Siahaan, L. Retrieving the National Main Commodity Maps in Indonesia Based on High-Resolution Remotely Sensed Data Using Cloud Computing Platform. *Land* **2020**, *9*, 377. [CrossRef]
29. Mithöfer, D.; van Noordwijk, M.; Leimona, B.; Cerutti, P.O. Certify and shift blame, or resolve issues? Environmentally and socially responsible global trade and production of timber and tree crops. *Int. J. Biodivers. Sci. Ecosyst. Serv. Manag.* **2017**, *13*, 72–85.
30. Leimona, B.; van Noordwijk, M.; Mithöfer, D.; Cerutti, P.O. Certifying Environmental Social Responsibility. Environmentally and socially responsible global production and trade of timber and tree crop commodities: Certification as a transient issue-attention cycle response to ecological and social issues. *Int. J. Biodivers. Sci. Ecosyst. Serv. Manag.* **2018**, *13*, 497–502. [CrossRef]
31. KLHK. *Peta Penutupan Lahan Indonesia, 2000, 2010, 2018*; Kementrian Lingkunan Hidup dan Kehutanan: Jakarta, Indonesia, 2018; Available online: http://geoportal.menlhk.go.id/arcgis/rest/services/ (accessed on 9 October 2020).
32. Pribadi, U.A.; Setiabudi Suryadi, I.; Laumonier, Y. West Kalimantan Ecological Vegetation Map 1:50000. 2020. Available online: https://data.cifor.org/dataset.xhtml?persistentId=doi:10.17528/CIFOR/DATA.00203 (accessed on 15 October 2020). [CrossRef]
33. Descals, A.; Wich, S.; Meijaard, E.; Gaveau, D.L.A.; Peedell, S.; Szantoi, Z. High-resolution global map of smallholder and industrial closed-canopy oil palm plantations. *Earth Syst. Sci. Data Discuss* **2020**. [CrossRef]
34. SK Menteri Kehutanan no 733/2014. Available online: https://fdokumen.com/document/sk-no733-tahun-2014-penunjukan-kalbar.html (accessed on 30 October 2020).
35. SK Menteri Kehutanan no 863/2014. Available online: https://fdokumen.com/document/surat-keputusan-menteri-kehutanan-nomor-sk-863menhut-ii2014-tentang-kawasan.html (accessed on 30 October 2020).
36. Galudra, G.; Sirait, M. A discourse on Dutch colonial forest policy and science in Indonesia at the beginning of the 20th century. *Int. For. Rev.* **2009**, *11*, 524–533. [CrossRef]
37. Rachman, N.F.; Siscawati, M. Forestry Law, Masyarakat Adat and Struggles for Inclusive Citizenship in Indonesia. *Routledge Handb. Asian Law* **2016**, 224–249. [CrossRef]
38. Bakker, L.; Moniaga, S. The space between: Land claims and the law in Indonesia. *Asian J. Soc. Sci.* **2010**, *38*, 187–203. [CrossRef]
39. Hermosilla, A.C.; Fay, C. *Strengthening Forest Management in Indonesia through Land Tenure Reform: Issues and Framework for Action*; The World Bank: Washington, DC, USA, 2005; p. 55.

40. Ministry of Environment and Forestry. *The State of Indonesia's Forest*; Ministry of Environment and Forestry, Republic of Indonesia: Jakarta, Indonesia, 2018.
41. Suwarno, A.; van Noordwijk, M.; Weikard, H.P.; Suyamto, D. Indonesia's forest conversion moratorium assessed with an agent-based model of Land-Use Change and Ecosystem Services (LUCES). *Mitig. Adapt. Strateg. Glob. Chang.* **2018**, *23*, 211–229. [CrossRef] [PubMed]
42. Directorate of Peat Degradation Control. Penetapan Fungsi Ekosistem Gambut (Peatland Ecosystem Function Establishment). 2020. Available online: http://pkgppkl.menlhk.go.id/v0/penetapan-fungsi-ekosistem-gambut/ (accessed on 15 October 2020).
43. Decree of Minister of Forestry No. 47/Kpts-II/1998 Concerning the Designation of +/− 29,000 Hectares of the Protective Forest and Limited Production Forest Zone, From the Pesisir Forest Grouping, in West Lampung District, Province of Lampung, Which Are Covered by Damar Agroforests ('Repong Damar') Managed by Communities under Customary Law (Masyarakat Hukum Adat), as a Zone with Distinct Purpose (Kawasan Dengan Tujuan Istimewa). Available online: https://peraturan.huma.or.id/pub/surat-keputusan-menteri-kehutanan-nomor-47-kpts-ii-1998 (accessed on 30 November 2020).
44. Kusters, K.; De Foresta, H.; Ekadinata, A.; van Noordwijk, M. Towards solutions for state vs. local community conflicts over forestland: The impact of formal recognition of user rights in Krui, Sumatra, Indonesia. *Hum. Ecol.* **2007**, *35*, 427–438. [CrossRef]
45. Michon, G. *Domesticating Forests—How Farmers Manage Forest Resources*; CIFOR: Bogor, Indonesia, 2005.
46. De Royer, S.; Van Noordwijk, M.; Roshetko, J.M. Does community-based forest management in Indonesia devolve social justice or social costs? *Int. For. Rev.* **2018**, *20*, 167–180. [CrossRef]
47. Rakatama, A.; Pandit, R. Reviewing social forestry schemes in Indonesia: Opportunities and challenges. *For. Policy Econ.* **2020**, *111*, 102052. [CrossRef]
48. Kelly, A.B.; Peluso, N.L. Frontiers of commodification: State lands and their formalization. *Soc. Nat. Resour.* **2015**, *28*, 473–495. [CrossRef]
49. Brown, D.W. *Addicted to Rent: Corporate and Spatial Distribution of Forest Resources in Indonesia: Implications for Forest Sustainability and Government Policy*; Indonesia-UK Tropical Forest Management Programme, Provincial Forest Management Programme: Jakarta, Indonesia, 1999.
50. Resosudarmo, B.P. (Ed.) *The Politics and Economics of Indonesia's Natural Resources*; Institute of Southeast Asian Studies: Singapore, 2005. [CrossRef]
51. Brown, T.; Leitmann, J.; Boccucci, M.; Jurgens, E. *Sustaining Economic Growth, Rural Livelihoods, and Environmental Benefits: Strategic Options for Forest Assistance in Indonesia*; Worldbank: Jakarta, Indonesia, 2006; Available online: http://documents1.worldbank.org/curated/en/986501468049447840/pdf/392450REVISED0IDWBForestOptions.pdf (accessed on 24 October 2020).
52. Mulyani, M.; Jepson, P. REDD+ and forest governance in Indonesia: A multistakeholder study of perceived challenges and opportunities. *J. Environ. Dev.* **2013**, *22*, 261–283. [CrossRef]
53. Van Noordwijk, M.; Agus, F.; Dewi, S.; Purnomo, H. Reducing emissions from land use in Indonesia: Motivation, policy instruments and expected funding streams. *Mitig. Adapt. Strateg. Glob. Chang.* **2014**, *19*, 677–692. [CrossRef]
54. Singer, B.; Giessen, L. Towards a donut regime? Domestic actors, climatization, and the hollowing-out of the international forests regime in the Anthropocene. *For. Policy Econ.* **2017**, *79*, 69–79. [CrossRef]
55. Hall, D. Land grabs, land control, and Southeast Asian crop booms. *J. Peasant Stud.* **2011**, *38*, 837–857. [CrossRef]
56. Gouyon, A.; De Foresta, H.; Levang, P. Does 'jungle rubber' deserve its name? An analysis of rubber agroforestry systems in southeast Sumatra. *Agrofor. Syst.* **1993**, *22*, 181–206. [CrossRef]
57. Verbist, B.; Putra, A.E.D.; Budidarsono, S. Factors driving land use change: Effects on watershed functions in a coffee agroforestry system in Lampung, Sumatra. *Agric. Syst.* **2005**, *85*, 254–270. [CrossRef]
58. Ruf, F.; Ehret, P. Smallholder cocoa in Indonesia: Why a cocoa boom in Sulawesi? In *Cocoa Pioneer Fronts Since 1800*; Palgrave Macmillan: London, UK, 1996; pp. 212–231. [CrossRef]
59. Li, T.M. Practices of Assemblage and Community Forest Management. *Econ. Soc.* **2007**, *36*, 264–294. [CrossRef]
60. Galudra, G.; van Noordwijk, M.; Agung, P.; Suyanto, S.; Pradhan, U. Migrants, land markets and carbon emissions in Jambi, Indonesia: Land tenure change and the prospect of emission reduction. *Mitig. Adapt. Strateg. Glob. Chang.* **2014**, *19*, 715–731. [CrossRef]

61. Pelzer, K.J. Pioneer Settlement in the Asiatic Tropics, Studies in land utilisation and agricultural colonization in southeastern Asia. In *Pioneer Settlement in the Asiatic Tropics. Special Publication No. 29*; American Geographical Society: New York, NY, USA, 1945.
62. De Koninck, R.; McTaggart, W.D. Land settlement processes in Southeast Asia: Historical foundations, discontinuities, and problems. *Asian Res. Serv.* **1987**, *15*, 341–356.
63. Angelsen, A. *Forest Cover Change in Space and Time: Combining the von Thunen and Forest Transition Theories*; The World Bank: Washington, DC, USA, 2007.
64. De Koninck, R. The challenges of the agrarian transition in Southeast Asia. *Labour Cap. Soc.* **2004**, *37*, 285–288.
65. Dove, M. *The Banana Tree at the Gate: A History of Marginal Peoples and Global Markets in Borneo*; Yale University Press: New Haven, CT, USA, 2011.
66. Geertz, C. *Agricultural Involution: The Processes of Ecological Change in Indonesia*; Univ. of California Press: Berkeley, CA, USA, 1963.
67. van Noordwijk, M.; Bizard, V.; Wangpakapattanawong, P.; Tata, H.L.; Villamor, G.B.; Leimona, B. Tree cover transitions and food security in Southeast Asia. *Glob. Food Secur.* **2014**, *3*, 200–208. [CrossRef]
68. Potter, L. How can the people's sovereignty be achieved in the oil palm sector? Is the plantation model shifting in favour of smallholders. In *Land and Development in Indonesia: Searching for the People's Sovereignty*; McCarthy, J.F., Robinson, K., Eds.; ISEAS-Yusof Ishak Institute: Singapore, 2016; pp. 315–342.
69. Colfer, C.J.P.; Resosudarmo, I.A.P. Which way forward. In *People, Forests and Policymaking in Indonesia*; Resources for the Future: Washington, DC, USA, 2002.
70. Berkes, F.; Colding, J.; Folke, C. Rediscovery of traditional ecological knowledge as adaptive management. *Ecol. Appl.* **2000**, *10*, 1251–1262. [CrossRef]
71. Gadgil, M.; Olsson, P.; Berkes, F.; Folke, C. Exploring the role of local ecological knowledge in ecosystem management: Three case studies. In *Navigating Social-Ecological Systems: Building Resilience for Complexity and Change*; Berkes, F., Colding, J., Folkes, C., Eds.; Cambridge University Press: Cambridge, UK, 2003. [CrossRef]
72. Villamor, G.B.; Pontius, R.G.; van Noordwijk, M. Agroforest's growing role in reducing carbon losses from Jambi (Sumatra), Indonesia. *Reg. Environ. Chang.* **2014**, *14*, 825–834. [CrossRef]
73. Santika, T.; Wilson, K.A.; Budiharta, S.; Kusworo, A.; Meijaard, E.; Law, E.A.; Friedman, R.; Hutabarat, J.A.; Indrawan, T.P.; St John, F.A.; et al. Heterogeneous impacts of community forestry on forest conservation and poverty alleviation: Evidence from Indonesia. *People Nat.* **2019**, *1*, 204–219. [CrossRef]
74. Directorate General of Estate Crops. *Tree Crop Estate Statistics of Indonesia 2018–2020—Palm Oil*; Directorate General of Estate, Ministry of Agriculture: Jakarta, Indonesia, 2019.
75. Bakhtiar, I.; Suradiredja, D.; Santoso, H.; Saputra, W. *Hutan Kita Bersawit–Gagasan Penyelesaian untuk Perkebunan Kelpa Sawit dalam Kawasan Hutan*; KEHATI: Jakarta, Indonesia, 2019.
76. Gaveau, D.; Sheil, D.; Husnayaen Salim, M.A.; Arjakusuma, S.; Ancrenaz, M.; Pacheco, P.; Meijaard, E. Rapid conversions and avoided deforestation: Examining four decades of industrial plantation expansion in Borneo. *Sci. Rep.* **2016**, *6*, 3201. [CrossRef]
77. Meijaard, E.; Garcia-Ulloa, J.; Sheil, D.; Wich, S.A.; Carlson, K.M.; Juffe-Bignoli, D.; Brooks, T.M. (Eds.) *Oil Palm and Biodiversity. A Situation Analysis by the IUCN Oil Palm Task Force*; IUCN Oil Palm Task Force: Gland, Switzerland, 2018; p. 116. [CrossRef]
78. Austin, K.G.; Mosnier, A.; Pirker, J.; McCallum, I.; Fritz, S.; Kasibhatla, P.S. Shifting patterns of oil palm driven deforestation in Indonesia and implications for zero-deforestation commitments. *Land Use Policy* **2017**, *69*, 41–48. [CrossRef]
79. DJP. *Statistik Perkebunan Indonesia; Kelapa Sawit 2018–2020*; Direktorat Jenderal Perkebunan: Jakarta, Indonesia, 2020; p. 68.
80. Badan Pusat Statistik, Sistem Informasi Administrasi Kependudukan, Kemendagri, 2019. Indonesia Administrative Level 0–4 Boundaries, Indonesia (IDN) Administrative Boundary Common Operational Database (COD-AB), Update 23 December 2019, Badan Pusat Statistik, Sistem Informasi Administrasi Kependudukan. Kemendagri (Ministry of Home Affairs). Available online: https://data.humdata.org/dataset/indonesia-administrative-level-0-4-boundariesx (accessed on 16 May 2020).
81. Potter, L. Where Are the Swidden Fallows Now? An Overview of Oil Palm and Dayak Agriculture across Kalimantan, with Case Studies from Sanggau, in West Kalimantan. In *Shifting Cultivation and Environmental Change: Indigenous People, Agriculture and Forest Conservation*; Routledge: Abingdon, UK, 2015.

82. Penot, E.; Chambon, B.; Wibawa, G. History of rubber agroforestry systems development in Indonesia and Thailand as alternatives for sustainable agriculture and income stability. *Int. Proc. IRC* **2017**, *1*, 497–532. [CrossRef]
83. Feintrenie, L.; Levang, P. Sumatra's rubber agroforests: Advent, rise and fall of a sustainable cropping system. *Small-Scale For.* **2009**, *8*, 323–335. [CrossRef]
84. Therville, C.; Feintrenie, L.; Levang, P. Farmers' perspectives about agroforests conversion to plantations in Sumatra. Lessons learnt from Bungo District (Jambi, Indonesia). *For. Trees Livelihoods* **2011**, *20*, 15–33. [CrossRef]
85. Koh, L.P.; Miettinen, J.; Liew, S.C.; Ghazoul, J. Remotely sensed evidence of tropical peatland conversion to oil palm. *Proc. Natl. Acad. Sci. USA* **2011**, *108*, 5127–5132. [CrossRef]
86. Afriyanti, D.; Hein, L.; Kroeze, C.; Zuhdi, M.; Saad, A. Scenarios for withdrawal of oil palm plantations from peatlands in Jambi Province, Sumatra, Indonesia. *Reg. Environ. Chang.* **2019**, *19*, 1201–1215. [CrossRef]
87. Global Forest Watch. Universal Mill List. 2020. Available online: http://data.globalforestwatch.org/datasets/universal-mill-list/data (accessed on 21 March 2020).
88. ten Kate, A.; Kuepper, B.; Piotrowski, M. NDPE Policies Cover 83% of Palm Oil Refineries; Implementation at 78%. Available online: https://chainreactionresearch.com/report/ndpe-policies-cover-83-of-palm-oil-refineries-implementation-at-75/ (accessed on 25 October 2020).
89. Budidarsono, S.; Susanti, A.; Zoomers, A. Oil palm plantations in Indonesia: The implications for migration, settlement/resettlement and local economic development. In *Biofuels—Economy, Environment and Sustainability*; INTECH: London, UK, 2013; pp. 173–193. [CrossRef]
90. DJP. *Statistik Perkebunan Indonesia*; Kelapa Sawit; Direktorat Jenderal Perkebunan: Jakarta, Indonesia, 1990.
91. Jelsma, I.; Schoneveld, G.C.; Zoomers, A.; van Westen, A.C.M. Unpacking Indonesia's independent oil palm smallholders: An actor-disaggregated approach to identifying environmental and social performance challenges. *Land Use Policy* **2017**, *69*, 281–297. [CrossRef]
92. Schoneveld, G.C.; van der Haar, S.; Ekowati, D.; Andrianto, A.; Komarudin, H.; Okarda, B.; Jelsma, I.; Pacheco, P. Certification, good agricultural practice and smallholder heterogeneity: Differentiated pathways for resolving compliance gaps in the Indonesian oil palm sector. *Glob. Environ. Chang.* **2019**, *57*, 101933. [CrossRef]
93. Hidayat, K.N.; Glasbergen, P.; Offermans, A. Sustainability Certification and Palm Oil Smallholders' Livelihood: A comparison between scheme smallholders and independent smallholders in Indonesia. *Int. Food Agribus. Manag. Rev.* **2015**, *18*, 25–48. [CrossRef]
94. Euler, M.; Schwarze, S.; Siregar, H.; Qaim, M. Oil Palm Expansion among Smallholder Farmers in Sumatra, Indonesia. *J. Agric. Econ.* **2016**, *67*, 658–676. [CrossRef]
95. IFC. *Diagnostic Study on Indonesian Palm Oil Smallholders: Developing a Better Understanding of Their Performane and Potential*; International Finance Corporation: Washington, DC, USA, 2013; p. 96.
96. Gillespie, P. How Does Legislation Affect Oil Palm Smallholders in the Sanggau District of Kalimantan, Indonesia? *Australas. J. Nat. Resour. Law Policy* **2011**, *14*, 1–37.
97. World Bank. Indonesia: Strategies for Sustained Development of Tree Crops. In *Agriculture Operations Division*; Asia Regional Office, Ed.; World Bank: Washington, DC, USA, 1989.
98. Badrun, M. *Milestone of Change: Developing a Nation through Oil Palm "PIR"*; Directorate General of Estate Crops: Jakarta, Indonesia, 2011; p. 229.
99. Mongabay. Indonesia Moves to End Smallholder Guarantee Meant to Empower Palm Oil Farmers. 2020. Available online: https://news.mongabay.com/2020/05/indonesia-palm-oil-plasma-plantation-farmers-smallholders/ (accessed on 30 November 2020).
100. Jelsma, I.; Slingerland, M.; Giller, K.E.; Bijman, J. Collective action in a smallholder oil palm production system in Indonesia: The key to sustainable and inclusive smallholder palm oil? *J. Rural Stud.* **2017**, *54*, 198–210. [CrossRef]
101. Daemeter. *Overview of Indonesian Smallholder Farmers: A Typology of Organizational Models, Needs, and Investment Opportunities*; Daemeter: Bogor, Indonesia, 2015.
102. Cramb, R.A. Palmed Off: Incentive Problems with Joint-Venture Schemes for Oil Palm Development on Customary Land. *World Dev.* **2013**, *43*, 84–99. [CrossRef]
103. Susanti, A.; Maryudi, A. Development narratives, notions of forest crisis, and boom of oil palm plantations in Indonesia. *For. Policy Econ.* **2016**, *73*, 130–139. [CrossRef]
104. Bissonnette, J.-F.; De Koninck, R. The return of the plantation? Historical and contemporary trends in the relation between plantations and smallholdings in Southeast Asia. *J. Peasant Stud.* **2017**, *44*, 918–938. [CrossRef]

105. Grass, I.; Kubitza, C.; Krishna, V.V.; Corre, M.D.; Mußhoff, O.; Pütz, P.; Drescher, J.; Rembold, K.; Ariyanti, E.S.; Barnes, A.D.; et al. Trade-offs between multifunctionality and profit in tropical smallholder landscapes. *Nat. Commun.* **2020**, *11*, 1186. [CrossRef] [PubMed]
106. Feintrenie, L.; Chong, W.; Levang, P. Why do Farmers Prefer Oil Palm? Lessons Learnt from Bungo District, Indonesia. *Small-Scale For.* **2010**, *9*, 379–396. [CrossRef]
107. Kubitza, C.; Krishna, V.V.; Alamsyah, Z.; Qaim, M. The economics behind an ecological crisis: Livelihood effects of oil palm expansion in Sumatra, Indonesia. *Hum. Ecol.* **2018**, *46*, 107–116. [CrossRef]
108. Jelsma, I.; Woittiez, L.S.; Ollivier, J.; Dharmawan, A.H. Do wealthy farmers implement better agricultural practices? An assessment of implementation of Good Agricultural Practices among different types of independent oil palm smallholders in Riau, Indonesia. *Agric. Syst.* **2019**, *170*, 63–76. [CrossRef]
109. Woittiez, L.S.; Slingerland, M.; Rafik, R.; Giller, K.E. Nutritional imbalance in smallholder oil palm plantations in Indonesia. *Nutr. Cycl. Agroecosyst.* **2018**, *111*, 73–86. [CrossRef]
110. McCarthy, J.F.; Zen, Z. Agribusiness, agrarian change, and the fate of oil palm smallholders in Jambi. In *The Oil Palm Complex: Smallholders, Agribusiness and the State in Indonesia and Malaysia*; Cramb, R., McCarthy, J.F., Eds.; NUS Press: Singapore, 2016; pp. 109–155. ISBN 978-981-4722-06-3.
111. Krishna, V.; Euler, M.; Siregar, H.; Qaim, M. Differential livelihood impacts of oil palm expansion in Indonesia. *Agric. Econ.* **2017**, *48*, 639–653. [CrossRef]
112. Krishna, V.V.; Kubitza, C.; Pascual, U.; Qaim, M. Land markets, Property rights, and Deforestation: Insights from Indonesia. *World Dev.* **2017**, *99*, 335–349. [CrossRef]
113. Purnomo, H.; Shantiko, B.; Sitorus, S.; Gunawan, H.; Achdiawan, R.; Kartodihardjo, H.; Dewayani, A.A. Fire economy and actor network of forest and land fires in Indonesia. *For. Policy Econ.* **2017**, *78*, 21–31. [CrossRef]
114. Zen, Z.; Barlow, C.; Gondowarsito, R.; McCarthy, J.F. Interventions to promote smallholder oil palm and socio-economic improvement in Indonesia. In *The Oil Palm Complex: Smallholders, Agribusiness and the State in Indonesia and Malaysia*; Cramb, R., McCarthy, J.F., Eds.; NUS Press: Singapore, 2016; pp. 78–108. ISBN 978-981-4722-06-3.
115. Aulia, A.; Sandhu, H.; Millington, A. Quantifying the Economic Value of Ecosystem Services in Oil Palm Dominated Landscapes in Riau Province in Sumatra, Indonesia. *Land* **2020**, *9*, 194. [CrossRef]
116. Ayompe, L.M.; Schaafsma, M.; Egoh, B.N. Towards sustainable palm oil production: The positive and negative impacts on ecosystem services and human wellbeing. *J. Clean. Prod.* **2021**, *278*, 123914. [CrossRef]
117. Manoli, G.; Meijide, A.; Huth, N.; Knohl, A.; Kosugi, Y.; Burlando, P.; Fatichi, S. Ecohydrological changes after tropical forest conversion to oil palm. *Environ. Res. Lett.* **2018**, *13*, 064035. [CrossRef]
118. Fitzherbert, E.B.; Struebig, M.J.; Morel, A.; Danielsen, F.; Brühl, C.A.; Donald, P.F.; Phalan, B. How will oil palm expansion affect biodiversity? *Trends Ecol. Evol.* **2008**, *23*, 538–545. [CrossRef] [PubMed]
119. Koh, L.P.; Wilcove, D.S. Is oil palm agriculture really destroying tropical biodiversity? *Conserv. Lett.* **2008**, *1*, 60–64. [CrossRef]
120. Booth, S.L. The Relationship between Carbon Emissions, Land Use Change and the Oil Palm Industry within Southeast Asia. Master's Thesis, University of San Fransisco, San Fransisco, CA, USA, 2017. Available online: https://repository.usfca.edu/capstone/562 (accessed on 30 November 2020).
121. Comte, I.; Colin, F.; Whalen, J.; Olivier, G.; Caliman, J.-P. Agricultural practices in oil palm plantations and their impact on hydrological changes, nutrient fluxes and water quality in Indonesia. A review. *Adv. Agron.* **2012**, *116*, 71–122. [CrossRef]
122. Merten, J.; Röll, A.; Guillaume, T.; Meijide, A.; Tarigan, S.; Agusta, H.; Hölscher, D. Water scarcity and oil palm expansion: Social views and environmental processes. *Ecol. Soc.* **2016**, *21*, 21. [CrossRef]
123. Obidzinski, K.; Andriani, R.; Komarudin, H.; Andrianto, A. Environmental and Social Impacts of Oil Palm Plantations and their Implications for Biofuel Production in Indonesia. *Ecol. Soc.* **2012**, *17*. [CrossRef]
124. Dislich, C.; Keyel, A.C.; Salecker, J.; Kisel, Y.; Meyer, K.M.; Auliya, M.; Wiegand, K. A review of the ecosystem functions in oil palm plantations, using forests as a reference system. *Biol. Rev. Camb. Philos. Soc.* **2017**, *92*, 1539–1569. [CrossRef]
125. Morana, L.S. *Oil Palm Plantations: Threats and Opportunities for Tropical Ecosystems. Thematic Focus: Ecosystem Management and Resource Efficiency*; United Nation environment Programme Global Environmental Alert Service (GEAS): Geneva, Switzerland, 2011.

126. Safitri, L.; Sastrohartono, H.; Purboseno, S.; Kautsar, V.; Saptomo, S.; Kurniawan, A. Water Footprint and Crop Water Usage of Oil Palm (*Eleasis guineensis*) in Central Kalimantan: Environmental Sustainability Indicators for Different Crop Age and Soil Conditions. *Water* **2018**, *11*, 35. [CrossRef]
127. Barthel, M.; Jennings, S.; Schreiber, W.; Sheane, R.; Royston, S.; Fry, J.; McGill, J. *Study on the Environmental Impact of Palm Oil Consumption and on Existing Sustainability Standards*; (07.0201/2016/743217/ETU/ENV.F3); European Commission: Brussels, Belgium, 2018; Available online: https://ec.europa.eu/environment/forests/pdf/palm_oil_study_kh0218208enn_new.pdf (accessed on 30 November 2020).
128. Azhar, B.; Lindenmayer, D.B.; Wood, J.; Fischer, J.; Manning, A.; McElhinny, C.; Zakaria, M. The conservation value of oil palm plantation estates, smallholdings and logged peat swamp forest for birds. *For. Ecol. Manag.* **2011**, *262*, 2306–2315. [CrossRef]
129. Petrenko, C.; Paltseva, J.; Searle, S. *Ecological Impacts of Palm Oil Expansion in Indonesia*; White Paper; International Council on Clean Transportation: Washington, DC, USA, 2016.
130. Fischer, J.; Lindenmayer, D. Landscape modification and habitat fragmentation: A synthesis. *Glob. Ecol. Biogeogr.* **2007**, *16*, 265–280. [CrossRef]
131. Gallmetzer, N.; Schulze, C.H. Impact of oil palm agriculture on understory amphibians and reptiles: A Mesoamerican perspective. *Glob. Ecol. Conserv.* **2015**, *4*, 95–109. [CrossRef]
132. Murdiyarso, D.; Van Noordwijk, M.; Wasrin, U.R.; Tomich, T.P.; Gillison, A.N. Environmental benefits and sustainable land-use options in the Jambi transect, Sumatra. *J. Veg. Sci.* **2002**, *13*, 429–438. [CrossRef]
133. Aratrakorn, S.; Thunhikorn, S. Changes in bird communities following conversion of lowland forest to oil palm and rubber plantations in Thailand. *Bird Conserv. Int.* **2006**, *16*, 71–82. [CrossRef]
134. Edwards, F.A.; Edwards, D.P.; Hamer, K.C.; Davies, R.G. Impacts of logging and conversion of rainforest to oil palm on the functional diversity of birds in Sundaland. *IBIS* **2013**, *155*, 313–326. [CrossRef]
135. Konopik, O.; Gray, C.L.; Grafe, T.U.; Steffan-Dewenter, I.; Fayle, T.M. From rainforest to oil palm plantations: Shifts in predator population and prey communities, but resistant interactions. *Glob. Ecol. Conserv.* **2014**, *2*, 385–394. [CrossRef]
136. Payán, E.; Boron, V. The Future of Wild Mammals in Oil Palm Landscapes in the Neotropics. *Front. For. Glob. Chang.* **2019**, *2*, 61. [CrossRef]
137. Danielsen, F.; Beukema, H.; Burgess, N.; Parish, F.; Brühl, C.; Donald, P.; Fitzherbert, E. Biofuel plantations on forested lands: Double jeopardy for biodiversity and climate. *IOP Conf. Ser. Earth Environ. Sci.* **2009**, *6*, 242014. [CrossRef]
138. Vijay, V.; Pimm, S.L.; Jenkins, C.N.; Smith, S.J. The Impacts of Oil Palm on Recent Deforestation and Biodiversity Loss. *PLoS ONE* **2016**, *11*, e0159668. [CrossRef]
139. Ancrenaz, M.; Oram, F.; Ambu, L.; Lackman, I.; Ahmad, E.; Elahan, H.; Meijaard, E. Of Pongo, palms and perceptions: A multidisciplinary assessment of Bornean orang-utans Pongo pygmaeus in an oil palm context. *Oryx* **2014**, *49*, 465–472. [CrossRef]
140. Baraniuk, C. Orangutans Are Hanging on in the Same Palm Oil Plantations That Displace Them, Conservationists Still Need to Work to Minimize Conflict between the Endangered Apes and Humans. *Sci. Am.* 2020. Available online: https://www.scientificamerican.com/article/orangutans-are-hanging-on-in-the-same-palm-oil-plantations-that-displace-them/ (accessed on 30 November 2020).
141. Jonas, H.; Abram, N.K.; Ancrenaz, M. *Addressing the Impact of Large-Scale Oil Palm Plantations on Orangutan Conservation in Borneo: A Spatial, Legal and Political Economy Analysis*; IIED: London, UK, 2017.
142. Nellemann, C.; Miles, L.; Kaltenborn, B.; Virtue, M.; Ahlenius, H. *Last Stand of the Orang-Utan*; A UNEP Rapid Response Assessment: Nairobi, Kenya, 2007.
143. Swarna Nantha, H.; Tisdell, C. The orangutan–oil palm conflict: Economic constraints and opportunities for conservation. *Biodivers. Conserv.* **2009**, *18*, 487–502. [CrossRef]
144. Suba, R.B.; van de Ploeg, J.; Zelfde, M.; Lau, Y.W.; Wissingh, T.F.; Kustiawan, W.; de Iongh, H.H. Rapid expansion of oil palm is leading to human–elephant conflicts in North Kalimantan province of Indonesia. *Trop. Conserv. Sci.* **2017**, *10*, 1940082917703508. [CrossRef]
145. Shuhada, S.N.; Salim, S.; Nobilly, F.; Lechner, A.M.; Azhar, B. Conversion of peat swamp forest to oil palm cultivation reduces the diversity and abundance of macrofungi. *Glob. Ecol. Conserv.* **2020**, *23*, e01122. [CrossRef]

146. Paoletti, A.; Darras, K.; Jayanto, H.; Grass, I.; Kusrini, M.; Tscharntke, T. Amphibian and reptile communities of upland and riparian sites across Indonesian oil palm, rubber and forest. *Glob. Ecol. Conserv.* **2018**, *16*, e00492. [CrossRef]
147. Gillison, A.N.; Bignell, D.E.; Brewer, K.R.; Fernandes, E.C.; Jones, D.T.; Sheil, D.; May, P.H.; Watt, A.D.; Constantino, R.; Couto, E.G.; et al. Plant functional types and traits as biodiversity indicators for tropical forests: Two biogeographically separated case studies including birds, mammals and termites. *Biodivers. Conserv.* **2013**, *22*, 1909–1930. [CrossRef]
148. Unjan, R.; Nissapa, A.; Phitthayaphinant, P. An Identification of Impacts of Area Expansion Policy of Oil Palm in Southern Thailand: A Case Study in Phatthalung and Nakhon Si Thammarat Provinces. *Procedia Soc. Behav. Sci.* **2013**, *91*, 489–496. [CrossRef]
149. Anamulai, S.; Sanusi, R.; Zubaid, A.; Lechner, A.M.; Ashton-Butt, A.; Azhar, B. Land use conversion from peat swamp forest to oil palm agriculture greatly modifies microclimate and soil conditions. *PeerJ* **2019**, *7*, e7656. [CrossRef]
150. Larsen, R.K.; Jiwan, N.; Rompas, A.; Jenito, J.; Osbeck, M.; Tarigan, A. Towards 'hybrid accountability' in EU biofuels policy? Community grievances and competing water claims in the Central Kalimantan oil palm sector. *Geoforum* **2014**, *54*, 295–305. [CrossRef]
151. Córdoba, D.; Juen, L.; Selfa, T.; Peredo, A.M.; Montag, L.F.d.A.; Sombra, D.; Santos, M.P.D. Understanding local perceptions of the impacts of large-scale oil palm plantations on ecosystem services in the Brazilian Amazon. *For. Policy Econ.* **2019**, *109*, 102007. [CrossRef]
152. Tarigan, S.D. Land Cover Change and its Impact on Flooding Frequency of Batanghari Watershed, Jambi Province, Indonesia. *Procedia Environ. Sci.* **2016**, *33*, 386–392. [CrossRef]
153. Heidari, A.; Mayer, A.; Watkins, D.; Castillo, M.M. Hydrologic impacts and trade-offs associated with developing oil palm for bioenergy in Tabasco, Mexico. *J. Hydrol. Reg. Stud.* **2020**, *31*, 100722. [CrossRef]
154. Röll, A.; Niu, F.; Meijide, A.; Hardanto, A.; Hendrayanto, H.; Knohl, A.; Hölscher, D. Transpiration in an oil palm landscape: Effects of palm age. *Biogeosciences* **2015**, *12*, 5619–5633. [CrossRef]
155. Carr, M. The water relations and irrigation requirements of oil palm (*Elaeis guineensis*): A review. *Exp. Agric.* **2011**, *47*, 629–652. [CrossRef]
156. Chong, S.Y.; Teh, C.B.S.; Ainuddin, A.N.; Philip, E. Simple net rainfall partitioning equations for nearly closed to fully closed canopy stands. *Pertanika J. Trop. Agric. Sci.* **2018**, *41*, 81–100.
157. Farmanta, Y.; Dedi, S. Rainfall Interception by Palm Plant Canopy. 2016. Available online: http://repository.unib.ac.id/11344/1/046%20Yong%20F%20%26%20Dedi%20S.pdf (accessed on 30 November 2020).
158. Hardanto, A.; Röll, A.; Niu, F.; Meijide, A.; Hendrayanto; Hölscher, D. Oil Palm and Rubber Tree Water Use Patterns: Effects of Topography and Flooding. *Front. Plant Sci.* **2017**, *8*, 452. [CrossRef]
159. Bruijnzeel, L.A. Hydrological functions of tropical forests: Not seeing the soil for the trees? *Agric. Ecosyst. Environ.* **2004**, *104*, 185–228. [CrossRef]
160. Tarigan, S.; Stiegler, C.; Wiegand, K.; Knohl, A.; Murtilaksono, K. Relative contribution of evapotranspiration and soil compaction to the fluctuation of catchment discharge: Case study from a plantation landscape. *Hydrol. Sci. J.* **2020**, *65*, 1239–1248. [CrossRef]
161. Sumarga, E.; Hein, L.; Hooijer, A.; Vernimmen, R. Hydrological and economic effects of oil palm cultivation in Indonesian peatlands. *Ecol. Soc.* **2016**, *21*, 19. [CrossRef]
162. Chellaiah, D.; Yule, C.M. Effect of riparian management on stream morphometry and water quality in oil palm plantations in Borneo. *Limnologica* **2018**, *69*, 72–80. [CrossRef]
163. Juen, L.; Cunha, E.J.; Carvalho, F.G.; Ferreira, M.C.; Begot, T.O.; Andrade, A.L. Effects of oil palm plantations on the habitat structure and biota of streams in eastern Amazon. *River Res. Appl.* **2016**, *32*, 2081–2094. [CrossRef]
164. Sheaves, M.; Johnston, R.; Miller, K.; Nelson, P.N. Impact of oil palm development on the integrity of riparian vegetation of a tropical coastal landscape. *Agric. Ecosyst. Environ.* **2018**, *262*, 1–10. [CrossRef]
165. Jamaludin, N.F.; Muis, Z.A.; Hashim, H. An integrated carbon footprint accounting and sustainability index for palm oil mills. *J. Clean. Prod.* **2019**, *225*, 496–509. [CrossRef]
166. Lewis, K.; Rumpang, E.; Kho, L.K.; McCalmont, J.; Teh, Y.A.; Gallego-Sala, A.; Hill, T.C. An assessment of oil palm plantation aboveground biomass stocks on tropical peat using destructive and non-destructive methods. *Sci. Rep.* **2020**, *10*, 2230. [CrossRef] [PubMed]

167. Khasanah, N.M.; van Noordwijk, M.; Ningsih, H. Aboveground carbon stocks in oil palm plantations and the threshold for carbon-neutral vegetation conversion on mineral soils. *Cogent Environ. Sci.* **2015**, *1*, 1119964. [CrossRef]
168. Kotowska, M.M.; Leuschner, C.; Triadiati, T.; Meriem, S.; Hertel, D. Quantifying above- and belowground biomass carbon loss with forest conversion in tropical lowlands of Sumatra (Indonesia). *Glob. Chang. Biol.* **2015**, *21*, 3620–3634. [CrossRef] [PubMed]
169. Paterson, R.R.M.; Lima, N. Climate change affecting oil palm agronomy, and oil palm cultivation increasing climate change, require amelioration. *Ecol. Evol.* **2018**, *8*, 452–461. [CrossRef]
170. Khasanah, N.M.; van Noordwijk, M.; Ningsih, H.; Rahayu, S. Carbon neutral? No change in mineral soil carbon stock under oil palm plantations derived from forest or non-forest in Indonesia. *Agric. Ecosyst. Environ.* **2015**, *211*, 195–206. [CrossRef]
171. van Noordwijk, M.; Khasanah, N.M.; Dewi, S. Can intensification reduce emission intensity of biofuel through optimized fertilizer use? Theory and the case of oil palm in Indonesia. *GCB Bioenergy* **2017**, *9*, 940–952. [CrossRef]
172. IPCC. 2013 Supplement to the 2006 IPCC guidelines for national greenhouse gas inventories: Wetlands. In *Intergovernmental Panel on Climate Change (IPCC)*; Hiraishi, T., Krug, T., Tanabe, K., Srivastava, N., Baasansuren, J., Fukuda, M., Troxler, T.G., Eds.; IPCC: Geneva, Switzerland, 2014.
173. Khasanah, N.M.; van Noordwijk, M. Subsidence and carbon dioxide emissions in a smallholder peatland mosaic in Sumatra, Indonesia. *Mitig. Adapt. Strateg. Glob. Chang.* **2019**, *24*, 147–163. [CrossRef]
174. Rival, A.; Montet, D.; Pioch, D. Certification, labelling and traceability of palm oil: Can we build confidence from trustworthy standards? *OCL* **2016**, *23*, D609. [CrossRef]
175. Tinhout, B.; van de Hombergh, H. *Setting the Biodiversity Bar for Palm Oil Certification; Assessing the Rigor of Biodiversity and Assurance Requirements of Palm Oil Standards*; IUCN: Amsterdam, The Netherlands, 2019.
176. Pye, O. Commodifying sustainability: Development, nature and politics in the palm oil industry. *World Dev.* **2019**, *121*, 218–228. [CrossRef]
177. Wiggs, C.; Kuepper, B.; Piotr, M. Spot Market Purchases Allow Deforestation-Linked Palm Oil to Enter NDPE Supply Chains. Chain Reaction Research. Available online: https://chainreactionresearch.com/report/spot-market-purchases-allow-deforestation-linked-palm-oil-to-enter-ndpe-supply-chains/ (accessed on 30 October 2020).
178. Santika, T.; Law, E.; Wilson, K.A.; John, F.A.S.; Carlson, K.; Gibbs, H.; Morgans, C.L.; Ancrenaz, M.; Meijaard, E. Impact of Palm Oil Sustainability Certification on Village Well-Being and Poverty in Indonesia. 2020. Available online: https://osf.io/preprints/socarxiv/5qk67/ (accessed on 30 November 2020).
179. RSPO. Impact Update 2019. In *Round Table on Sustainable Palm Oil*; RSPO: Kuala Lumpur, Malaysia, 2019; p. 52.
180. Astari, A.J.; Lovett, J.C. Does the rise of transnational governance 'hollow-out' the state? Discourse analysis of the mandatory Indonesian sustainable palm oil policy. *World Dev.* **2019**, *117*, 1–12. [CrossRef]
181. Schouten, G.; Bitzer, V. The emergence of Southern standards in agricultural value chains: A new trend in sustainability governance? *Ecol. Econ.* **2015**, *120*, 175–184. [CrossRef]
182. Higgins, V.; Richards, C. Framing sustainability: Alternative standards schemes for sustainable palm oil and South-South trade. *J. Rural Stud.* **2019**, *65*, 126–134. [CrossRef]
183. Hidayat, N.K.; Offermans, A.; Glasbergen, P. Sustainable palm oil as a public responsibility? On the governance capacity of Indonesian Standard for Sustainable Palm Oil (ISPO). *Agric. Hum. Values* **2018**, *35*, 223–242. [CrossRef]
184. Kaoem, T. A False Hope? In *An Analysis of the Draft New Indonesia Sustainable Palm Oil (ISPO) Regulation*; EIA: London, UK, 2020; p. 6.
185. Van Noordwijk, M.; Leimona, B.; Amaruzaman, S. Sumber Jaya from conflict to source of wealth in Indonesia: Reconciling coffee agroforestry and watershed functions. In *Sustainable Development through Trees on Farms: Agroforestry in Its Fifth Decade*; Van Noordwijk, M., Ed.; World Agroforestry (ICRAF): Bogor, Indonesia, 2019; pp. 177–192.
186. Erb, M.; Anggal, W. Conflict and the growth of democracy in Manggarai District. In *Deepening Democracy in Indonesia.*; Erb, M., Sulistiyanto, P., Eds.; Institute of Southeast Asian Studies: Singapore, 2009; pp. 283–302.
187. Setiawan, E.N.; Maryudi, A.; Purwanto, R.H.; Lele, G. Opposing interests in the legalization of non-procedural forest conversion to oil palm in Central Kalimantan, Indonesia. *Land Use Policy* **2016**, *58*, 472–481. [CrossRef]

188. TEMPO English. *Held Captive at Rokan Hulu*; Tempo English: Jakarta, Indoenisa, 2016; pp. 36–37.
189. Gaveau, D.L.A.; Salim, M.A.; Hergoualc'h, K.; Locatelli, B.; Sloan, S.; Wooster, M.; Marlier, M.E.; Molidena, E.; Yaen, H.; DeFries, R.; et al. Major atmospheric emissions from peat fires in Southeast Asia during non-drought years: Evidence from the 2013 Sumatran fires. *Sci. Rep.* **2014**, *4*, 6112. [CrossRef] [PubMed]
190. Suryanata, K. Fruit trees under contract: Tenure and land use change in upland Java, Indonesia. *World Dev.* **1994**, *22*, 1567–1578. [CrossRef]
191. Robiglio, V.; Reyes, M. Restoration through formalization? Assessing the potential of Peru's Agroforestry Concessions scheme to contribute to restoration in agricultural frontiers in the Amazon region. *World Dev. Perspect.* **2016**, *3*, 42–46. [CrossRef]
192. Slingerland, M.A.; Khasanah, N.M.; van Noordwijk, M.; Susanti, A.; Meilantina, M. Improving smallholder inclusivity through integrating oil palm with crops. In *Exploring Inclusive Palm Oil Production*; ETFRN and Tropenbos International: Wageningen, The Netherland, 2019; pp. 147–154.
193. Khasanah, N.; Van Noordwijk, M.; Slingerland, M.; Sofiyudin, M.; Stomph, D.; Migeon, A.F.; Hairiah, K. Oil Palm Agroforestry Can Achieve Economic and Environmental Gains as Indicated by Multifunctional Land Equivalent Ratios. *Front. Sustain. Food Syst.* **2020**, *3*, 122. [CrossRef]
194. Woittiez, L.S.; van Wijk, M.T.; Slingerland, M.; van Noordwijk, M.; Giller, K.E. Yield gaps in oil palm: A quantitative review of contributing factors. *Eur. J. Agron.* **2017**, *83*, 57–77. [CrossRef]
195. Kehati. *Solusi Sawitt Salam Kawasn Hutan [Oilmpalm Solutins in the Forest Zone]. SymSPOSia, Episode 10*; Kehati: Jakarta, Indonesia, 2020; Available online: https://www.youtube.com/watch?v=0fsEfK5jsGU&feature=youtu.be (accessed on 29 November 2020).
196. Nurrochmat, D.R.; Boer, R.; Ardiansyah, M.; Immanuel, G.; Purwawangsa, H. Policy forum: Reconciling palm oil targets and reduced deforestation: Landswap and agrarian reform in Indonesia. *For. Policy Econ.* **2020**, *119*, 102291. [CrossRef]
197. Iiyama, M.; Neufeldt, H.; Dobie, P.; Hagen, R.; Njenga, M.; Ndegwa, G.; Mowo, J.G.; Kisoyan, P.; Jamnadass, R. Opportunities and challenges of landscape approaches for sustainable charcoal production and use. In *Climate-Smart Landscapes: Multifunctionality in Practice*; World Agroforestry Centre (ICRAF): Nairobi, Kenya, 2015; pp. 195–209.

7

Traditional Pollarding Practices for Dimorphic Ash Tree (*Fraxinus dimorpha*) Support Soil Fertility in the Moroccan High Atlas

Abdessamad Fakhech [1], Didier Genin [2,*], Mohamed Ait-El-Mokhtar [3], El Mustapha Outamamat [1], Soufiane M'Sou [1], Mohamed Alifriqui [1], Abdelilah Meddich [3] and Mohamed Hafidi [1,4]

1 Laboratory of Microbiological Biotechnology, Agricultural Sciences & Environment (BioMAgE), Faculty of Sciences Semlalia, Cadi Ayyad University, Marrakech 40000, Morocco; abdessamad.fakhech@edu.uca.ma (A.F.); elmostapha.outamamat@ced.uca.ma (E.M.O.); soufiane.msou@ced.uca.ma (S.M.); Alifriqui@uca.ma (M.A.); Hafidi@uca.ma (M.H.)

2 IRD, Laboratory Population, Environment, Dévelopment, UMR151 AMU-IRD, 1331 Marseille, France

3 Laboratory of Agri-food,, Biotechnologies & Bioresource Valorisation (AGROBIOVAL), Faculté des Sciences Semlalia, Université Cadi Ayyad, Marrakech 40000, Morocco; mohamed.aitelmokhtar@ced.uca.ma (M.A.-E.-M.); a.meddich@uca.ma (A.M.)

4 AgroBioSciences Program, Mohammed IV Polytechnic University (UM6P), Benguerir 43150, Morocco

* Correspondence: didier.genin@univ-amu.fr

Abstract: Shaping and pollarding of dimorphic ash tree (*Fraxinus dimorpha*) are two traditional practices used by the local inhabitants in agropastoral parklands of the Moroccan High Atlas to secure their production systems and increase tree production and strength. This study focused on assessing the impact of these practices on soil quality. Abiotic parameters and mycorrhizal attributes of the samples of four soil types related to different ash tree morphotypes were assessed and compared. Rhizospheric soils (Rs) of three *F. dimorpha* morphotypes were sampled: trees regularly pollarded and shaped for stem anastomosis (An), regularly pollarded multistemmed trees (Na), and multistemmed trees belonging to a public forest under national forestry service management and sporadically illegally pollarded (Fo). The fourth soil was a non-Rs found in bare soils, which represented the control (Nr). Results showed a sizable difference between An soil properties and the other soil types ones, with significantly higher phosphorus (×6), nitrogen (×5), and carbon (×2) levels and higher mycorrhizal (×6) status than Nr soil, and showed 37% more mycorrhization intensity than Fo. Na showed intermediary levels between An and Fo. Fo had ×2 P, ×3 Total Kjeldahl Nitrogen (TKN), 58% more Total Organic Carbon (TOC) content, and twice the spore density compared with Nr. It is concluded that shaping and pollarding have a positive impact on the soil characteristics of the studied species and could make a useful contribution to sound agroforest management schemes.

Keywords: *Fraxinus dimorpha*; soil chemical characteristics; mycorrhizal attributes; traditional ecological knowledge; anastomosis; agroforest

1. Introduction

Traditional ecological knowledge (TEK) is defined as "a cumulative body of knowledge, practices and beliefs, handed down through generations by cultural transmission and evolving by adaptive processes, about the relationship between living beings with one another and with their forest environment" [1]. It refers to the knowledge developed by native or local people over generations through direct interactions with their environment. While TEK remains under-recognized, efforts to preserve it are increasing [2]. Over the last years, TEK has become an alternative approach to better understand and adapt to climate and biodiversity change [3,4]. Examples of how TEK provides insights into balanced forest management have been reported and discussed [5]. In Morocco, this approach was particularly documented by Genin et al. [6] in the forested landscapes of the Aït M'hamed region in Azilal province, situated in the Central High Atlas. A unique traditional practice relative to dimorphic ash tree stands was described, consisting of strict periodic tree pollarding and favoring tree stem coalescence and fusion in order to improve the productivity of foliar fodder [6,7].

The dimorphic ash tree, *Fraxinus dimorpha*, is a native tree growing spontaneously in some parts of the Moroccan High Atlas. It is called "imts" in the local Amazigh language, and it grows primarily on rocky inclines and valleys, at altitudes between 1200 and 2000 m. For the local inhabitants, it is a multifunctional tree, which constitutes a keystone species in the functioning of their agro-silvo-pastoral systems and livelihoods [8]. It supplies them with firewood, poles, and beams for houses and agricultural tools, spices, tinctorial, and medicinal products [6]. But its most essential use is as fodder, in times when pasture is scarce and insufficient. From August until November, *F. dimorpha* trees are relied upon for providing fodder for small stocks. For a single tree, branches are pollarded every 4 years to allow enough regrowth time. Sheep and goats graze directly on the ground chopped leafy branches [7].

F. dimorpha in this area is characterized by trunk heterogeneity with four reported morphotypes: (1) large anastomosed trunks located mainly in privately owned, cultivated, and/or grazed parklands; (2) multistemmed trunks with multiple 10–20 cm diameter stems located both in parklands and in the adjacent public forest; (3) single-stemmed trunks, which are the least frequently occurring morphotype; and (4) shrublike trees, which dominate the public forest (Figure 1) [6].

Figure 1. Four different *F. dimorpha* morphotypes: (**a**) anastomosed (background) and multistemmed (foreground) (image © Genin), (**b**) single-stemmed, (**c**) shrublike (images © Fakhech).

The most widespread morphotype within the privately owned lands is the anastomosed trunks. This morphotype is the fruit of one of the most remarkable features of traditional tree management of this region. In the local Amazigh language, this is called "Tahboucht", which translates roughly as "educating" or "mothering." It consists in building rock walls around small or overgrazed trees until they are out of the reach of sheep and goats. Only the best-developed and straightest stems are then attached together as close as possible to favor the process of "anastomosis," by which they can fuse and form a single large trunk. Anastomosis promotes anatomical changes, such as increased proportion of parenchymatous cells and cross section surface area [9]. This singular practice is viewed by Genin et al. [6] as an original and effective option for resource scarcity management in this region.

As perennial plants, woody trees and shrubs represent an important ecosystem stabilization component by ensuring fundamental ecosystem functions, such as soil organic matter improvement and mutual symbiosis [10,11], while producing biomass, services, and goods to support local populations. The case of *F. dimorpha* could help to develop a more global approach to ensure both the preservation and valorization of efficient TEK practices, and the conservation of ecosystemic functionality. Genin et al. [6] showed that the practice of trunk anastomosis allowed a 36% increase in foliage production after a 4-year cycle of exploitation, compared with non-anastomosed trees, and promoted the resilience and longevity of the trees. The effects of this practice on belowground properties have still to be investigated. Here we focused on mycorrhizal status since the arbuscular mycorrhizal fungi (AMF) are a major rhizospheric component of most plant roots [12].

The aim of this study was thus to investigate the response of some of *F. dimorpha* rhizospheric soil (Rs) characteristics found below different tree ports as a result of contrasted tree exploitation practices. We hypothesized that aerial improvement of *F. dimorpha* should be linked to an improvement of its belowground characteristics since the improvement of the tree's aerial parts requires a mobilization of all the rhizospheric components.

2. Material and Methods

2.1. *Study Site and Sampling*

Field sampling was conducted in Aït M'hamed rural commune, located in the Central High Atlas, Azilal province, Morocco (31°49′N; 6°35′W) (Figure 2). Dimorphic ash is the dominantly distributed spontaneous tree species in the landscape, forming tree parklands fully integrated within local agro-silvo-pastoral systems. Altitudes range from 1300 to 1700 m. The climate is mountain Mediterranean with annual rainfall between 450 and 600 mm. The mean minimum temperature is 5 °C, and the mean maximum temperature is 28 °C. Soils are relatively homogeneous, largely dominated by a sandy-loamy calcareous skeletal soil. Sampling was carried out in May on rhizospheric soil (Rs) found below three *F. dimorpha* morphotypes: anastomosed (An), multistemmed (Na) (both tree types located in occasionally cultivated parklands), and public forest (Fo) tree types. Ten 1 kg soil samples (Rs) were randomly collected from between 10 and 40 cm depth under the tree cover for each morphological type of trees, within a homogeneous edaphotopographic slope. Ten additional samples of non-Rs, situated 50 m away from any dimorphic ash tree cover, were also randomly collected as control (Nr). Root samples of *F. dimorpha* were collected for mycorrhizal colonization assessment for An, Na, and Fo trees.

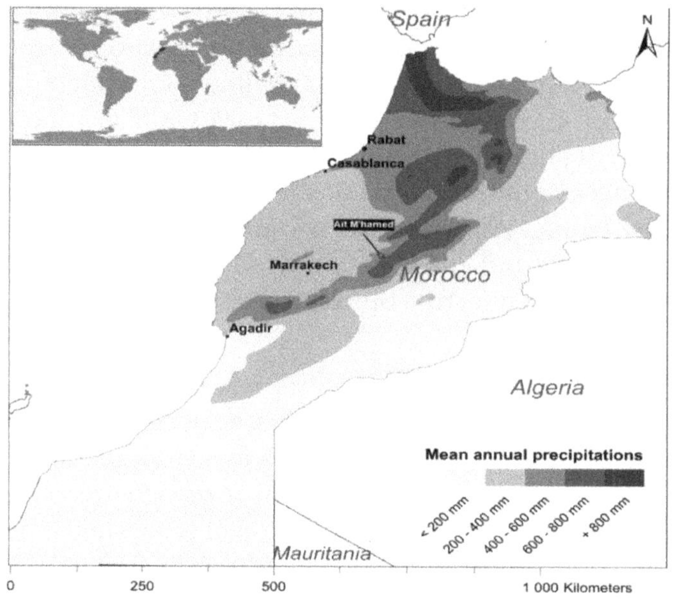

Figure 2. Localization of the study area.

2.2. *Soil Chemical and Physical Analysis*

Soil chemical and physical characteristics were determined using the following methods: Total Kjeldahl nitrogen (TKN) was determined using the Kjeldahl method [13], which consists in distilling the transformed organic nitrogen in a soil sample in boric acid (H_3BO_3) as ammonium ($NH^+{}_4$) and titrating it with dilute sulfuric acid. Available phosphorus (P) was determined using the Olsen method [14]; soil phosphoric acid was extracted using sodium bicarbonates ($NaHCO_3$), and the resulting proportional blue color of the phosphomolybdic complex was evaluated using spectrophotometry at 820 nm. Total organic carbon (TOC) was determined with the potassium dichromate method [13], soil sample organic matter was completely oxidized using potassium dichromate ($K_2Cr_2O_7$), and residual potassium dichromate was dosed with Mohr salt ($(NH_4)_2Fe(SO_4)_2(H_2O)_6$). Carbon-to-nitrogen (C/N) ratio was calculated using the last two parameters. Soil texture was determined using a Robinson pipette [13]; soil fine particles (sands, silt, and clay) were separated by diameter and gravity after different time periods and at different depths using the Stokes equation. Electrical conductivity and pH were measured using a conductometer (Basic 30) and a pH meter (Basic 20), respectively, for a 1:5 soil-to-water ratio.

2.3. *Mycorrhizal Attributes*

2.3.1. Root Mycorrhization

Arbuscular mycorrhizal fungi colonization assessment was performed according to the Phillips and Hayman [15] clearing and coloring method. Fine roots were cleared in 10% potassium hydroxide (KOH). They were then colored with 5 mL Trypan blue with repeated washing between and after. Next, samples were cut into 1 cm fragments, and 20 were placed on each slide, with three repetitions for each sample. Glycerol was added for conservation, as well as a cover slip before visualization under microscope. Mycorrhizal structures appear in dark blue. Relative mycorrhization intensity was measured by assigning a colonization index from 0 to 5: 0, no colonization; 1, colonization percentage less than 1%; 2, between 1% and 10%; 3, between 11% and 50%; 4, between 51% and 90%; 5, greater than 91% [16]:

$$M\,\% = \frac{(95 \times n5 + 70 \times n4 + 30 \times n3 + 5 \times n2 + n1)}{Tf} \quad (1)$$

where n5 is the fragments number that represents the colonization degree corresponding to index 5, the same for n4, n3, n2, and n1, respectively; and Tf is total fragments number.

2.3.2. Spore Enumeration

Arbuscular mycorrhizal fungi spores isolation was conducted using the Gerdemann and Nicolson method [17]: wet sieving and decanting, in combination with the Walker sucrose gradient centrifugation method [18]. Soil samples were passed through a set of sieves (800 and 50 μm) under running water. The trapped fraction in between was processed with two-step centrifugation at 3000 rpm, first with distilled water, then with two sucrose solutions (40% and 60%), with filtration being carried out on Whatman paper. Spores were observed and counted under a stereoscopic microscope at ×40.

2.4. Statistical Analysis

Results were tested using one-way analysis of variance (ANOVA) after dismissing heteroscedasticity and confirming normality using the Levene and Shapiro–Wilk tests, respectively [18]. Tukey's honest significant difference (HSD) was performed to describe significance. The Pearson product-moment coefficient was used to measure the linear correlation between the studied variables, for which p-values were approximated using the F-distributions. Values of p lower than 0.05 were considered significant. Principal component analysis (PCA) was also performed on all the variables to emphasize their variation and visualize how they evolve [19]. Statistical analyses were performed using the LibreOffice Calc v6.4.4 and R v4.0 software under a rolling ArchLabs distribution.

3. Results

3.1. Soil Chemical and Physical Analyses

Variability in Rs' abiotic variables is represented with box and whisker plots (Figure 3). An soil had significantly higher P (Figure 3a), TKN (Figure 3b), and TOC (Figure 3c) levels than the other soils (Table 1). Na had higher levels of the same parameters than Fo with no statistical significance. Nr had significantly the lowest levels of these parameters. The C/N ratio was 8 for An and Na and 9 for Fo. The Nr C/N ratio, on the other hand, was 19 and significantly higher than the other soil ratios (Figure 3d and Table 1). Texture did not change significantly for all the soil types and ranged from loam to loamy sand; pH was neutral, and Electrical Conductivity (EC) was around 3 μS/cm for all soils with no statistical difference. Thus these were left out of the box and whisker plots.

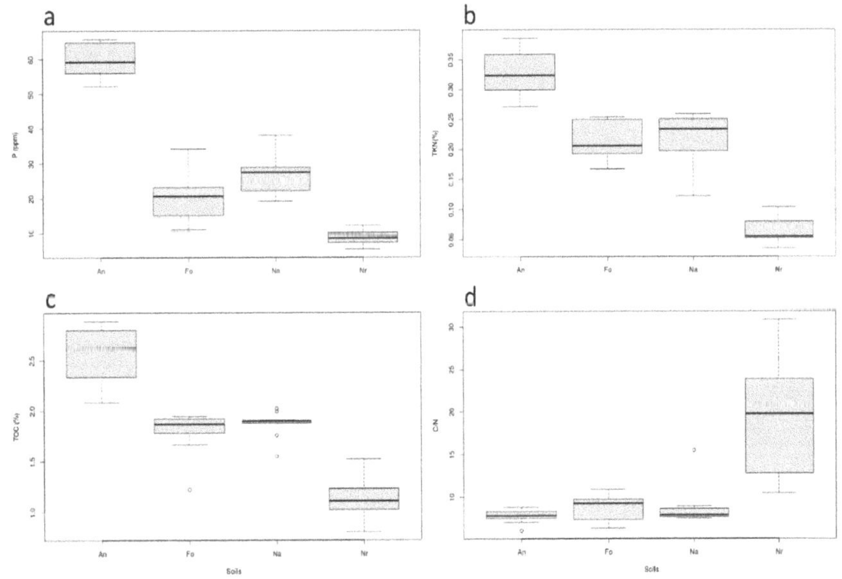

Figure 3. Variability of (**a**) available phosphorus (P), (**b**) total Kjeldahl nitrogen (TKN), (**c**) total organic carbon (TOC), and (**d**) carbon-to-nitrogen (C/N) ratios between the studied soils. The box and whisker diagrams include median value (dark rectangle), range of 50% of the samples (large rectangle), maximum and minimum values (cross bars), and outliers (circle).

Table 1. Analysis of variance and Tukey's honest significant difference (HSD) *p*-values of the studied variables and soil types.

	ANOVA *p*-Values		Tukey's HSD *p*-Values					
	***F*-Value**	**Pr (>F)**	**An–Na**	**An–Fo**	**An–Nr**	**Na–Fo**	**Na–Nr**	**Fo–Nr**
M	321.9	$<2 \times 10^{-16}$ ***	3.13×10^{-5}	0	0	0.06	0	0
Sn	92.38	$<2 \times 10^{-16}$ ***	0	0	0	0.09	2.4×10^{-6}	0.003
P	142.4	$<2 \times 10^{-16}$ ***	0	0	0	0.07	1×10^{-7}	0.0001
TKN	95.32	$<2 \times 10^{-16}$ ***	4×10^{-7}	0	0	0.86	0	0
TOC	68.31	6.24×10^{-15} ***	4×10^{-7}	0	0	0.86	0	5×10^{-7}
C/N	20.36	6.9×10^{-8} ***	0.9	0.9	4×10^{-7}	0.9	2.6×10^{-6}	2.1×10^{-6}
pH	0.82	0.49	-	-	-	-	-	-
EC	0.65	0.59	-	-	-	-	-	-

Significance codes: ***, 0.001; P, available phosphorus; TKN, total Kjeldahl nitrogen; TOC, total organic carbon; M, relative mycorrhization intensity; Sn, spore number; EC, electrical conductivity. An, Anastomosed trees; Na: Non-anastomosed trees; Fo: Public forest; Nr: non-rhyzospheric soil.

3.2. Soil Mycorrhizal Attributes

An significantly had the highest spore density (Figure 4a) and mycorrhization intensity (Figure 4b), followed by Na, Fo, and Nr (Table 1). All root samples were mycorrhized and consequently had 100% mycorrhization frequency. The mycorrhized fragments number was equal to the total observed fragments number (M% = m%). That is the reason why only the relative mycorrhization intensity (M%) was used to evaluate the difference in colonization levels.

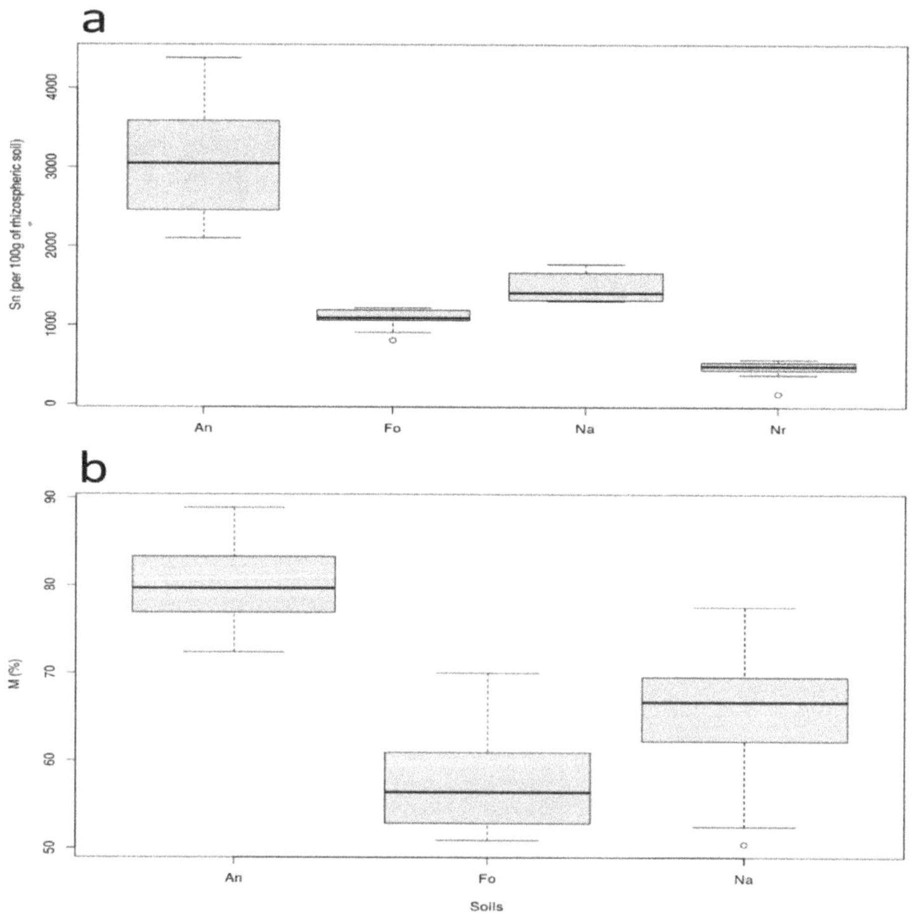

Figure 4. Variability of (**a**) spore number (Sn) and (**b**) mycorrhization intensity (M) between the studied soils. The box and whisker diagrams include median value (dark rectangle), range of 50% of the samples (large rectangle), maximum and minimum values (cross bars), and outliers (circle).

Compared with the control, An had six times higher P levels, five times higher TKN levels, and twice higher TOC levels. It also showed 37% higher mycorrhization intensity than Fo and six times higher spore density than Nr. Compared with Na, An had twice higher P levels, 47% higher TKN levels, 35% higher TOC levels, 22% higher M, and twice higher spore density. Compared with Nr, Na had three times higher P levels, three times higher TKN levels, and 65% times higher TOC levels. It also showed 12% higher mycorrhization intensity than Fo and three times higher spore density than Nr. Na and Fo did not differ much, and all differences were insignificant. Lastly, Fo had twice higher P content than Nr, three times higher TKN content, 58% higher TOC content, and twice higher spore density than Nr.

The studied variables' correlations were grouped by soil type in Tables 2–5. For An, a strong and significant positive correlation was found between P and M, P and Sn, M and Sn, Sn and TKN, and Sn and TOC and a moderate positive correlation between P and TKN (Table 2). For Na, a strong and significant positive correlation was found between P and M and between M and TKN, a moderate positive correlation between P and TKN and between M and TOC) and a strong negative one between TKN and C/N (Table 3). For Fo, a strong and significant positive correlation was noted between P and M and between M and Sn and a strong negative one between TKN and C/N) (Table 4). Lastly, the Nr soil type had one strong significant negative correlation between TKN and C/N (Table 5).

Figure 5a shows PCA compressed 56.5% of the data in the first component and 15.7% in the second. Figure 5b shows the biplot of the PCA with samples and variables represented in the plot. The PCA biplot shows the intimate co-evolution of P, TKN, and Sn, followed by M and TOC, highly driven by the An soil, while Nr had no influence. An showed a small overlap with Na and an even smaller one with Fo. Nr did not show any overlap with the other soil types.

Table 2. An soil parameters' correlation coefficients (left down) and corresponding *p*-values (right up); significant cases are highlighted.

	M	Sn	P	TKN	TOC	C/N	pH	EC
M		0.01	0.01	0.07	0.05	0.99	0.3	0.19
Sn	0.76		0.01	0.02	0	0.72	0.59	0.28
P	0.77	0.78		0.04	0.09	0.63	0.71	0.34
TKN	0.59	0.73	0.66		0.06	0.14	0.53	0.35
TOC	0.63	0.91	0.56	0.61		0.28	0.44	0.47
C/N	0.01	0.13	−0.17	−0.5	0.38		0.07	0.8
pH	−0.36	−0.2	0.14	0.22	−0.28	−0.59		0.91
EC	0.45	0.38	0.34	0.33	0.26	−0.09	0.04	

P, available phosphorus; TKN, total Kjeldahl nitrogen; TOC, total organic carbon; M, relative mycorrhization intensity; Sn, spore number; EC, electrical conductivity.

Table 3. Na soil parameters' correlation coefficients (left down) and corresponding *p*-values (right up); significant cases are highlighted.

	M	Sn	P	TKN	TOC	C/N	pH	EC
M		0.13	0	0.01	0.04	0.07	0.49	0.25
Sn	0.51		0.22	0.26	0.18	0.44	0.28	0.26
P	0.94	0.43		0.04	0.07	0.19	0.63	0.4
TKN	0.79	0.39	0.66		0.25	0	0.44	0.18
TOC	0.66	0.46	0.6	0.4		0.95	0.66	0.29
C/N	−0.59	−0.28	−0.46	−0.9	0.02		0.28	0.4
pH	0.25	0.38	0.17	0.28	−0.16	−0.38		0.95
EC	0.4	−0.39	0.3	0.46	0.37	−0.3	−0.02	

P, available phosphorus; TKN, total Kjeldahl nitrogen; TOC, total organic carbon; M, relative mycorrhization intensity; Sn, spore number; EC, electrical conductivity.

Table 4. Fo soil parameters' correlation coefficients (left down) and corresponding *p*-values (right up); significant cases are highlighted.

	M	Sn	P	TKN	TOC	C/N	pH	EC
M		0	0.01	0.53	0.13	0.69	0.53	0.89
Sn	0.82		0.29	0.62	0.2	0.61	0.58	0.99
P	0.77	0.37		0.2	0.23	0.65	0.15	0.73
TKN	0.23	0.18	0.44		0.4	0.03	0.84	0.52
TOC	0.51	0.44	0.42	0.3		0.17	0.24	0.09
C/N	0.14	0.18	−0.16	−0.69	0.47		0.61	0.04
pH	−0.22	0.2	−0.49	−0.08	−0.41	−0.18		0.86
EC	0.05	0	0.13	0.23	−0.57	−0.66	0.06	

P, available phosphorus; TKN, total Kjeldahl nitrogen; TOC, total organic carbon; M, relative mycorrhization intensity; Sn, spore number; EC, electrical conductivity.

Table 5. Nr soil parameters' correlation coefficients (left down) and corresponding *p*-values (right up); significant cases are highlighted.

	Sn	P	TKN	TOC	C/N	pH	EC
Sn		0.1	0.54	0.42	0.85	0.52	0.81
P	0.55		0.66	0.29	0.86	0.26	0.25
TKN	0.22	−0.16		0.92	0	0.92	0.23
TOC	0.29	−0.37	−0.04		0.14	0.33	0.41
C/N	0.07	0.06	−0.85	0.5		0.54	0.58
pH	−0.23	0.4	0.04	−0.35	−0.22		0.41
EC	−0.09	−0.4	0.41	0.3	−0.2	0.29	

P, available phosphorus; TKN, total Kjeldahl nitrogen; TOC, total organic carbon; Sn, spore number; EC, electrical conductivity.

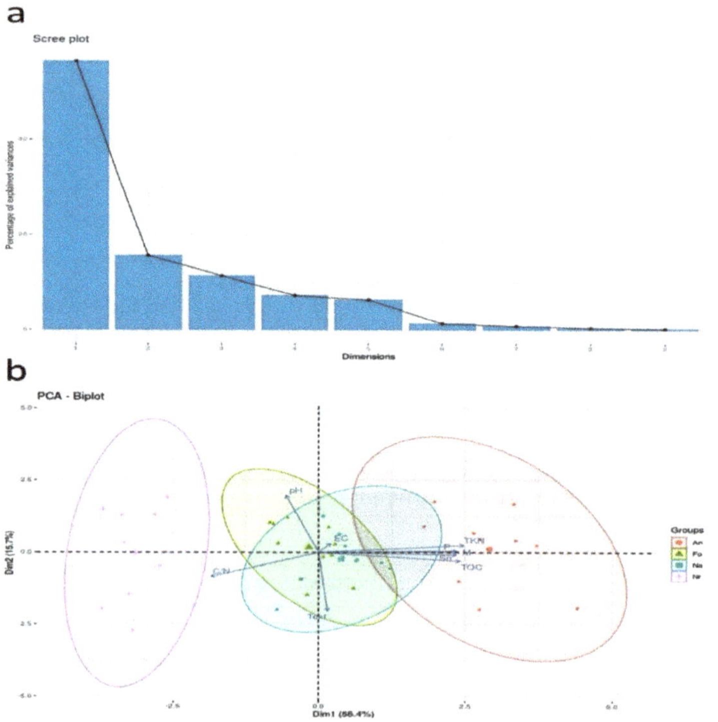

Figure 5. (**a**) Percentage of variances explained by each principal component; (**b**) PCA biplot of samples and variables, obtained from the analyzed variables for the different soil types. P, available phosphorus; TKN, total Kjeldahl nitrogen; TOC, total organic carbon; M, relative mycorrhization intensity; Sn, spore number; EC, electrical conductivity.

4. Discussion

The differences between the studied soil types are very considerable, considering that they are under the same climatic, topographic, and edaphic conditions. Since their only differences rely on whether or not they are located under specific tree morphotypes or not submitted to tree management (Fo and Nr), we consider that these differences are to be related to human intervention in tree shaping. Tree shaping seems to lead to an increase in biological activity and nutrient fluxes, and thus to differences in soil properties, as found in a Himalayan context [20]. For the An and Na soil types, the only difference is the presence/absence of anastomosis. Fo trees are subjected to higher but still moderate browsing [7]. This could explain the Fo soil's low recorded values for the studied variables and high C/N ratio compared with the other soil types. Heavy grazing has been proved to have a negative impact on soil fertility and AMF attributes [21–26]. In the present case, slight grazing could explain the relatively elevated N levels in soils [27,28], as well as the technique used to deliver forage material to flocks. Grazing in moderation was also recorded to have an increasing effect on AMF [29]. The only difference between An and Na samples was the shaping applied on trees. Some samples were no more than 10 m away from each other, but presented completely different soil profiles (Figure 1a). Most Na and Fo parameters were statistically similar and had almost the same profile. This suggests that the major explanation for the remarkably high An soil parameters is linked to the tree shaping status. The biomass produced by anastomosed trees is higher [6], which enriches the soil more than elsewhere, leading to increased soil biological activity.

The C/N ratio is used to characterize the patterns of change in organic matter in soil. It can represent an important indicator describing the organic matter state and its influence on soil microbial activity and identity and soil productivity [30,31]. Carbon is lost by mineralization faster than nitrogen, and the C/N ratio decreases over time. Low C/N ratios generally indicate that plants provide soil biological activity with necessary resources so that it releases, in return, the elements contained in the organic matter faster, thus helping to nourish the trees more actively [32]. In this study, An, Na, and Fo C/N ratios point to rapid decomposition of soil organic matter and release of N into the soil, which can also be an alternative explanation for the soils' elevated N levels. The higher Nr C/N ratio was still within the range, favoring good organic matter decomposition with little to no release of N into the soil [33].

Correlating the studied variables showed some similarities and clear differences between the behaviors of these variables in the different studied soil types. Starting with the similarities, all soil types had a significant positive correlation between P and M. An and Fo had one similarity as both had a strong and significant positive correlation between M and Sn. An and Na also had one similarity, a strong and significant positive correlation between P and TKN. The differences were that Nr showed no distinctive correlations between almost all of its parameters, and only An had a strong and significant positive correlation between Sn and TOC and between Sn and TKN. An had a distinctive new pattern never previously mentioned in the literature, where three variables (P, M, and Sn) positively correlated strongly with each other. To our knowledge, this is the first time such a pattern emerged. In the available literature, one or two of these variables commonly correlate negatively with the others, often being spore density and available phosphorus [34–37], although some studies sometimes showed positive correlations [38–40]. One study even showed that P can correlate with either M or Sn [41], but we could not find one that cites such case where the three variables simultaneously correlate positively with each other. This newly observed pattern could be interpreted as the result of an indirect effect of stems shaping on the discussed soil variables.

Arbuscular mycorrhizal fungi are a fungi group that forms mutualistic symbiosis with most land plants [42]. They are well known for increasing plant nutrient uptake and productivity [43]. Their impacts on a wide range of ecosystem processes and several species are very well documented [44–46]. In high concentrations, phosphorus is known to suppress AMF infection and spore density [47], whereas in low concentrations, AMF take on the task of P prospectors and make it more available to the host at the hyphosphere and mycorrhizosphere [48]. Arbuscular Mycorrhizal Fungi's positive correlation with TKN has also been reported [49,50], while different nitrogen responses have also been mentioned, ranging from positive correlation to P [51] to even opposing levels of P [52,53].

Principal Component analysiss confirmed the noted correlations and showed that Na and Fo shared more similarities with high variance overlap, while for Nr, the absence of variance overlap singled it out, not having any significant similarities with any of the other soil types. M and Sn co-evolution has also been noted in another ecosystem on a different tree species, *Juniperus phoenicea* [36]. It also showed that pH, EC, and texture evolved independently of the rest of the variables not having any drive effect on them.

Agroforestry is undergoing a revival of interest nowadays, particularly to manage soil fertility and in the context of ongoing climate change [54]. Pollarding is a secular technique found historically in many countries [55]. If it almost disappeared in many of them due to the eviction of trees in croplands, it is still alive in various developing countries and supports local livelihoods [20,56]. It is considered to be an efficient technique to delay tree ageing [57], as well as to modify the allocation of resources, especially in the short-term shift in shoot–root balance induced by pruning or pollarding, which can lead to altered root distribution and fine-root turnover enriching soil biota [58]. This knowledge has to be enriched, by both observation of traditional practices and experimentation, and could be useful to develop alternative schemes for agricultural production.

Finally, these results provide arguments in support of the intermediate disturbance hypothesis [59,60], which argues that biological diversity and ecological functioning are enhanced when environments are subjected to intermediate levels of disturbance. Trunk anastomosis, by enhancing tree productivity, seems to promote organic restitutions to the soil and favor its biological activity. These results have to be confirmed in other situations, particularly in zones where pollarding has been used for long time.

5. Conclusions

Traditional tree shaping by pollarding and favoring trunk anastomosis have a clear and distinct positive impact on the dimorphic ash tree's rhizospheric soil variables found in agroforestral systems in the Central Moroccan High Atlas, particularly on phosphorus, nitrogen, and carbon content, and on mycorrhizal activity, compared with uncultivated or poorly managed soils. This illustrates that, in some cases, human practices can indirectly improve the delivery of ecosystem services from spontaneous forested stands. Our results confirm the initial hypothesis that soil characteristics respond to *F. dimorpha* shaping, as such changes in the aerial parts of the tree require recruitment of all the belowground actors, both abiotic and biotic. They tend to show that soil biological activity can be boosted by modifying natural tree port, and by a management that allows higher organic restitutions. However, these results suggest that abiotic factors alone are insufficient to fully explain the spectacular improvement of leaf productivity on anastomosed tree, and physiological, histological, and biomolecular experiments have to be conducted to better understand the impact of shaping dimorphic ash tree at these different levels, and their mutual interactions.

Author Contributions: Conceptualization, A.F., D.G., M.A., M.H.; writing—original draft, A.F., D.G.; methodology, M.A.-E.-M., A.M., A.F., M.H.; investigation, A.F., E.M.O., S.M.; writing—review and editing, A.F., M.A.-E.-M., D.G., M.A., A.M., M.H. All authors have read and agreed to the published version of the manuscript.

References

1. Parrotta, J.; Agnoletti, M. Traditional forest knowledge: Challenges and opportunities. *Forest Ecol. Manag.* **2007**, *249*, 1–4. [CrossRef]
2. Maffi, L.; Woodley, E. *Biocultural Diversity Conservation: A Global Sourcebook*; Routledge: London, UK, 2012.

3. De Freitas, J.G.; Bastos, M.R.; Dias, J.A. Traditional Ecological Knowledge as a Contribution to Climate Change Mitigation and Adaptation: The Case of the Portuguese Coastal Populations. In *Handbook of Climate Change Communication: Case Studies in Climate Change Communication*; Filho, W.L., Manolas, E., Azul, A.M., Azeiteiro, U.M., McGhie, H., Eds.; Springer International Publishing: Cham, Switzerland, 2018; pp. 257–269. [CrossRef]
4. Wyllie de Echeverria, V.R.; Thornton, T.F. Using traditional ecological knowledge to understand and adapt to climate and biodiversity change on the Pacific coast of North America. *Ambio* **2019**, *48*, 1447–1469. [CrossRef] [PubMed]
5. Parrotta, J.; Yeo-Chang, Y.; Camacho, L.D. Traditional knowledge for sustainable forest management and provision of ecosystem services. *Int. J. Biodivers. Sci. Ecosyst. Serv. Manag.* **2016**, *12*, 1–4. [CrossRef]
6. Genin, D.; M'Sou, S.; Ferradous, A.; Alifriqui, M. Another vision of sound tree and forest management: Insights from traditional ash shaping in the Moroccan Berber mountains. *For. Ecol. Manag.* **2018**, *429*, 180–188. [CrossRef]
7. Genin, D.; Crochot, C.; MSou, S.; Araba, A.; Alifriqui, M. Meadow up a tree: Feeding flocks with a native ash tree in the Moroccan mountains. *Pastoralism* **2016**, *6*. [CrossRef]
8. Genin, D.; Alifriqui, M. The agroforestry parklands with dimorphic ash trees of Aït M'Hamed. A well-guarded local specificity. In *The Emergence of Local Specificities in the Mediterranean Hinterlands*; Aderghal, M., Genin, D., Hanafi, A., Landel, P.A., Michon, G., Eds.; Laboratoire Population-Environnement-Développement: Marseille, France, 2019; pp. 30–51. ISBN 979-10-96763-08-5.
9. Schweingruber, F.H. *Wood Structure and Environment*; Springer Science & Business Media: Berlin/Heidelberg, Germany, 2007.
10. Howard, K.S.C.; Eldridge, D.J.; Soliveres, S. Positive effects of shrubs on plant species diversity do not change along a gradient in grazing pressure in an arid shrubland. *Basic Appl. Ecol.* **2012**, *13*, 159–168. [CrossRef]
11. Birhane, E.; Ahmed, S.; Hailemariam, M.; Negash, M.; Rannestad, M.M.; Norgrove, L. Carbon stock and woody species diversity in homegarden agroforestry along an elevation gradient in southern Ethiopia. *Agrofor. Syst.* **2020**, *94*, 1099–1110. [CrossRef]
12. Rillig, M.C.; Mummey, D.L. Mycorrhizas and soil structure. *New Phytol.* **2006**, *171*, 41–53. [CrossRef]
13. Schinner, F.; Öhlinger, R.; Kandeler, E.; Margesin, R. *Methods in Soil Biology*; Springer Science & Business Media: Berlin/Heidelberg, Germany, 2012.
14. Olsen, S.R. Estimation of available phosphorus in soils by extraction with sodium bicarbonate. *USDA Circular* **1954**, *24*. [CrossRef]
15. Phillips, J.M.; Hayman, D.S. Improved procedures for clearing roots and staining parasitic and vesicular-arbuscular mycorrhizal fungi for rapid assessment of infection. *Trans. Br. Mycol. Soc.* **1970**, *55*, 158. [CrossRef]
16. Trouvelot, A.; Kough, J.L.; Gianinazzi-Pearson, V. Measurement of the mycorrhization rate VA of a root system. In search for an estimation method with functional significance Presented at the Physiological and genetical aspects of mycorrhizae. In Proceedings of the 1st European Symposium on Mycorrhizae, Dijon, France, 1–5 July 1985; pp. 217–221.
17. Gerdemann, J.W.; Nicolson, T.H. Spores of mycorrhizal Endogone species extracted from soil by wet sieving and decanting. *Trans. Br. Mycol. Soc.* **1963**, *46*, 235–244. [CrossRef]
18. Scheffe, H. *The Analysis of Variance*; John Wiley & Sons, Inc.: Hoboken, NJ, USA, 1959; pp. 351–358.
19. Jolliffe, I.T. Principal Component Analysis. In *Series in Statistics*; Springer: New York, NY, USA, 2012. [CrossRef]
20. Singh, R.K.; Singh, A.; Garnett, S.T.; Zander, D.K.; Lobsang, D.K.; Tsering, D. Paisang (Quercus griffithii); a keystone tree species in sustainable agroecosystem management and livelihoods in Arunachal Pradesh, India. *Environ. Manag.* **2015**, *55*, 187–204. [CrossRef] [PubMed]
21. Walker, C.; Mize, C.W.; McNabb, H.S. Populations of endogonaceous fungi at two locations in central Iowa. *Can. J. Bot.* **1982**, *60*, 2518–2529. [CrossRef]
22. Liu, J.; Zhang, Q.; Li, Y.; Di, H.; Xu, J.; Li, J.; Pan, H. Effects of pasture management on soil fertility and microbial communities in the semi-arid grasslands of Inner Mongolia. *J. Soils Sediments* **2016**, *16*, 235–242. [CrossRef]
23. Guangyu, Z.; Zhuangsheng, T.; Lei, C.; Zhouping, S.; Lei, D. Overgrazing depresses soil carbon stock through changing plant diversity in temperate grassland of the Loess Plateau. *Plant Soil Environ.* **2018**, *64*, 1–6. [CrossRef]
24. Cai, Y.; Yan, Y.; Xu, D.; Xu, X.; Wang, C.; Wang, X.; Eldridge, D.J. The fertile island effect collapses under extreme overgrazing: Evidence from a shrub-encroached grassland. *Plant Soil* **2020**, *448*, 201–212. [CrossRef]
25. Faghihinia, M.; Zou, Y.; Chen, Z.; Bai, Y.; Li, W.; Marrs, R.; Staddon, P.L. The response of grassland mycorrhizal fungal abundance to a range of long-term grazing intensities. *Rhizosphere* **2020**, *13*, 100178. [CrossRef]

26. Yang, X.; Chen, J.; Shen, Y.; Dong, F.; Chen, J. Global negative effects of livestock grazing on arbuscular mycorrhizas: A meta-analysis. *Sci. Total Environ.* **2020**, *708*, 134553. [CrossRef]
27. Li, Y.; Wu, J.; Shen, J.; Liu, S.; Wang, C.; Chen, D.; Zhang, J. Soil microbial C:N ratio is a robust indicator of soil productivity for paddy fields. *Sci. Rep.* **2016**, *6*, 35266. [CrossRef]
28. Shen, H.; Dong, S.; Li, S.; Xiao, J.; Han, Y.; Yang, M.; Yeomans, J.C. Grazing enhances plant photosynthetic capacity by altering soil nitrogen in alpine grasslands on the Qinghai-Tibetan plateau. *Agric. Ecosyst. Environ.* **2019**, *280*, 161–168. [CrossRef]
29. Wan, X.; Huang, Z.; He, Z.; Yu, Z.; Wang, M.; Davis, M.R.; Yang, Y. Soil C:N ratio is the major determinant of soil microbial community structure in subtropical coniferous and broadleaf forest plantations. *Plant Soil* **2015**, *387*, 103–116. [CrossRef]
30. Eom, A.-H.; Wilson, G.W.T.; Hartnett, D.C. Effects of ungulate grazers on arbuscular mycorrhizal symbiosis and fungal community structure in tallgrass prairie. *Mycologia* **2001**, *93*, 233–242. [CrossRef]
31. Li, W.; Cao, W.; Wang, J.; Li, X.; Xu, C.; Shi, S. Effects of grazing regime on vegetation structure, productivity, soil quality, carbon and nitrogen storage of alpine meadow on the Qinghai-Tibetan Plateau. *Ecol. Eng.* **2017**, *98*, 123–133. [CrossRef]
32. Brust, G.E. Chapter 9-Management Strategies for Organic Vegetable Fertility. In *Safety and Practice for Organic Food*; Biswas, D., Micallef, S.A., Eds.; Academic Press: Cambridge, MA, USA, 2019; pp. 193–212. [CrossRef]
33. Zhang, S.; Zheng, Q.; Noll, L.; Hu, Y.; Wanek, W. Drivers of microbial nitrogen use efficiency and soil inorganic N processes at the continental scale. *Geophys. Res. Abstr.* **2019**, *21*, 1–4.
34. Khakpour, O.; Khara, J. Spore density and root colonization by arbuscular mycorrhizal fungi in some species in the northwest of Iran. *Int. Res. J. Appl. Basic Sci.* **2012**, *3*, 977–982.
35. Tamil Pradha, K.; Sivakumar, K. Effects of am fungi on spores density and root colonization of chilli (*Capsicum annum*) at different levels of rock phosphate. *Int. J. Curr. Res. Life Sci.* **2018**, *7*, 1415–1419.
36. Fakhech, A.; Ouahmane, L.; Hafidi, M. Seasonality of mycorrhizal attributes, soil phosphorus and nitrogen of Juniperus phoenicea and Retama monosperma boiss. in an Atlantic sand dunes forest. *J. Sustain. For.* **2019**, *38*, 1–17. [CrossRef]
37. Gao, D.; Pan, X.; Zhou, X.; Wei, Z.; Li, N.; Wu, F. Phosphorus fertilization and intercropping interactively affect tomato and potato onion growth and rhizosphere arbuscular mycorrhizal fungal community. *Arch. Agron. Soil Sci.* **2020**. [CrossRef]
38. Rubio, R.; Borie, F.; Schalchli, C.; Castillo, C.; Azcón, R. Occurrence and effect of arbuscular mycorrhizal propagules in wheat as affected by the source and amount of phosphorus fertilizer and fungal inoculation. *Appl. Soil Ecol.* **2003**, *23*, 245–255. [CrossRef]
39. Cuéllar, A.E.; Martinez, L.R.; Espinosa, R.R.; Cuéllar, E.E. The Inoculation with an Effecient AMF Strain Decreases the Phosphoric Fertilizer Requirements in Ipomea Batata (L.), Lam in Dry Period. *J. Chem. Environ. Biol. Eng.* **2019**, *3*, 13. [CrossRef]
40. Effendy, M.; Wijayani, B.W. Estimation of Available Phosphorus in Soil Using the Population of Arbuscular Mycorrhizal Fungi Spores. *J. Trop. Soils* **2019**, *16*, 225–232. [CrossRef]
41. Abdel-Fattah, G.M.; El-Dohlob, A.F.; El-Dohlob, G.M.; El-Haddad, S.M.; Hafez, S.A.; Rashad, Y.M. An ecological view of arbuscular mycorrhizal status in some Egyptian plants. *J. Environ. Sci.* **2010**, *37*, 123–136.
42. Brundrett, M.C.; Tedersoo, L. Evolutionary history of mycorrhizal symbioses and global host plant diversity. *New Phytol.* **2018**, *220*, 1108–1115. [CrossRef] [PubMed]
43. Smith, S.E.; Smith, F.A. Roles of Arbuscular Mycorrhizas in Plant Nutrition and Growth: New Paradigms from Cellular to Ecosystem Scales. *Annu. Rev. Plant Biol.* **2011**, *62*, 227–250. [CrossRef] [PubMed]
44. Powell, J.R.; Rillig, M.C. Biodiversity of arbuscular mycorrhizal fungi and ecosystem function. *New Phytol.* **2018**, *220*, 1059–1075. [CrossRef] [PubMed]
45. Fakhech, A.; Ouahmane, L.; Hafidi, M. Analysis of symbiotic microbial status of Atlantic sand dunes forest and its effects on Acacia gummifera and Retama monosperma (Fabaceae) to be used in reforestation. *J. For. Res.* **2019**. [CrossRef]
46. Ait-El-Mokhtar, M.; Fakhech, A.; Anli, M.; Ben-Laouane, R.; Boutasknit, A.; Wahbi, S.; Meddich, A. Infectivity of the palm groves arbuscular mycorrhizal fungi under arid and semi-arid climate and its edaphic determinants towards efficient ecological restoration. *Rhizosphere* **2020**, 100220. [CrossRef]
47. Tucker, M. Effects of Phosphorus on Mycorrhizal Ccolonization of Wetland Plants under Natural and Controlled Conditions. Master's Thesis, Wilfrid Laurier University, Waterloo, Belgium, 2020.

48. Linderman, R.G. The mycorrhizosphere phenomenon. In Proceedings of the International Symposium "Mycorrhiza for Plant Vitality" and the Joint Meeting of Working Groups 1–4 of COST Action 870, Hannover, Germany, 3–5 October 2007; pp. 341–355.
49. Nuccio, E.E.; Hodge, A.; Pett-Ridge, J.; Herman, D.J.; Weber, P.K.; Firestone, M.K. An arbuscular mycorrhizal fungus significantly modifies the soil bacterial community and nitrogen cycling during litter decomposition: AMF alters soil bacterial community and N cycling. *Environ. Microbiol.* **2013**, *15*, 1870–1881. [CrossRef]
50. Lin, C.; Wang, Y.; Liu, M.; Li, Q.; Xiao, W.; Song, X. Effects of nitrogen deposition and phosphorus addition on arbuscular mycorrhizal fungi of Chinese fir (Cunninghamia lanceolata). *Sci Rep.* **2020**, *10*, 12260. [CrossRef]
51. Liu, C.; Ravnskov, S.; Liu, F.; Rubæk, G.H.; Andersen, M.N. Arbuscular mycorrhizal fungi alleviate abiotic stresses in potato plants caused by low phosphorus and deficit irrigation/partial root-zone drying. *J. Agric. Sci.* **2018**, *156*, 46–58. [CrossRef]
52. Camenzind, T.; Homeier, J.; Dietrich, K.; Hempel, S.; Hertel, D.; Krohn, A.; Rillig, M.C. Opposing effects of nitrogen versus phosphorus additions on mycorrhizal fungal abundance along an elevational gradient in tropical montane forests. *Soil Biol. Biochem.* **2016**, *94*, 37–47. [CrossRef]
53. Xiao, D.; Che, R.; Liu, X.; Tan, Y.; Yang, R.; Zhang, W.; Wang, K. Arbuscular mycorrhizal fungi abundance was sensitive to nitrogen addition but diversity was sensitive to phosphorus addition in karst ecosystems. *Biol. Fertil. Soils* **2019**, *55*, 457–469. [CrossRef]
54. Jose, S. Agroforestry for ecosystem services and environmental benefits: An overview. *Agrofor. Syst.* **2009**, *76*, 1–10. [CrossRef]
55. Petit, S.; Watkins, C. Forgotten peasant practices: Tree pollarding in Great Britain. *Etudes Rural.* **2004**, *169*, 197–214. [CrossRef]
56. Genin, D.; Aumeerudy-Thomas, Y.; Balent, G.; Nasi, R. The multiple dimensionsof rural forests: Lessons from a comparative analysis. *Ecol. Soc.* **2013**, *18*, 27. [CrossRef]
57. Ferrini, F. Pollarding and its effects on tree physiology: A look to mature and senescent tree management in Italy. In *Pollard Trees in Europe, European Meeting on Pollard Trees*; Dumont, E., Ed.; Maison Botanique: Boursay, France, 2006; pp. 71–79.
58. Van Noordwijk, M.; Purnomosidhi, P. Root architecture in relation to tree-soil-crop interactions and shoot pruning in agroforestry. *Agrofor. Syst.* **1995**, *30*, 161–173. [CrossRef]
59. Connell, J. Diversity in tropical rain forests and coral reefs. *Science* **1978**, *199*, 1302–1310. [CrossRef]
60. Siebert, S.F.; Belsky, J.M. Historic livelihoods and land uses as ecological disturbances and their role in enhancing biodiversity: An example from Bhutan. *Biol. Conserv.* **2014**, *177*, 82–89. [CrossRef]

8

Agroforestry Innovation through Planned Farmer Behavior: Trimming in Pine–Coffee Systems

Edi Dwi Cahyono [1,*], Salsabila Fairuzzana [1], Deltanti Willianto [1], Eka Pradesti [1], Niall P. McNamara [2], Rebecca L. Rowe [2] and Meine van Noordwijk [3]

1 Faculty of Agriculture, Universitas Brawijaya, Malang 65145, Indonesia; salsabilazatira@gmail.com (S.F.); deltantiwww@gmail.com (D.W.); eka_pradesti@student.ub.ac.id (E.P.)
2 UK Centre of Ecology & Hydrology, Lancaster LA14AP, UK; nmcn@ceh.ac.uk (N.P.M.); rebrow@ceh.ac.uk (R.L.R.)
3 World Agroforestry (ICRAF), Bogor 16155, Indonesia; m.vannoordwijk@cgiar.org
* Correspondence: edidwi.fp@ub.ac.id

Abstract: Knowledge transfer depends on the motivations of the target users. A case study of the intention of Indonesian coffee farmers to use a tree canopy trimming technique in pine–based agroforestry highlights path-dependency and complexity of social-ecological relationships. Farmers have contracts permitting coffee cultivation under pine trees owned by the state forestry company but have no right to fell trees. A multidisciplinary international team of scientists supported farmers at the University of Brawijaya Forest in East Java to trial canopy trimming to improve light for coffee production while maintaining tree density. Data were collected using surveys through interviews, case study analysis using in-depth interviews, focus group discussions and nonparticipant observations. Using the Theory of Planned Behavior, we found that though farmer attitudes toward trimming techniques were positive, several factors needed to be scrutinized: perceived limited socio-policy support and resources. While there is hope that canopy trimming can improve coffee production and local ecosystem services, a participatory and integrative extension and communication strategy will be needed. In the relationship between farmers as agents and forest authorities as principals, any agroforestry innovation needs to incorporate knowledge and concerns in the triangle of farmers, policymakers and empirical science.

Keywords: agroforestry; innovation transfer; trimming; intention; participatory and integrative research-extension; stakeholders

1. Introduction

Innovators—and especially developers of innovations that others are supposed to use—may assume that once innovations are transferred they are inherently beneficial and accessible to the potential adopters [1–4]. In other words, they are subject to a "pro-innovation bias". By starting from the needs of the targeted adopters a different design of an innovation process may lead to alternate outcomes [5–8]. How benefits and costs of any innovation are distributed is key to the acceptance of innovations, but groups, rituals, affiliation, status and power are major drivers of human decision-making in a social context [9]. In a context of agroforestry, trees and farmers interact, at the plot, landscape and policy scales [10]. Ecological interactions between trees and other agroforestry components, both complementary and antagonistic, can be a mirror for the social interactions between forest institutions and farmers. This is especially true in a specific form of agroforestry where forestry institutions own the land and are interested in permanent tree cover and timber harvests. While farmers may benefit indirectly or directly from timber production, they are primarily benefiting from and interested in understory crops leading to potentially antagonistic relationships. The principal–agent theory, as developed in institutional economics, analyses the strategic interactions between principals who delegate, often limited in time or scope, a conditional grant of authority to an agent [11,12]. An agent (e.g., farmers or farmer groups) is an individual or entity that has the authority to act on behalf of the principal (party or parties, e.g., forestry institution) who has power over the agent. A dispute—namely the principal–agent problem—arises when the actions of agents pursue different interests or even are in conflict with those of the principal [12]. Forms of "social forestry" (locally known as "tumpangsari" and internationally as "taungya") as implemented on densely populated Java (Indonesia) have such "principal–agent" interactions between the forestry principals, who under the social forestry program have granted increased authority to farmer to act as "agents" for the "principal" [13]. This principle–agent interaction faces the inherent challenges of this type of relationship, with little harmony in management objectives expected and strict enforcement of forestry rules essential, but not always feasible leading to the risk of agent shirking and slippage. Analysis [14] suggests that other types of benefit-sharing arrangements are needed before a much wider range of locally adapted solutions can be found, which more fully exploit the opportunities of tree–crop synergy. Social forestry programs in Indonesia have tried to bring more balance to the stakes of the forestry principals and farmer agents, including timber sharing arrangements [15]. However, technical innovation may still be constrained by the institutional context and its history (path dependency), as our case study explores.

For many decades, the interaction between Indonesian forest authorities and forest-edge farmers who have tried to expand their coffee into existing forests has been strongly antagonistic in Indonesia, as also evident elsewhere [16–18]. Only after the 1997 "reformasi" period and the 1998 forestry law, the "tumpangsari"—contracts that allow farmers to grow annual food crops among newly planted trees for a few years—were extended to allow longer-term shade-tolerant perennials such as coffee or fodder for a cut-and-carry system to be grown [19–21]. For the state forest company authorized to manage these lands, the survival and growth of the tree plantation has been the priority and farmers risked being left out of future contracts if trees were found to have been damaged. The intercrops were contractually treated as a "share-crop", with around one-third of the yield belonging to the "landlord," (in this case, the state forest company). These rules gave the state forest company a stake in the productivity of intercrops (different from when a fixed "land rent" would have been charged) but reduced the profitability for the farmer. Any intervention, such as social innovations from forest authorities or technical innovations from development agencies can destabilize the agent–farmer economic structure and cause disputes with the existing authority-principal policies, which cause resistance towards agroforestry innovation [22,23]. A balance needs to be found between agents (or principals) who may be profiteers, prove-it accountants or visionary prophets [24,25]. Profiteers are profit agents (entrepreneurs) as part of the national economy; prove-it accountants are intermediary agents that encourage transparent processes for mutual benefit. Prophets see themselves as representative of the local community to voice their interests, needs and perspectives in the complexity of the external environment. The roles of NGOs (prove-it accountants) as an intermediary for marginal groups to negotiate with the authorities (profiteers) have been identified [26,27]. On the other hand, efforts to put forward local forest community interest ("the prophet") form a challenge [28] that open spaces for more a deeper study.

Our case study focuses on an area of forest on the slopes of the Arjuna volcano in Malang regency, Java where principal–agent interactions had been ongoing for two decades. The forest management was handed over to Brawijaya University (UB) to be operated as an "education forest" and the new university management team proposed a change in forest-farmer relations. Rather than trying to collect the coffee share crop revenue directly, they asked all farmers to sell their coffee (purportedly for going market prices) to the new university UB-Forest Management company. It was anticipated that through better coffee processing and targeting other market segments enough profit would be made to offset the foregone sharecropping payments. UB-Forest management was also looking to return profit or benefit to the farmers in other ways, e.g., community electricity provision which may not be fully recognized or thought valuable by the farmers. In practice, farmers quickly reverted to selling most of their coffee to outside traders, as they could achieve a better price than selling to the UB-Forest Management company.

As part of the wider interaction between UB and the forest community, UB researchers independent of the UB-Forest management team proposed to address technical constraints around farmer yield and income (Figure 1). One such constraint was the low levels of light for the promotion of healthy understory coffee growth due to the overlying dense pine forest canopy. Our overarching hypothesis was that while farmers perceived the introduced innovation positively, some individual and external factors may hamper their intention to practice it.

Figure 1. Coffee plants under the pine forest canopy and the interdisciplinary science experts.

Our specific research questions were (1) what farmers thought about innovative efforts to provide more light to understory crops, (2) how this relates to their knowledge and expectations and (3) under what conditions it could be sufficiently profitable for them to undertake this innovation. We explored these questions with reference the theory of planned behavior, as we will describe here, before providing details of the site and specific research methods.

2. Methods

2.1. *Theory of Planned Behavior*

The theory of planned behavior (TPB) (Figure 2) can reveal the cognitive, affective and motivational dynamics of farmers facing new techniques. TPB is commonly employed in the fields of health and economics [29,30], but rarely in the field of agroforestry, agriculture or environmental management. A recent study using TPB in such a field was carried out in Serbia [31]. Some researchers argue that self-efficacy is interrelated with the three antecedent variables to measure intentions [32,33]. For this reason, we included self-efficacy as part of the variable of perceived behavioral control.

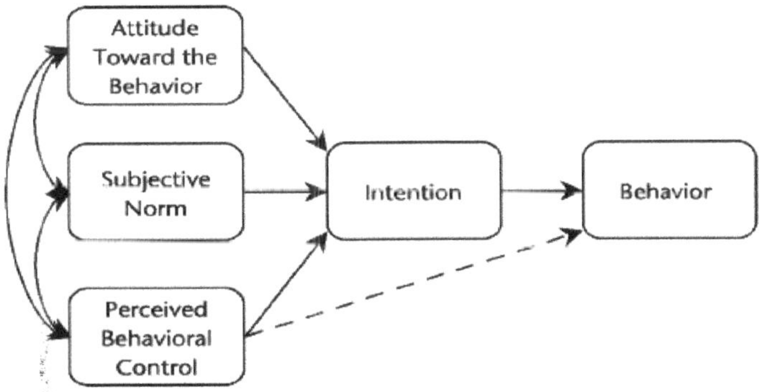

Figure 2. Theory of planned behavior, adapted from [32], showing the three antecedent interacting aspects: (1) attitude towards behavior/behavioral beliefs; (2) subjective norms/normative beliefs; and (3) perceived behavioral control, intention and behavior/action.

The single construct of intention may be compared with the three antecedent variables of intention. In addition, as can be seen in the model, Perceived Behavioral Control (PBC) is a unique variable that acts as an antecedent for intention and actual behavior as well. Intention is interpreted as volitional control or willingness to take certain actions [34–36]. In this study, intention variables were treated as the consequence or impact of attitudes, subjective norms and perceived behavioral control. We also measure intention solely (as a single variable), to be exemplified farmer intentions in applying trimming techniques; the score can be compared with the three antecedent variables of intention.

Typically, previous studies used "perception" and/or "attitude "as a determinant of adoption [37–40]. The TBP allows the variables of attitudes, subjective norms, and, perceived behavioral control to be combined as determinants of intention [41,42]. Furthermore, the self-efficacy variable (self-confidence) may be added to TBP to predict intentions. Hence, TBP can describe the behavior that is influenced by external factors (nonvolitional behavior) and self-efficacy as the individual's internal factor [43].

In the context of agroforestry, we expect that an intention to use pine trimming is associated with a positive attitude (expectations of benefit), strong social support and perceived low constraints of farmers to implement such techniques. In practice, the possible low scores of intention variables can lead to the rearrangement of social and communicative interventions to increase the adoption of trimming techniques. Given the current challenges and opportunities in the UB-Forest, there are simultaneous efforts to explore: (1) techno-ecological options for pine–coffee agroforestry revitalization (such as increasing light intensity, improving crop and soil management (trimming leaves used as litter/organic fertilizer); (2) socioeconomic options (the integration of stakeholders, such as to the financing of production factors); and (3) renewing the existing Indonesia's forestry policy framework to overcome past conflicts (sharing systems, alternative marketing channels and land management regulations) and achieving effective joint management (involving farmers, local governments and forest authority).

2.2. Place, Location and Research Participants

Currently, a variety of collaborative schemes on community forest management exist in Java. The collaboration is between forest management and members of the surrounding communities. UB-Forest management inherited contracts made by the state forest company in response to massive forest land annexation by the community at the end of the 20th century due to political reforms. Cultivation of coffee under pine (or mahogany) trees is common in UB-Forest. At the end of 2015, The UB-Forest management through the University of Brawijaya (UB) received the mandate to administer the forest on behalf of the Indonesian Minister of Forestry and Environment. This production forest—which is located on the slopes of Mount Arjuno (altitude 1200–1900 m in the Karangploso District, Malang regency) has three specific objectives: education, research and ecotourism. Historically, the pioneer farmers settled around the UB Forest in the 1920s and the present farmers are the descendants from the early farmers. Their livelihoods are highly dependent on the agroforestry system, in which they also cultivate food crops and vegetables along with the coffee under the pine trees.

The research location was the Sumbersari Hamlet in the enclave area of UB-Forest (Figure 3).

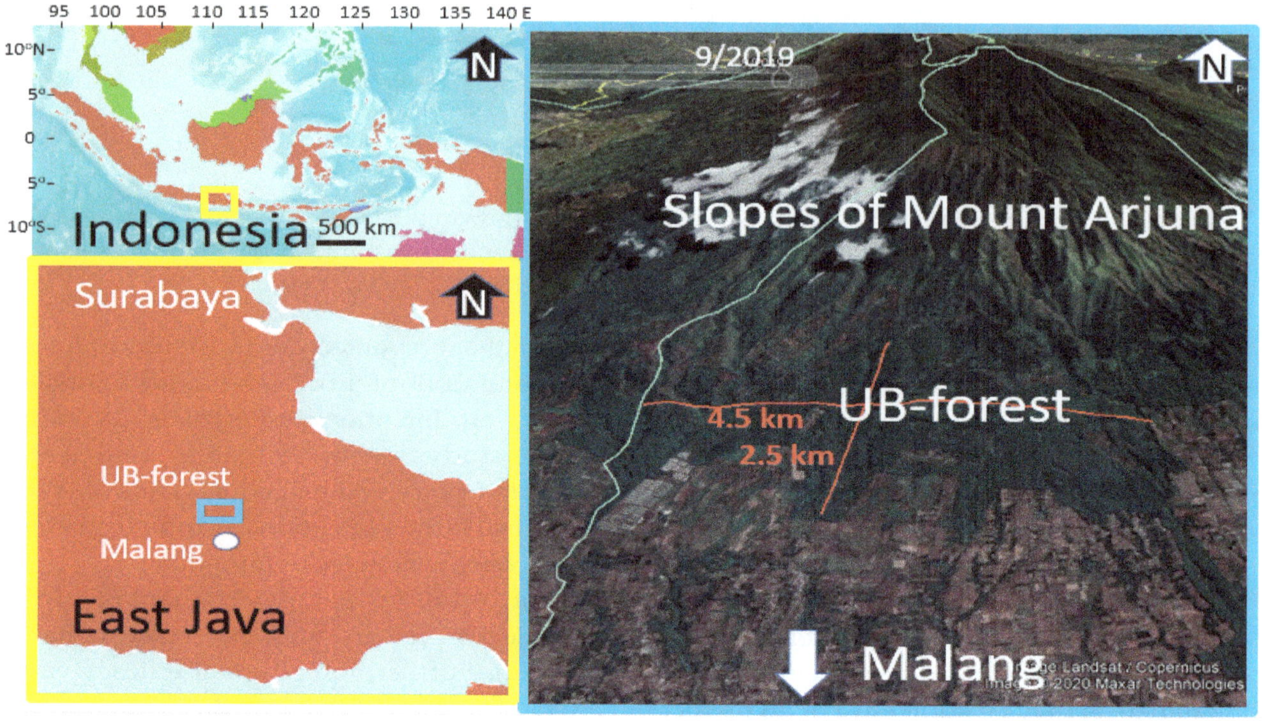

Figure 3. Research location UB Forest in East Java, Indonesia.

The hamlet is located in the Kalisari Watershed of Malang Regency, East Java. Data were collected in the period of March–May 2019. The senior researchers were those from the fields of agricultural extension/communication, soil science and agroforestry. Communication was established with a local key person to get wider access to the sites and respondents. Approval for this study was obtained from the management of UB-Forest and local regional authorities. Anonymity and voluntary participation of the participants were ensured. The research participants for the survey comprised 22 farmers out of an overall 30 within the hamlet, who were cultivating coffee under the pine. The research participants were verbally informed about the purpose, use and possible impacts of the research; in the local context written consent was not deemed to be required. Several members of the hamlet, including its leader, were involved in testing the innovation in the forest.

2.3. Research Design and Methods

The research design used the four stages of the action research approach (Figure 4): plan, act, observe and reflect [44]. It used mixed methods, during the observation stage combining quantitative (through a survey) and qualitative data collection (case study).

Plan—Preparations, site visit. Researchers from UB-UKCEH-ICRAF conducted site visits and a series of discussions under the umbrella of the agroforestry research group at UB to identify issues to be solved through action research. UB and UKCEH accessed funding through different research funding schemes to support the project.

Act—Intervention on-site and communicate with farmers. To overcome low coffee productivity under the pine trees, canopy pine trimming techniques were disseminated. This technique was considered new, as it had never been applied by local farmers. Strategies to communicate this innovation was by (1) "advocacy" through personal communication to a key farmer as the gatekeeper to gain trust; (2) "behavior development" through group meetings with farmers followed by a demonstration of pine trimming techniques on the site. The communication for the advocacy strategy was designed to ensure the key farmer understood the intended benefits of using the technique. Then, he acted as the "innovation endorser"—influencing other farmers to adopt the technique. For this reason, he was also asked to demonstrate the techniques to his peers.

Figure 4. Stages of action research for transferring the trimming techniques.

Observe—Collection data on sites/evaluate evidence. Farmer responses were measured using a mixed-methods sequential explanatory design [45]. This design covers two phases: (1) quantitative data collection through a "survey" of the coffee growers (executed March—April 2019); (2) qualitative data collection using a single instrument case study (May 2019) regarding the grower perceptions of the trimming technique in the context of UB Forest management. The qualitative data collection helped to get a deeper understanding of the quantitative database. This was especially important when dealing with extreme cases or demographic factors that may be associated with farmer intentional behaviors.

Reflect—Reflection was based on the research findings/evidence, which was applied through (1) individual/researcher evaluation; (2) discussions between researchers; and (3) stakeholder views through research finding seminar (involving resource person; a representative from university, governmental agents, NGOs, etc.); this is also a way to increase external validity. Subsequently, assertions/affirmation was conducted by clarifying the findings using the theory of planned behavior and other references and followed by giving options for further study and intervention approach.

2.4. Research Instruments

For the survey, a closed questionnaire (quantitative database) was administered to all participants (22 farmers), covering variables of intention that were (1) attitude toward trimming (6 indicators, e.g., covering statements such as "I believe that pine trimming ... is necessary because the litters can become mulch"; (2) subjective norms (8 indicators, e.g., "In my opinion ... local community supports trimming"; and (3) perceived behavioral control (9 indicators, e.g., "I am confident to trim, because ... it is easy to trim at 10 m high". The five-point Likert scale technique was employed: "strongly disagree" (score 1), "disagree" (2), "neutral" (3), "agree" (4) and "strongly agree" (5). The statements were based on theoretical concepts and adjusted with pine–coffee agroforestry issues as exposed by a key farmer (Appendix A); Additional questions were about farmers" demographic status (age, education, farming systems and experience, source of coffee farming information). The related questions for in-depth interviews can be seen in Appendix B. To reduce nonresponse error, respondents were interviewed face-to-face to overcome their unfamiliarity to fill out the questionnaire.

2.5. Data Validity

For the quantitative data, content validity and reliability tests were managed by experts, discussions with farmers and statistical tests. The steps to develop indicators for intentional aspects: (1) the variables were derived from published theories [46–48] and indicators arranged based on discussions with peers and leading researchers and (2) from extracting key concepts from key farmers during initial interview sessions. The validity of the statement items in the questionnaire was justified by members of the research team. Questions were then tested for clarity in a pilot study with 10 farmers following which sentences within the questionnaire were revised to increase participants" understanding. Statistically, item validity is measured by the correlation coefficient between an item's score and the total score of all items at a certain significant level (10%). The validity of the intention related items was tested with the SPSS statistical package resulting in a Pearson's correlation value (r) of > 0.36, indicating the items have high validity.

As a measure of the reliability of items, internal consistency was also tested. This technique examines the correlation between multiple items to measure a construct. The calculation of Cronbach's alpha Coefficient was rendered for item sections in the TBP questionnaire. The high instrument reliability was indicated by a coefficient of more than 0.80; any item coefficient of less than 0.30 need to be eliminated to avoid an unreliable indicator [49]. The Cronbach's alpha value was 0.85, indicating high reliability of the items.

2.6. Data Collection Methods

As a follow-up, after analyzing quantitative data, the researchers interviewed key farmers again to cross-check noteworthy quantitative data (such as the mode or extreme data) as a case for presentation (embedded analysis) [50]. Other techniques were used to triangulate: (1) nonparticipant observation (by observing the agroforestry setting/location); (2) member-checks" that is discussions with the key farmer and other farmers regarding the findings; (3) focus group discussions (FGD). The first FGD was conducted with 12 coffee farmers at the local meeting house to discuss the technical and social aspects of pine–based agroforestry cultivation. The second FGD was carried out in the local village meeting hall (*Balai desa*), by inviting key farmers (those who were the first to grow coffee or "the innovators") and coffee farmer representatives (those who planted coffee afterward or "the followers") and the stakeholders (representatives of the neighborhood and citizen associations, village officer, rural forest community member, village and sub-district officers, UB Forest employee, Perhutani/state-own forestry company agent and UB agroforestry research group members). In this activity, farmers were welcome to express their views on various issues, such as the relationship between the local community and the forest management authority, the history and quality of partnership, farmer economy and policy. These steps were useful to explore and describe sensitive issues and complexity of farmer livelihoods as the context which may influence their intentional behaviors. Field notes, camera imagery, and audio recordings were used to record the data. The data were recorded on a CD for storage only while useful for further intended research. Data are presented in an anonymized way to protect respondents.

2.7. Data Analysis

For the data analysis, descriptive statistics (frequencies, percentages, mean scores and standard deviations) were calculated to summarize the survey data. Separate tabulation was applied to capture qualitative data. The qualitative findings (statements, observation, photos) being used to validate and provide context to the qualitative results. Research findings were also presented in a series of seminars with members of the research team and stakeholders as part of the reflection and interpretation process.

3. Results

3.1. Social Context and the Demographic Factors

The agroforestry land in the UB-Forest area managed by the respondents of this study totaled 33.04 ha and the total area of UB-Forest is 544 ha. All respondent-farmers managed crops both in UB Forest 13.5 ha (pine-coffee agroforestry), but also in the state-owned forest (Perhutani) 18 ha (pine-vegetable agroforestry) and other forested areas 1.5 ha (pine-vegetables). Although, the survey data shows that the average of land areas managed by the respondent was relatively limited, at 1.2 ha (in Perhutani land); 0.7 ha (in UB Forest) and 0.5 ha (in farmer home gardens). The demographic data (Figure 5) show that 15 of 22 farmers (68%) have been farming within the agroforestry scheme for more than 20 years. Most respondents (41%) were in the >50-year age class and 77% of the farmers have had limited access to education being educated only to the elementary school level.

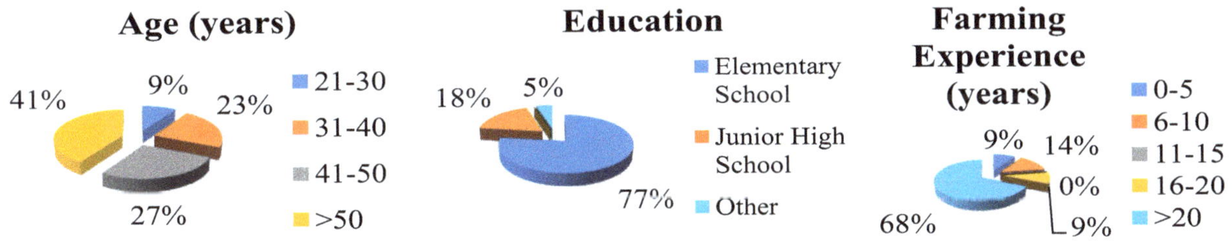

Figure 5. Demographic characteristics of respondents (n = 22).

Moreover, living in relatively remote areas (enclave) had become a constraint for their economic development. The results of an interview with a farmer revealed that in the past (especially before the reform era) the forest authorities did not allow farmers to grow coffee under the pine canopies, arguing that "forests cannot be turned into gardens." The coffee planted by farmers as an agroforestry system was the result of farmers" struggles to survive in these enclaved forest areas, despite causing conflicts with the forest authority at that time. In a lower level of preference, farmers also cultivate other types of crops, such as horticulture and food staples in sufficient quantities for their daily needs. The current use of chemical fertilizers from outside is still difficult as before, though they need it. One farmer stated, "apart from being far away from obtaining fertilizer, it is also expensive," indicating their high dependence on local efforts to obtain adequate coffee yields.

We asked the respondents about their experience in accessing the information on coffee cultivation. The data on Table 1 shows that 19 of 22 respondents (86%) had participated in extension activities discussing coffee cultivation. The sources of information were from public agricultural workers and/or from Perhutani personnel or experts from UB. Only 4 of 22 farmers (18%) received information on coffee cultivation from farmer groups and 7 of 22 (32%) received the information from other sources.

Table 1. Transfer of information on coffee cultivation (n = 22).

No	Source of Information	Number of Respondents Receiving the Information
1	Office of agriculture	19
2	State forest Company/Perhutani	19
3	University staff	19
4	Forum/farmer Group	4
5	Others	7

3.2. Attitudes toward the Trimming

As has been noted, the primary aim of this study was to measure the intention levels of farmers to adopt pine trimming techniques. We calculated the supporting variables: attitudes toward the trimming, subjective norms (social supports) and perceived behavioral control (obstacles in doing trimming) using a Likert scale from one (strongly disagree) to five (strongly agree). The study showed that 79% of respondents agree or strongly agree that pine trimming "provides benefits" for farmers (item #5; Table 2). The perceived benefits primarily related to the pine litters being transformed into mulch (item #1); and to a lesser extent benefits were expected to arise from fertilizing the soil, increasing light at the level of the coffee canopy or use of branches as firewood (items #1, #3 and #6). Moreover 83% of respondents agreed or strongly agreed that trimming can "increase coffee production" (#2).

Table 2. Attitudes towards pine trimming (n = 22), value based on a Likert scale from one (strongly disagree) to five (strongly agree).

No	Statements: I Believe that Pine Trimming ...	% "Agree" or "Strongly Agree"	Means (M) of Attitudes	Standard Deviation (Sd)
1	■ ... is necessary because the leaf litters can be made into mulch.	84.55	4.23	0.53
2	■ ... increases coffee production.	82.73	4.14	0.64
3	■ ... fertilizes the soil.	81.82	4.09	0.68
4	■ ... increases the incoming light.	81.82	4.09	0.87
5	■ ... provides benefits.	79.09	3.95	0.72
6	■ ... is useful because the branches can be used as firewood	78.18	3.91	0.68

It is worth noting that the mean score of "provides benefits" (#5) is slightly lower than most of the other indicators (except benefits of getting firewood), indicating the possibility that the farmers are not just considering the direct "benefits" to coffee yields, as is the objective of the question, but are instead considering the balance between benefits and risks (Table 2).

As seen in Table 2, there is a considerable variation in the perception scores. An in-depth interview with a farmer revealed some uncertainty as to the level of benefit of pine trimming, "If it is only trimmed, then it does not have much effect ... only the number of leaves on the coffee plants increases." This indicates that although considered generally useful, there is an uncertain view of the trimming technique.

3.3. *Subjective Norms of Trimming*

Farmers were asked about the level of social support related to pine trimming, which reflects the subjective norm variable (Table 3). When confronted with the statement "experts help farmers know how to trim through demonstration" (item #2), 81% of respondents said agree or strongly agree. This is consistent with the response to the statement "I understand the discussion about trimming (#1): 83% expressed agree or strongly agree. To a lesser extent, there were only 75% of respondents who agreed or strongly agreed that "hamlet personnel support trimming" (indicator #8). It is interesting to note that a relatively moderate response was given regarding the perceived support by UB-Forest management, both in terms of support for trimming and especially for land maintenance (indicators #5 and #7).

The findings show that respondents felt that they received support not only from UB-UKCEH-ICRAF personnel, but also from the local community and their families. On the other hand, the support from the UB-Forest management and local officers (hamlet personnel) were moderate or even considered limited. This is likely because of perceived problems of the newly revised coffee sharing mechanism and also possible latent conflicts in the community. Moreover, a complicated relationship was experienced by farmers with *Perhutani*, the former state forest management entity.

Table 3. Subjective norms/social supports of pine trimming (n = 22).

No	Statements: In My Opinion ...	%"Agree" and "Strongly Agree"	Means/M of Subjective Norms	Standard Deviation/Sd
1	■ ... I understand the discussion about trimming.	82.73	4.14	0.35
2	■ ... experts help farmers know how to trim through demonstration.	81.82	4.09	0.53
3	■ ... the local community supports trimming.	80.91	4.05	0.38
4	■ ... my family supports trimming.	80.91	4.05	0.58
5	■ ... UB Forest supports trimming.	79.09	3.95	0.58
6	■ ... discussions with friends encourage me to look for additional information about trimming.	78.18	3.91	0.81
7	■ ... UB Forest takes care farmers to take care of the agroforestry lands.	76.36	3.82	0.40
8	■ ... hamlet personnel support trimming	74.55	3.73	0.70

A key farmer expressed the historical context plainly:

"These coffee plants ... in [the year of] 96 was very difficult because at that time the Perhutani did not allow us to plant coffee under the shade ... under the stands of pine."

"The reason was that the forest cannot be used as a garden. However, I think because we live in a forest area ... the impression is that it destroys the forest, but we must earn a livelihood for our survival."

"For this reason, I tried to plant whatever crops I planted at that time. The plant that was able to grow and produce just coffee. It means that the people here are still looking for income in the forest area."

Hence, this study shows that farmers receive various levels of support from different parties. Relatively high support from the UB-UKCEH-ICRAF is probably because of the action of disseminating the trimming techniques through site demonstrations and discussions with farmers on some occasions.

3.4. Perceived Behavioral Control of Trimming

We noted that perceived behavioral control influences farmer intentions to undertake to trim. We use two primary variables to measure this self-efficacy (individual factors) and non-self-efficacy (other factors). The indicators of self-efficacy covering aspects of farmer confidence to trim because of expected improvements in their livelihoods, easiness of trimming techniques or suitability with local conditions, etc. Meanwhile, the indicators of non-self-efficacy include the availability of finance, time, information or other resources. The results are presented in Table 4.

When confronted with the statement "if funds are available" (item #1) and "the demonstration by experts encourages me to trim" (#2), 83% and 80% of respondents stated "agree"/A and "strongly agree"/SA, respectively. Furthermore, the number of respondents who strongly agreed or agreed however is reduced when responding to the statement "it helps improve my life" (#3); and "it is easy to understand" (#4).

Table 4. Perceived behavioral control/PBC of trimming (n = 22).

No	Statements "I Am Confident to Trim . . . "	%"Agree" and "Strongly Agree"	Means/M of PBC	Standard Deviation/ Sd
1	■ . . . if funds are available.	82.73	4.14	0.35
2	■ . . . because the demonstration by experts encourage me to trim.	80.00	4.00	0.76
3	■ . . . because it helps improve my life.	77.27	3.86	0.64
4	■ . . . because it is easy to understand.	76.36	3.82	0.85
5	■ . . . because I will do trimming in the farming land.	76.36	3.82	0.91
6	■ . . . because trimming is suitable with local condition.	75.45	3.77	0.69
7	■ . . . because it is easy to do.	70.00	3.50	1.10
8	■ . . . because it is cheap.	67.27	3.36	1.05
9	■ . . . it is easy to trim at 10 m high.	56.36	2.82	1.05

The more statements cover aspects of follow-up or practicality, the less the farmer's intention to do the trimming ("I will do trimming in the farming land": agree or strongly agree = 76). Less than 77% of respondents agree or strongly agree to practice the trimming. To exemplify suitability to local conditions (#6); availability of funds (#8) and time (#7). These responses are also consistent with a finding in our study, in which 75.45% of respondents stated "my time is limited to trim. "The lowest agreement is for the ease of trimming at a height of 10 m (#9), indicating the potential safety risk of this work (Figure 6). The key farmer highlighted this case:

"As you get older . . . you only dare [to go up] in part . . . If you go up maybe up to three to five trees you can still be brave, but not for the whole day . . . If more than that [five trees] I start to tremble . . . often slides. It means that the conditions are not possible to carry out all day . . . even though one hectare has more than 600 trees."

Figure 6. Demonstration of pine trimming by a key farmer.

Only a moderate number of farmers (76%) "will do trimming in the farming land" (#5, M = 3.82; Sd = 0.91). However, further analysis shows that proper extension methods may increase the farmer's intention to seek additional information—or apply trimming. Among these methods were interpersonal communication, such as discussions with friends on pine trimming (#6); especially discussions with the experts; and provision of demonstration plots (#2).

Last, but not least, there are strong indications that external factors were also key points to raise the level of farmer intention. The statement "if funds are available", then they will be confident to trim (A/SA = 83%). Similarly, when the trimming is associated with other external resources, such as local condition (#6), easiness to do (#7), money (#8) and potential risk (#9) (agree or strongly agree < 76%). These indicators suggest the conditions that may hamper the application of trimming, as expressed by a farmer: "To be honest, if I trim it, I'm not happy ... the pine remains small, but high. Aside from work safety, the risk is falling ... "*suloyo*" [local term for "sluggish, weak or helpless"]. Don't let it happen."

4. Discussion

The process to change individual behavior, commonly seen in an innovation/technological transfer, tends to pro-innovation bias [3]. The assumption is that new technology is better than the previous one. One of the problems is that for targeted individuals the nature of technology is perceived as complicated, as is evident in this study. Nevertheless, our findings have compiled evidence that shows trimming is complex from a farmer's perspective. An individual's intentions are also influenced by considerations of individual capacity, including technical complexity and resource availability to operate. This is particularly evident when such innovation is interfaced with marginal farmers, as in the case of pine–based agroforestry community in Java. We have used the theory of planned behavior as the lens to understand farmer intention to undertake the innovation of pine canopy trimming. We found that the value of the three factors of intention varied.

The attitude/behavioral beliefs—Measurements of the theme or aspect of attitude indicate that, to some extent, the economic benefit is an important element in forming farmers" attitudes towards pine trimming. However, income stability and production costs are also significant factors, regardless of plant productivity. The additional cost for trimming could be a sensitive factor as limitations of family labor and the age of farmers would necessitate the employment of outside labor to complete the trimming. The sensitivity of the extra costs for a small enterprise has long been identified as a sensitive factor for marginal farmers [50].

From the ecological perspective, trimming is perceived useful for (1) increasing light intensity for coffee plants; and (2) fertilizing and mulching the soil to retaining moisture. The literature shows the significant roles of mulch for improving microclimates for plant growth while suppressing weeds [51,52]. Retaining soil moisture is particularly valuable for dry land agroforestry such as in the UB Forest. Therefore, from the techno-ecological views, farmers are responsive to the innovation; their motivation is to increase productivity under the pine trees—vegetables, food crops, and especially coffee plants. The current pine canopy has suppressed coffee growth; the optimum light intensity should be 60–80% [53] but is lower in UB due to over-shading by the pine. This can result in the mortality of coffee tree branches and problems of shortening the harvesting period [54,55].

The subjective norms—We have noted that subjective norms are a belief to fulfill the social expectation. The support from the UB-UKCEH-ICRAF is shown in the trimming demonstration site to help farmers' awareness of and interest in the trimming. However, limited assistance from the UB-Forest management, possibly because of limited involvement at the early phase of the project, is a potential challenge although engagement with this UB-UKCEH-ICRAF project is now ongoing. Moreover, farmers are unhappy as their previous proposals to the forest management to get agricultural inputs were neglected, as expressed during an interview session. Farmers have never practiced trimming, perhaps because of the perceived strict rules on the treatment of the pine tree by the former and current forest management. Hence, the trimming technique itself may be simple in practice, but there are socio-psychological barriers due to the considerations of past rules.

Perceived behavioral control—As mentioned, self-efficacy is a belief to do something. We use this concept to measure farmer confidence to apply the trimming technique to predict the adoption in the future [56]. Generally, farmers are confident to trim because it may improve their livelihoods and due to the availability of information on trimming [57]. In our case, the information brought by the UB-UKCEH-ICRAF personnel is understood to dictate that trimming is only for the lower dry branches on the trees to a height of 10–12 m. However, in practice, farmers are uncomfortable trimming to 12 m. The data show that most farmers were seniors, which may have physical and health constraints. The old farmers were only able to trim three to five trees per day roughly, as revealed in the interview sessions; and only one to three trees for those above 7 m. Furthermore, farmers were unsure of their ability to perform trimming over a large area, with an area of one hectare containing 600 trees, regardless of their long experience of farming (over 20 years). In summary, it is easy for farmers to understand the trimming procedures, but not to practice it. Furthermore, the non-self-efficacy factor—that is the combination of externality aspects [33,58]—is also a significant consideration for intentions. We found that trimming is perceived to be expensive (when done by somebody else) and time-consuming; though having ecological benefits.

Hence, the trimming technique is perceived as complicated because of sound technical problems and limited supports from the stakeholders, regardless of the plant productivity and ecological benefits. Hence, the question is with the farmers" resource scarcity and social supports, will they be willing to apply the technique in the future? Discussions with the farmers revealed suggestions for the option of loans (from the forest management) to pay workers to trim. The other alternative is to using machinery to fasten the trimming process, which is probably cheaper, faster and safer. The use of machinery is something that researchers will investigate as part of any future forest management strategy, for example, telescopic pruners can reach 8 m in height. In-depth discussions reveal an interesting thing; farmers had proposed alternative techniques beyond trimming—such as pine thinning by removal of selected trees. The advantages proposed by farmers were (a) increasing light intensity for coffee plants (just as the trimming); (b) no need to add fertilizers; (c) no need to climb pine trees at extended heights; and (d) cheap and easy to do. A leading farmer experienced that without thinning, the soil fertility and strength of pine trees would decrease with time, as exemplified at a certain forest location (identified as plot #84); plant growth as stunted and at the risk of falling during heavy rains. He argued that it would be dangerous for individuals under the trees and considering that the forest is an also educational and recreational destination. He recommends that thinning needs to be accomplished every 10–15 years. The research team also heard of other farmers" innovative techniques on agroforestry management that are beyond our study at this stage.

5. Implications

Based on our experience at UB Forest, the trimming innovation has a prospect to be adopted, but with caution. On one side there is a sound theoretical basis that trimming can increase the productivity of pine–coffee agroforestry and improve the microclimate, suggesting the innovation has productivity-ecological benefits. However, it may be hampered by limited support and scarcity of farmer resources and conditions (such as old age), indicating issues on psycho-social burdens and practicality. The theory of planned behavior is a useful theoretical lens to identify the opportunity and obstacles of applying trimming techniques from the farmer's point of view. Furthermore, we have shared these issues with stakeholders (researchers, local government officers, UB forest management, key farmers, etc.) through a series of seminars. As a reflection, the attendance and engagement of the stakeholders including the government agents were high at every seminar and showing support to the research team and the findings of this study.

For the implications, the trimming program, which has been employed over two years, we have highlighted options: keep the innovation by (1) providing subsidy or loans for the farmers to hire workers to trim or buy trimming machinery, and/or (2) integrate with the farmer techniques. The first option has conditional acceptance because of perceived financial and technical burdens. The second option has a higher chance of acceptance and implementation by farmers, first, for one reason, it would better align with the farmer's needs and capabilities. For other reasons, it would also enhance collaboration between researchers and farmers. Both options, however, have a challenge of how to cooperate with the stakeholders. For this reason, the plan and knowledge learned from action research need to be shared with the stakeholder from the beginning of any agroforestry innovation development. To reflect, this experience has improved the understanding of researchers and the collaborative options that would enhance the capacities for all the social actors in the agroforestry system. Based on this research we would encourage the application of the second option in the future.

We learn that technical constraints (the majority were senior farmers and low income) and lack of optimal external support may be important limitations in the future if there are no specific interventions. The new intervention should be not only on technical assistance, but also the sharing of information or knowledge and collaborations between stakeholders. To sustain, the communicative intervention processes require a participatory and integrative approach. Co-management of agroforestry innovations may be developed throughout the stages, from the creation, pilot/local implementation and scaling up. We need to explore any aspirations and local knowledge, contemplating the specific demographic conditions and needs of the marginal farmers at the UB Forest and the demand for stakeholder integration. Our experience may be useful as a lesson learned for policymakers and executives to comprehend the social-ecological dynamic of agroforestry management, let alone the Javanese farmers who reside close to forest lands.

As a reflection, this action research is useful, but needs improvement. We notice that the process of transferring agroforestry innovation cannot only consider innovation and farmers, but also the forest management authority. The key to accelerating the innovation transfer process is the use of an integrated communication and extension strategy. For this reason, it is necessary to establish connections between external scientists as initiators of innovation, farmers as practitioners and relevant authorities [59,60], in the case of our study is the UB Forest management. Moreover, a study shows that farmers" knowledge is often overlooked in the transfer of innovation [61,62]. In our case, farmers can observe biophysical problems in the context of their socioeconomic complexity. A farmer revealed:

> "*Regarding the soil ... the farmers should fertilize it every six months ... the nutrients have been reduced because [it has been absorbed by] pine roots. I have tried like that [fertilizing], and then I calculated the maintenance costs. I calculated and the results were even. Finally, I let it go, I did not care—'simalakama'!*" (emphasis added).

Simalakama is the name of a bitter fruit for traditional medicinal ingredients. People usually say, "as if eating simalakama fruits," a local parable to express a dilemma when someone wants to do something or not.

More than this, farmers—through various considerations—have their rationality and solution preference. The results of in-depth interviews and FGDs show that outside of trimming, farmers have crop management preferences through thinning techniques. After 10 to 15 years, it is believed that pine plants need to be sparsely spaced, with a spacing of twice the current one (currently in 2 × 3 m). Farmers recommend a 2 × 6 m spacing for pine trees because with more incoming sunlight, then—according to farmer experience—there are simultaneous benefits. The advantages for productivity and ecology are: "soil fertility is increasing"; "reduced acidity"; "better coffee plant growth"; "bigger plant"; "sturdy, when exposed to the wind the plants will be stronger" and "groundwater storage is better". For social benefits, farmers framed it as: "both [parties] can get [benefits]; forestry gets, the community gets too." Other techniques, vice versa, jargonized as "*Sana untung, sini buntung*" [the party over there gets profits, the community is in disadvantaged]" (results of in-depth interviews and an FGD with farmers).

Consequently, the transfer of innovation does not only come from scientists focused on productivity-ecological aspects [63], but also those embracing the aspirations of farmers and the interest of powerholders. Therefore, knowledge of practical and political interests is needed simultaneously. This means that various modes of innovation transfer can be combined: from scientists (linear, top-down), farmers" needs (bottom-up) and interests from politics/UB Forest (lateral). Since its inception, bidirectional extension and communication strategies [64] or participatory strategies need to be developed although previous studies have shown that it is not easy for change agents to do so [6]. Included in the method action research is the need to apply transdisciplinary research, where farmers" local knowledge needs to be examined, confirmed and incorporated with those from scientific knowledge [65]—as an agroforestry knowledge pool. To sum up, efforts to balance the distinct roles of the tripartite agents for profit—prove-it—prophets may be more easily achieved through the participatory-collaborative approach [56,66].

Specific types of land management contracts between forest authorities and farmers looking for land to grow their food crops became known in 19th Century southeast Asian teak growing traditions under the name "taungya"—originally referring to indigenous swidden-fallow practices. The current pine–coffee management contracts represent considerable development since the earlier schemes [14,67], but the unequal power distribution of the past is still loading the dice against schemes that are both fair and efficient.

6. Conclusions

Based on the study, the adoption of the trimming innovation is prospective, though it needs adaptations. On one hand, on some levels trimming is believed to increase the productivity of pine–coffee agroforestry and maintain the microclimate, suggesting the productivity-ecological benefits of the innovation. However, this may be hampered by limited support and scarcity of farmer resources and conditions, indicating issues on psychosocial burdens and practicality. We have shared these issues with stakeholders (researchers, local government officers, UB forest management, key farmers, etc.) through a series of seminars. Practically, the trimming program that has been managed for two years, has two main options: keep the innovation by providing subsidy or loans for the farmers, and/or integrate with farmer techniques. Further options for innovation are necessary, including direct pruning of coffee and crop fertilization strategies, some of which are being tested now.

We learned that technical constraints (most senior farmers and low income) and lack of optimal external support may be an important limitation in the future if there are no specific interventions. a new intervention should be not only on technical assistance, but also the sharing of information or knowledge and collaborations between stakeholders. Sustaining the communicative intervention processes requires a participatory and integrative approach. Co-management of agroforestry innovations may be developed throughout the stages, from the creation, local implementation and scaling-up. This approach is crucial because stakeholders have roles and specific rights or interest in making decisions. Moreover, we had limited initiatives from farmers to be integrated with agroforestry innovation development. For these reasons, we need to explore more any local knowledge and practices to meet with the specific conditions and needs of the marginal farmers and their unique relations with forest management. Considering that this study had only a relatively small number of participants, a larger number of participants is needed to increase the generalizability of the results. However, our experience may be useful as a lesson learned for policymakers and executives to comprehend the social-ecological dynamic of agroforestry management, as well as the Javanese farmers who reside adjacent to forest lands.

Author Contributions: E.D.C.; M.v.N.; N.M. and R.R. conceptualized the study. E.D.C. designed the study, while S.F. and D.W. collected quantitative data; E.D.C. gathered qualitative data as well as acted as their undergraduate supervisor. Data curation and graph was prepared by E.P., D.W. and S.F. Then, E.D.C.; and M.v.N. assisted by E.P.; S.F.; and D.W. wrote the original manuscript preparation. M.v.N.; N.P.M.; R.L.R. and E.D.C. reviewed, added and edited the final manuscript. All authors have read and agreed to the published version of the manuscript.

Acknowledgments: Acknowledgment is given to the agroforestry group members from the Sumbersari community at UB Forest for their generosity in participating in this study and to the stakeholders of various institutions in their involvement during research result seminars.

Appendix A. Survey Instruments Used for Measuring Farmer Intention

NO.	Could you please respond (V) to the following statements with one of the answer options?	Answer options: 1 = strongly disagree; 2 = disagree; 3 = do not know (neutral); 4 = agree; 5 = strongly agree				
		1	2	3	4	5
	I believe that pine trimming . . .					
1.	. . . provides benefits.					
2.	. . . increases the incoming light.					
3.	. . . fertilizes the soil.					
4.	. . . is necessary because the leaf litters can be made into mulch.					
5.	. . . is useful because the branches can be used as firewood.					
6.	. . . increases coffee production.					
	In my opinion . . .					
7.	. . . I understand the discussion about trimming.					
8.	. . . experts help farmers know how to trim through demonstration.					
9.	. . . my family supports trimming.					
10.	. . . discussions with friends encourage me to look for additional information about trimming.					
11.	. . . the local community supports trimming.					
12.	. . . hamlet personnel support trimming.					
13.	UB Forest takes care farmers to take care of the agroforestry lands.					
14.	UB Forest supports trimming.					
	I am confident to trim, because . . .					
15.	. . . it is easy to understand.					
16.	. . . it is easy to do.					
17.	. . . it is easy to trim at 10 m high.					
18.	. . . it is cheap.					
19.	. . .					
20.	. . .					
21.	. . . it helps improve my life.					
22.	. . . the demonstration by experts encourage me to trim.					
23.	I will do trimming in the farming land.					

Appendix B. Guideline Questions for In-Depth Interviews

1. What types of plants or crops do you grow under the pine shade?
2. What are the responses from the forest management when you grow coffee under this shade?
3. Give your thoughts on the pine trimming technique recommended by the experts.
4. Please describe your experience when trimming pine.
5. How do you feel when trimming?
6. What obstacles do you face when you want to do trimming?
7. Do you have any experiences other than trimming to increase your coffee production?
8. What are the advantages of these alternative methods?

References

1. Padel, S. Conversion to organic farming: A typical example of the diffusion of an innovation? *Sociologiaruralis* **2001**, *41*, 40–61. [CrossRef]
2. Sveiby, K.E.; Gripenberg, P.; Segercrantz, B.; Eriksson, A.; Aminoff, A. Unintended and undesirable consequences of innovation. In Proceedings of the XX ISPIM Conference, The Future of Innovation, Vienna, Austria, 21–24 June 2009.
3. Rogers, E.M. *Diffusion of Innovations*; Simon and Schuster: New York, NY, USA, 2010.
4. Costantini, V.; Liberati, P. Technology transfer, institutions and development. *Technol. Forecast. Soc. Chang.* **2014**, *88*, 26–48. [CrossRef]
5. Pannell, D.J.; Marshall, G.R.; Barr, N.; Curtis, A.; Vanclay, F.; Wilkinson, R. Understanding and promoting adoption of conservation practices by rural landholders. *Aust. J. Exp. Agric.* **2006**, *46*, 1407–1424. [CrossRef]
6. Cahyono, E.D.; Agunga, R. Policy and practice of participatory extension in Indonesia: A case study of extension agents in Malang District, East Java Province. *J. Int. Agric. Ext. Educ. (JIAEE)* **2016**, *23*, 38–57. [CrossRef]
7. Agunga, A.; Cahyono, E.D.; Buck, E.; Scheer, S. Challenges of Implementing Participatory Extension in Indonesia. *J. Commun. Media Res.* **2016**, *8*, 20–45.
8. Cahyono, E.D. Persepsi Dan Tipe Keputusan Petani Madura Terhadap Inovasi Agens Hayati-PGPR. In *Proceeding Konferensi Nasional Penyuluhan&Komunikasi Pembangunan*; Cahyono, E.D., Ed.; Fakultas Pertanian Universitas Brawijaya: Malang, Indonesia, 2016; pp. 284–292.
9. Hofstede, G.J. GRASP agents: Social first, intelligent later. *AI Soc.* **2019**, *34*, 535–543. [CrossRef]
10. van Noordwijk, M. *Sustainable Development through Trees on Farms: Agroforestry in Its Fifth Decade*; World Agroforestry (ICRAF): Bogor, Indonesia, 2019.
11. Hawkins, D.G.; Lake, D.A.; Nielson, D.L.; Tierney, M.J. Delegation Under Anarchy: States, International Organizations, and Principal-Agent Theory. In *Delegation and Agency in International Organizations*; Cambridge University Press: New York, NY, USA, 2006; p. 21.
12. Eisenhardt, K.M. Agency theory: An assessment and review. *Acad. Manag. Rev.* **1989**, *14*, 57–74. [CrossRef]
13. Peluso, N.L. 'Traditions' of forest control in Java: Implications for social forestry and sustainability. *Nat. Resour. J.* **1992**, *32*, 883–918. [CrossRef]
14. Van Noordwijk, M.; Tomich, T.P. Agroforestry Technologies for Social Forestry: Tree-Crop Interactions and Forestry-Farmer Conflicts. In *Social Forestry and Sustainable Forest Management*; Hasanu, S.H., Sabarnurdin, S., Sumardi, HeruIswantoro, Eds.; PerumPerhutani: Jakarta, Indonesia, 1995; pp. 168–193.
15. Djamhuri, T.L. Community participation in a social forestry program in Central Java, Indonesia: The effect of incentive structure and social capital. *Agrofor. Syst.* **2008**, *74*, 83–96. [CrossRef]
16. Sunderland, T.C.H.; Deakin, E.L.; Kshatriya, M. Conclusions: Agrarian Change—A Change for the Better? In *Agrarian Change in Tropical Landscapes*; Center for International Forestry Research (CIFOR): Bogor, Indonesia, 2016.
17. McCarthy, J.F.; Cramb, R.A. Policy narratives, landholder engagement, and oil palm expansion on the Malaysian and Indonesian frontiers. *Geogr. J.* **2009**, *175*, 112–123. [CrossRef]
18. Sanders, A.J.; Ford, R.M.; Mulyani, L.; Larson, A.M.; Jagau, Y.; Keenan, R.J. Unrelenting games: Multiple negotiations and landscape transformations in the tropical peatlands of Central Kalimantan, Indonesia. *World Dev.* **2019**, *117*, 196–210. [CrossRef]
19. Potter, L.; Lee, J. *Tree planting in Indonesia: Trends, Impacts and Directions*; CIFOR Occasional Paper No. 18; CIFOR: Bogor, Indonesia, 1998; 85p.
20. Bachriadi, D.; Sardjono, M.A. Conversion or Occupation?: The Possibility of Returning Local Communities' Control Over Forest Lands in Indonesia. In *Luce Project on Green Governance Fellowship Paper*; University of California: Berkeley, CA, USA, 2005.
21. Siscawati, M.; Banjade, M.R.; Liswanti, N.; Herawati, T.; Mwangi, E.; Wulandari, C.; Tjoa, M.; Silaya, T. *Overview of Forest Tenure Reforms in Indonesia*; Working Paper 223; CIFOR: Bogor, Indonesia, 2017. [CrossRef]
22. Jerneck, A.; Olsson, L. More than trees! Understanding the agroforestry adoption gap in subsistence agriculture: Insights from narrative walks in Kenya. *J. Rural Stud.* **2013**, *32*, 114–125. [CrossRef]
23. Place, F.; Ajayi, O.C.; Torquebiau, E.; Detlefsen Rivera, G.; Gauthier, M.; Buttoud, G. Improved Policies for Facilitating the Adoption of Agroforestry. In *Agroforestry for Biodiversity and Ecosystem Services—Science and Practice*; Kaonga, M.L., Ed.; Intech Open: Rijeka, Croatia, 2013; pp. 113–128. [CrossRef]

24. Sahide, M.A.K.; Fisher, M.R.; Supratman, S.; Yusran, Y.; Pratama, A.A.; Maryudi, A.; Kim, Y.S. Prophets and profits in Indonesia's social forestry partnership schemes: Introducing a sequential power analysis. *For. Policy Econ.* **2020**, *115*, 102160. [CrossRef]
25. van Noordwijk, M. Prophets, Profits, Prove It: Social Forestry under Pressure. *One Earth* **2020**, *2*, 394–397. [CrossRef]
26. Rival, L. From Global Forest Governance to Privatised Social Forestry: Company–Community Partnerships in the Ecuadorian Choco. In *Privatising Development*; Martinus Nijhoff Publishers: Leiden, The Netherlands, 2005; pp. 253–270.
27. Pham, T.T.; Campbell, B.M.; Garnett, S.; Aslin, H.; Hoang, M.H. Importance and impacts of intermediary boundary organizations in facilitating payment for environmental services in Vietnam. *Environ. Conserv.* **2010**, *37*, 64–72. [CrossRef]
28. Taylor, P.L. Conservation, community, and culture? New organizational challenges of community forest concessions in the Maya Biosphere Reserve of Guatemala. *J. Rural Stud.* **2010**, *26*, 173–184. [CrossRef]
29. Wang, J.; Chu, M.; Deng, Y.; Lam, H.; Tang, J. Determinants of pesticide application: An empirical analysis with theory of planned behavior. *China Agric. Econ. Rev.* **2017**, *10*, 608–625. [CrossRef]
30. Satsios, N.; Hadjidakis, S. Applying the Theory of Planned Behaviour (TPB) in Saving Behaviour of Pomak Households. *Int. J. Financ. Res.* **2018**, *9*, 122–133. [CrossRef]
31. Despotović, J.; Rodić, V.; Caracciolo, F. Factors affecting farmers' adoption of integrated pest management in Serbia: An application of the theory of planned behavior. *J. Clean. Prod.* **2019**, *228*, 1196–1205. [CrossRef]
32. Ajzen, I. *Attitudes, Personality, and Behavior*; Open University Press: Maidenhead, UK, 2005.
33. Ramdhani, N.; Carver, C.S.; Scheier, M.F.; Segerstrom, S.C.; Solberg Nes, L.; Evans, D.R.; Khatimah, H. Penyusunanalatpengukurberbasis theory of planned behavior. *BuletinPsikologi* **2010**, *4*, 55–69. [CrossRef]
34. Shin, Y.H.; Hancer, M. The role of attitude, subjective norm, perceived behavioral control, and moral norm in the intention to purchase local food products. *J. Foodserv. Bus. Res.* **2016**, *19*, 338–351. [CrossRef]
35. Zeweld, W.; Van Huylenbroeck, G.; Tesfay, G.; Speelman, S. Smallholder farmers' behavioural intentions towards sustainable agricultural practices. *J. Environ. Manag.* **2017**, *187*, 71–81. [CrossRef] [PubMed]
36. Popa, B.; Niță, M.D.; Hălălișan, A.F. Intentions to engage in forest law enforcement in Romania: An application of the theory of planned behavior. *Forest Policy Econ.* **2019**, *100*, 33–43. [CrossRef]
37. Kahmir, H.R.; Muharam, A. Sifat Inovasi dan Aplikasi Teknologi Pengelolaan Terpadu Kebun Jeruk Sehat dalam Pengembangan Agribisnis Jeruk di Kabupaten Sambas, Kalimantan Barat. *J. Hortik.* **2008**, *18*, 477–490. Available online: http://ejurnal.litbang.pertanian.go.id/index.php/jhort/article/viewFile/899/743 (accessed on 22 September 2020).
38. Meijer, S.S.; Catacutan, D.; Ajayi, O.C.; Sileshi, G.W.; Nieuwenhuis, M. The role of knowledge, attitudes and perceptions in the uptake of agricultural and agroforestry innovations among smallholder farmers in sub-Saharan Africa. *Int. J. Agric. Sustain.* **2015**, *13*, 40–54. [CrossRef]
39. Zolkepli, I.A.; Kamarulzaman, Y. Social media adoption: The role of media needs and innovation characteristics. *Comput. Human Behav.* **2015**, *43*, 189–209. [CrossRef]
40. Malahayatin, D.M.; Cahyono, E.D. Faktor kesesuaian dengan kebutuhan petani dalam keputusan adopsi inovasi pola tanam jajar legowo (Studi kasus petani padi di Kecamatan Widang, Kabupaten Tuban). *J. Ekon. Pertan. Agribisnis* **2017**, *1*, 56–61. [CrossRef]
41. Fishbein, M.; Ajzen, I. *Intention and Behavior: An Introduction to Theory and Research*; Addison-Wesley: Reading, MA, USA, 1975.
42. Rezaei, R.; Safa, L.; Damalas, C.A.; Ganjkhanloo, M.M. Drivers of farmers' intention to use integrated pest management: Integrating theory of planned behavior and norm activation model. *J. Environ. Manag.* **2019**, *236*, 328–339. [CrossRef]
43. Sheppard, B.H.; Hartwick, J.; Warshaw, P.R. The theory of reasoned action: A meta-analysis of past research with recommendations for modifications and future research. *J. Consum. Res.* **1988**, *15*, 325–343. [CrossRef]
44. Burns, A. Action Research. In *Qualitative Research in Applied Linguistics*; Palgrave Macmillan: London, UK, 2009; pp. 112–134.
45. Creswell, J.W.; Creswell, J.D. *Research Design: Qualitative, Quantitative, and Mixed Methods Approaches*; Sage Publications: Newbury Park, CA, USA, 2017.
46. Ajzen, I. From Intentions to Actions: A Theory of Planned Behavior. In *Action Control*; Springer: Berlin/Heidelberg, Germany, 1982; pp. 11–39.

47. Notani, A.S. Moderators of perceived behavioral control's predictiveness in the theory of planned behavior: A meta-analysis. *J. Consum. Psychol.* **1998**, *7*, 247–271. [CrossRef]
48. Sussman, R.; Gifford, R. Causality in the theory of planned behavior. *Personal. Soc. Psychol. Bull.* **2019**, *45*, 920–933. [CrossRef] [PubMed]
49. De Vaus, D. *Surveys in Social Research*; Routledge: London, UK, 2013.
50. Taneo, J.M.; Cahyono, E.D. Analisis factor social ekonomis adopsi teknologi pupuk urea tablet. *Wawasan* **1999**, *7*, 8–16.
51. Mayun, I.A. Efek mulsa jerami padi dan pupuk kendang sapi terhadap pertumbuhan dan hasil bawang merah di daerah pesisir. *Agritrop* **2007**, *26*, 33–40.
52. Kader, M.A.; Senge, M.; Mojid, M.A.; Ito, K. Recent advances in mulching materials and methods for modifying soil environment. *Soil Tillage Res.* **2017**, *168*, 155–166. [CrossRef]
53. Utomo, S.B. *Dinamika Suhu Udara Siang Malam Terhadap Foto Respirasi Fase Generative Kopi Robusta di Bawah Naungan Yang Berbeda Pada Sistem Agroforestry*; Tesis: Universitas Jember. 2011. Available online: http://digilib.unej.ac.id (accessed on 3 November 2019).
54. Bote, A.D.; Struik, P.C. Effects of shade on growth, production and quality of coffee (*Coffea arabica*) in Ethiopia. *J. Hortic. For.* **2011**, *3*, 336–341.
55. Daras, U.; Sobari, I. *Pemangkasan Tanaman Kopi Dan Pemeliharaan Pohon Penaung*; Balai Penelitian Tanaman Industri dan Penyegar: Sukabumi, Indonesia, 2018.
56. McGinty, M.M.; Swisher, M.E.; Alavalapati, J. Agroforestry adoption and maintenance: Self-efficacy, attitudes and socio-economic factors. *Agroforest. Syst.* **2008**, *73*, 99–108. [CrossRef]
57. Binam, J.N.; Place, F.; Djalal, A.A.; Kalinganire, A. Effects of local institutions on the adoption of agroforestry innovations: Evidence of farmer managed natural regeneration and its implications for rural livelihoods in the Sahel. *Agric. Food Econ.* **2017**, *5*, 2. [CrossRef]
58. Sparks, P.; Guthrie, C.A.; Shepherd, R. The dimensional structure of the perceived behavioral control construct. *J. Appl. Soc. Psychol.* **1997**, *27*, 418–438. [CrossRef]
59. Sokolovska, N.; Fecher, B.; Wagner, G.G. Communication on the Science-Policy Interface: An Overview of Conceptual Models. *Publications* **2019**, *7*, 64. [CrossRef]
60. Böcher, M. Advanced approaches for a better understanding of scientific knowledge transfer in forest and forest-related policy. *For. Policy Econ.* **2020**, *114*, 102165. [CrossRef]
61. Cahyono, E.D. Participatory Communication and Extension for Indigenous Farmers: Empowering Local Paddy Rice Growers in East Java. In *Communicating for Social Change*; Palgrave Macmillan: Singapore, 2019; pp. 213–233.
62. Kankeu, R.S.; Demaze, M.T.; Krott, M.; Sonwa, D.J.; Ongolo, S. Governing knowledge transfer for deforestation monitoring: Insights from REDD+ projects in the Congo Basin region. *For. Policy Econ.* **2020**, *111*, 102081. [CrossRef]
63. Do, T.H.; Krott, M.; Böcher, M. Multiple traps of scientific knowledge transfer: Comparative case studies based on the RIU model from Vietnam, Germany, Indonesia, Japan, and Sweden. *For. Policy Econ.* **2020**, *114*, 102134. [CrossRef]
64. Roux, D.J.; Rogers, K.H.; Biggs, H.C.; Ashton, P.J.; Sergeant, A. Bridging the science–management divide: Moving from unidirectional knowledge transfer to knowledge interfacing and sharing. *Ecol. Soc.* **2006**, *11*, 4. [CrossRef]
65. Cahyono, E.D. *Inovasi Budidaya Agroforestri Kopi-Pinus di UB Forest: Intensi Petani Untuk Mengadopsi Teknik Trimming (Pemangkasan) Cabang Pinus Yang Diintroduksikan Oleh Kolaborasi UB-CEH*; Fakultas Pertanian Universitas Brawijaya: Malang, Indonesia, 2019.
66. Haggar, J.; Ayala, A.; Díaz, B.; Reyes, C.U. Participatory design of agroforestry systems: Developing farmer participatory research methods in Mexico. *Dev. Pract.* **2001**, *11*, 417–424. [CrossRef]
67. Smiet, A.C. Forest ecology on Java: Conversion and usage in a historical perspective. *J. Trop. For. Sci.* **1990**, *2*, 286–302.

9

Cost-Benefit Analysis of Landscape Restoration: A Stocktake

Priscilla Wainaina *, Peter A. Minang, Eunice Gituku and Lalisa Duguma

World Agroforestry Centre, UN Avenue, Gigiri, Nairobi 00100, Kenya; p.minang@cgiar.org (P.A.M.); E.Gituku@cgiar.org (E.G.); l.duguma@cgiar.org (L.D.)

* Correspondence: P.Wainaina@cgiar.org

Abstract: With the increase in demand for landscape restoration and the limited resources available, there is need for economic analysis of landscape restoration to help prioritize investment of the resources. Cost-benefit analysis (CBA) is a commonly applied tool in the economic analysis of landscape restoration, yet its application seems limited and varied. We undertake a review of CBA applications to understand the breadth, depth, and gaps. Of the 2056 studies identified in literature search, only 31 met our predefined criteria. Three studies offered a global perspective, while more than half were conducted in Africa. Only six countries benefit from at least 2 CBA studies, including Brazil, Ethiopia, Kenya, Vietnam, South Africa, and Tanzania. About 60% focus on agroforestry, afforestation, reforestation, and assisted natural regeneration practices. Only 16% covered all cost categories, with opportunity costs being the least covered. Eighty-four percent apply direct use values, while only 16% captured the non-use values. Similarly, lack of reliable data due to predictions and assumptions involved in data generation influenced CBA results. The limited number of eligible studies and the weaknesses identified hereinabove suggest strong need for improvements in both the quantity and quality of CBA to better inform planning, policies, and investments in landscape restoration.

Keywords: cost-benefit analysis; landscape restoration; global; stocktake

---

1. Introduction

There is a growing demand for restoration globally aimed at stopping further degradation and reversing degradation. Forest degradation, soil erosion, peatland, wetland drainage, and salinization have been the leading causes of land degradation globally over the past 50 years. According to the Intergovernmental Science-Policy Platform on Biodiversity and Ecosystem Services [1], 75% of the global land surface is significantly degraded, 66% of the ocean area is experiencing increasing cumulative impacts, and over 85% of wetlands have been lost. This has affected almost a quarter of the world's total land area, and the damage is felt through the loss of ecosystem goods and services. The damage costs the world an estimated USD 6.3 trillion a year in lost ecosystem service value, which includes climate regulation, clean air, recreational opportunities, freshwater, and fertile soils [2]. Besides, due to land degradation, the livelihoods of about half a billion people, especially poor people who depend on agricultural and forestlands, are jeopardized. Decreasing land productivity threatens water and food security, destabilizes sustainable development, and results in civil conflicts and human migration.

Global efforts have been put in place to stop further degradation and reverse degradation through restoration of forest and landscapes ecosystems. Over the last decade, there have been global restoration initiatives, including 'The Bonn Challenge' and 'The New York Declaration on Forests (NYDF)'. The Bonn Challenge is a global effort to bring 150 million hectares of the world's deforested and degraded land into restoration by 2020, and 350 million hectares by 2030[2], while the NYDF is a political declaration among governments, companies, indigenous peoples, and civil society to take action to halve the loss of natural forests by 2020 and halt it by 2030[1]. Within Africa, the African Forest Landscape Restoration Initiative, AFR100 aims to restore 100 million hectares of land in Africa by 2030[3]. Specific countries have also set definite restoration goals; for example, Kenya has committed to restoring 5.1 million hectares of degraded land[4], and Malawi committed to about 4.5 million hectares of land by 2030 [3].

Restoration is generally a costly undertaking, partly because it often begins after the environmental degradation is well-advanced and expensive to reverse, and is often labor and resource-intensive [4]. Funding sources for investment in landscape restoration include (1) Private finance, where capital is managed mainly to earn a financial return for the investor. (2) Public finance where funding comes from the government bodies. Public finance can further be divided into international donor support and domestic public expenditure. In public finance, public investments are largely made to generate economic, environmental and social benefits for the public. However, there may be a return to the government. (3) Philanthropic finance, which is charitable giving by individuals or organizations, typically with no intention of earning a financial return [5].

Numerous environmental, economic and social benefits are generated when degraded lands are restored. These benefits may range from conservation of biodiversity, creation of jobs, improvement in agricultural productivity, and so on. Despite the numerous benefits accruing from restoration, funding for landscape restoration falls short by about USD 300 billion a year [2]. Investment is inadequate for several key reasons, among them: (1) many of the benefits are public goods, which are difficult to monetize; (2) the long-term nature of investments does not match investors' desire for liquidity, i.e., restoration usually requires large investments upfront and has long lags before generating benefits, and (3) landscape projects are perceived to be risky. Overall, landscape restoration activities are often misunderstood as involving high up-front costs and low rates of return, and these ideas persist because few evaluations of restoration activities include a comprehensive and objective accounting of restoration's ecological and economic impacts. Economic analysis can encourage investment in restoration by clearly laying out the benefits and costs of restoration projects and their distribution among stakeholders. It is also instrumental in prioritizing scarce resources accordingly.

Cost-benefit analysis (CBA) is the commonly applied approach in the economic analysis of landscape restoration. However, compared to the significant number of landscape restoration projects and studies globally, relatively few CBA studies have been conducted on restoration projects. Reference [6] conducted a review of the rehabilitation of degraded dryland ecosystems. They found that, while numerous studies have been undertaken on restoration and rehabilitation of degraded ecosystems, there remains a gap in cost-benefit analysis of these interventions. They largely attributed this to missing data on the benefits and costs accruing from the restoration projects. Similarly, in a review of more than 2000 restoration case studies, the "The Economics of Ecosystems and Biodiversity (TEEB)[5] "2009 study found that less than 5% of the case studies provided meaningful cost data, and none provided an analysis of both costs and benefits [7].

While CBA continues to be a primary approach of economic assessment, adequate use of the tool requires a clear understanding of its limitations and pitfalls. Thus, the paper seeks to understand the breadth and depth of existing CBA studies. It also highlights gaps in the existing CBA studies on landscape restoration and how further studies can address these gaps.

---

2 www.nydfglobalplatform.org
1 www.bonnchallenge.org
3 http://afr100.org/
4 https://afr100.org/content/kenya
5 http://teebweb.org/

2. Background

2.1. Brief Background on Landscape Restoration

Ecological restoration is the process of assisting the recovery of an ecosystem that has been degraded, damaged, or destroyed [8]. The goal is to repair the ecosystem to its integrity and its health and to reverse land degradation, increase the resilience of biodiversity, and deliver important ecosystem services [9,10]. Ecological restoration also includes forest and landscapes restoration, which is a process that aims to regain ecological integrity and enhance human well-being in deforested or degraded forest landscapes. A key feature of forest and landscapes restoration is that a combination of forest and non-forest ecosystems, land uses, and restoration approaches can be accommodated within a landscape to achieve sustainable food production, ecosystem services provisioning, and biodiversity conservation [11].

Landscape restoration is not just a matter of planting trees but also involves assisting the recovery of a damaged or destroyed ecosystem. The restoration activities range from small local activities carried out by individuals or community groups through to regional, country, and even global-scale activities involving multiple agencies and large numbers of people [12]. Restoration can either be achieved through natural regeneration or active restoration. For example, natural forest regeneration is the spontaneous recovery of native tree species that colonize and establish in abandoned fields or natural disturbances. This process can also be assisted natural regeneration, whereby it is assisted through human interventions, such as fencing to control livestock grazing, weed control, and fire protection [13]. Active restoration, on the other hand, involves human involvement in a range of ways and involves a considerable cost in terms of labor and time [14]. It may require planting of nursery-grown seedlings, direct seeding, and/or the manipulation of disturbance regimes (for example, thinning and burning) to speed up the recovery process. This is often at a high cost to establish the vegetation structure, reassemble local species composition, and/or catalyze ecological succession.

Effective restoration should aim at the reestablishment of fully functioning ecosystems. To ensure effective restoration and achieve sustainability and resilience into the future, Reference [15] advocates for four principles of restoration; (1) restoration should increase ecological integrity (2) restoration should be sustainable in the long term (3) restoration ought to be informed by the past and future, and (4) the restoration benefits and engages society through direct participation. Ecological integrity has been defined as the "ability of an ecosystem to support and maintain a balanced, adaptive community of organisms having a species composition, diversity, and functional organization comparable to that of natural habitat [16].

Several factors ought to be considered while deciding on the restoration strategy to be employed; the specific ecosystem resilience, the land-use history, the landscape context, the goal for the restoration, and available resources [14]. In addition, depending on the level of land degradation, and the intended impact of restoration on land use and land-use changes [17] distinguishes four intensities/levels of restoration; ecological intensification, recovery/regeneration, reparation/recuperation, and remediation. Table 1 presents a summary of some of the commonly practiced landscape restoration options/strategies in different land use or ecosystem types adapted from Reference [18,19]. A detailed explanation of most of these landscape restoration practices and their restoration goals is provided by Reference [19].

Table 1. Landscape restoration options for different land-use types.

Land Use/Ecosystem Type	Landscape Restoration Options/Strategies
Forest land	- Afforestation and reforestation - Planted forests and woodlots - Natural regeneration - Silviculture - Assisted natural regeneration/Reclamation/Rehabilitation

Table 1. *Cont.*

Land Use/Ecosystem Type	Landscape Restoration Options/Strategies
Agricultural land	- Agroforestry - Integrated soil fertility management - Climate-smart agriculture - Improved fallow - Extended rotations in plantations - Farmer managed natural regeneration (FMNR)
Protective land and buffers	- Mangrove restoration - Watershed protection - Erosion control - Bamboo planting along water bodies and wetlands
Urban areas	- Green and blue infrastructure in urban areas
Wetlands	- Wetlands restoration and conservation
Freshwater (Rivers/lakes)	- River and lake restoration - Sediment management - Pound restoration - Integrated watershed management
Grasslands and Shrublands	- Assisted natural regeneration

Adapted from Reference [18,19].

2.2. *Economic Cost-Benefit Analysis of Landscape Restoration*

Generally, cost-benefit analysis can be either financial or economic; there are similarities and differences in both. In both analyses, the net benefits of a project investment are estimated, and the estimation is based on the difference between with-project and without-project situations. While conducting both analyses, the assumption of constant prices is made and, the techniques of evaluating costs and benefits through the discounting method remain the same. However, in financial CBA of projects, benefits and costs are compared to the enterprise, while, in economic CBA, benefits and costs are compared to the whole economy.

The true value that a project holds for society is highly considered in economic analysis. It considers all members of society and measures the project's positive and negative impacts in terms of willingness to pay (WTP) for units of increased consumption and to accept compensation for foregone units of consumption. Importantly, the economic analysis covers even the costs and benefits of goods and services which have no market price. In financial analysis, the project's sustainability and balance of investment is checked using market prices. In economic analysis, the legitimacy of using national resources to a certain project is measured using economic price, which has been converted from the market price by excluding tax, profit, subsidy, etc. In financial analysis, the taxes and subsidies included in the price of goods and services are integral parts of financial prices, but they are treated differently in economic analysis.

There is also a significant difference between financial and economic analysis in the way they treat their external effects (costs and benefits). Such externalities, health effects and non-technical losses tend to be valued in economic analysis. Both financial and economic analyses are supposed to include such externalities (side effects). In addition, economic and financial returns in both analyses do not converge. This is because what counts as a benefit or a cost to the project operator does not necessarily count as a benefit or cost to the economy. When restoration is viewed through a financial accounting lens that ignores public values and the inter-generational nature of restoration, the conclusions that are drawn tend to favor investing in less restoration than society would prefer [20]. For this review, we consider those studies that conducted an economic CBA.

There are nine steps for conducting an economic cost-benefit analysis for landscape restoration as outlined by Reference [21]: (1) Specify the set of restoration transitions: Define which degraded land uses will be restored and the activities that will be used to restore them, (2) Define the stakeholders who will be impacted by restoration, (3) Catalogue the impacts and define how they will be measured: Which impacts matter most to the stakeholders, who will be impacted by restoration and what units of measurement are most useful for measuring them? (4) Predict the impacts quantitatively over the time horizon of the project: Use ecosystem service models, household surveys, stakeholder engagement, and other estimation methods to quantify the expected impacts of restoration activities, (5) Monetize all of the impacts: Use appropriate direct and indirect methods to value the estimated impacts, (6) Discount benefits and costs to obtain present values: Select appropriate discount rates to make streams of future benefits and costs comparable at the present moment, (7) Calculate the Net Present Value (NPV) of each alternative: Subtract the discounted stream of implementation, transaction, and opportunity costs from the discounted stream of benefits as shown in the equation.

$$\textit{Net Present Value} = \sum_{t=0}^{n} \frac{B_t}{(1+r)^t} - \sum_{t=0}^{n} \frac{C_t}{(1+r)^t},$$

where B_t is the benefits from restoration at time t; r is the discount rate, and C_t is the total restoration costs at time t.

(8) Perform sensitivity analysis: The results of the CBA depend on assumptions, and the sensitivity of the results to changes in the underlying assumptions should be evaluated, (9) Make policy recommendations: From a Pareto-efficiency perspective, the restoration activities with the largest NPV should be recommended.

Overall, several categories of costs and benefits ought to be accounted for in a comprehensive CBA. The total cost of landscape restoration depends on how degraded a site is and how difficult it is to restore. Additionally, costs vary according to geography, degradation category, the objectives and contexts of specific restoration activities, and the types of restoration methods that are used. There are three investment phases involved in forest and landscape restoration: (1) Phase 1 is the initial readiness investment or up-front investment. During this phase, investments flow towards designing projects, planning, stakeholder engagement and participation, developing safeguards and capacity building. (2) Phase 2 is the investment for actual implementation. During this phase, it may involve policy reforms, implementation of the restoration of degraded lands, educational activities, land-use zoning and strengthening of capacities. (3) Phase 3 focuses on sustained financing for landscape ecosystem services and product services, for self-sustaining funding of the project's long-term running costs [5,18].

Similarly, these costs can be broadly categorized into three: (1) Implementation costs (usually very high); these costs include costs of raw materials, such as tree seedling, fencing, labor costs, transport costs, and other costs. In addition, this includes costs incurred in capacity building and training the local stakeholders. They are mostly the direct costs incurred in the project. Land users usually incur these costs, or they can be covered by the project. (2) Opportunity cost: this represents the cost of foregone opportunities and, represents the tangible goods and services that were foregone to make restoration possible. To capture the opportunity cost, it is necessary to conduct a baseline before implementing the project. (3) Transaction costs; transaction costs represent the cost for landowners and implementing agencies to identify viable land to restore and negotiate over terms that ensure restoration meets both local and national priorities [21]. These may include monitoring costs, as well.

Additionally, for a comprehensive economic CBA on restoration, the benefits to be considered ought to include both use and non-use values, as well as private and public benefits arising from the restoration, i.e., the total economic value of the restoration. The total economic value (TEV) of an investment attempts to estimate and monetize all economic impacts of the investment [22]. TEV recognizes that benefits and costs radiate far beyond the landowner or investor to the global effects. The TEV approach accounts for both the use and non-use benefits values [22].

The use-values are categorized into:

(a) Direct use values: These relate to the benefits obtained from the direct use of an ecosystem. Most of the direct products have market values, such as timber, poles, charcoal, gum arabica, and medicine, as well as other non-timber forest products (NTFPs), such as wild fruits, honey, fodder, crop harvests, recreation value, and others. They are the most straightforward benefit category to capture and account for since the data on quantities and value is available for most of the restoration projects.
(b) Indirect use values: These are usually associated with regulating services. These may include services, such as carbon sequestration, water treatment and regulation, soil erosion control, pollination, and so on.
(c) Option values: These include valuing ecosystem services for the option of future use, such as medicinal purposes.

Non-use values, on the hand, are categorized into:

(a) Bequest value: This captures the value arising from the satisfaction of knowing that future generations will access nature's benefits. This value is concerned with intergenerational equity.
(b) Altruist value: This value concerns intragenerational equity, i.e., the satisfaction of knowing that other people can also access nature's benefits.
(c) Existence value: This value is derived from the satisfaction of knowing a certain species exists. For example, indigenous trees, endangered species, and medical trees.

Direct use values are relatively easy to identify and value since they are tangible and usually have market values. On the other hand, indirect use and non-use value pose a challenge in valuation, and the valuation methods employed are often time and resource intensive. Several studies, such as Reference [23–25], detail how to adequately value these benefits through a wide range of valuation methods, including: production methods, choice experiments, contingent valuation methods (CVM), hedonic pricing, travel cost methods, cost-based approaches, benefit transfer, and mean-variance analysis.

3. Materials and Methods

Reviews aim to identify the most reliable research on a given question in a manner that minimizes selection biases in the literature search and screening process [26]. A comprehensive review of studies was conducted on cost-benefit analysis for landscape restoration by following the Preferred Reporting Items for Systematic Reviews and Meta-Analyses (PRISMA)[6] guidelines. Figure 1 shows the process of the review followed. The first step was a keyword search using the "publish or perish"[7] software. We searched for a combination of three phrases (1) "Cost-benefit analysis" and "Landscape restoration" (2) "Economics" and "Landscape restoration" (3) "Cost-benefit analysis" and "Land restoration". The search for "Cost-benefit analysis" and "Landscape restoration" generated 522 studies, while the search for "Economics" and "Landscape restoration" generated 922 studies and that of "Cost-benefit analysis" and "Land restoration" generated 612 studies.

For step two, the total 2056 publications were filtered down by titles and abstracts, resulting in 102 relevant publications. For the 102 publications, after reading through the full text, 31 publications were found that either entirely focused on CBA of landscape restoration or had a component of CBA of landscape restoration. The 31 publications are the ones that have been considered in the review, and a summary of these studies is presented in the Appendix A (Table A1). The summary is based on restoration options/strategies, country of focus, the year the study was conducted, data sources, the time considered in the CBA analysis, benefits and costs components, sensitivity analysis conducted, and the reported NPV. These variables are discussed in detail in Section 4 under results.

---

6 http://www.prisma-statement.org/
7 https://harzing.com/resources/publish-or-perish/windows

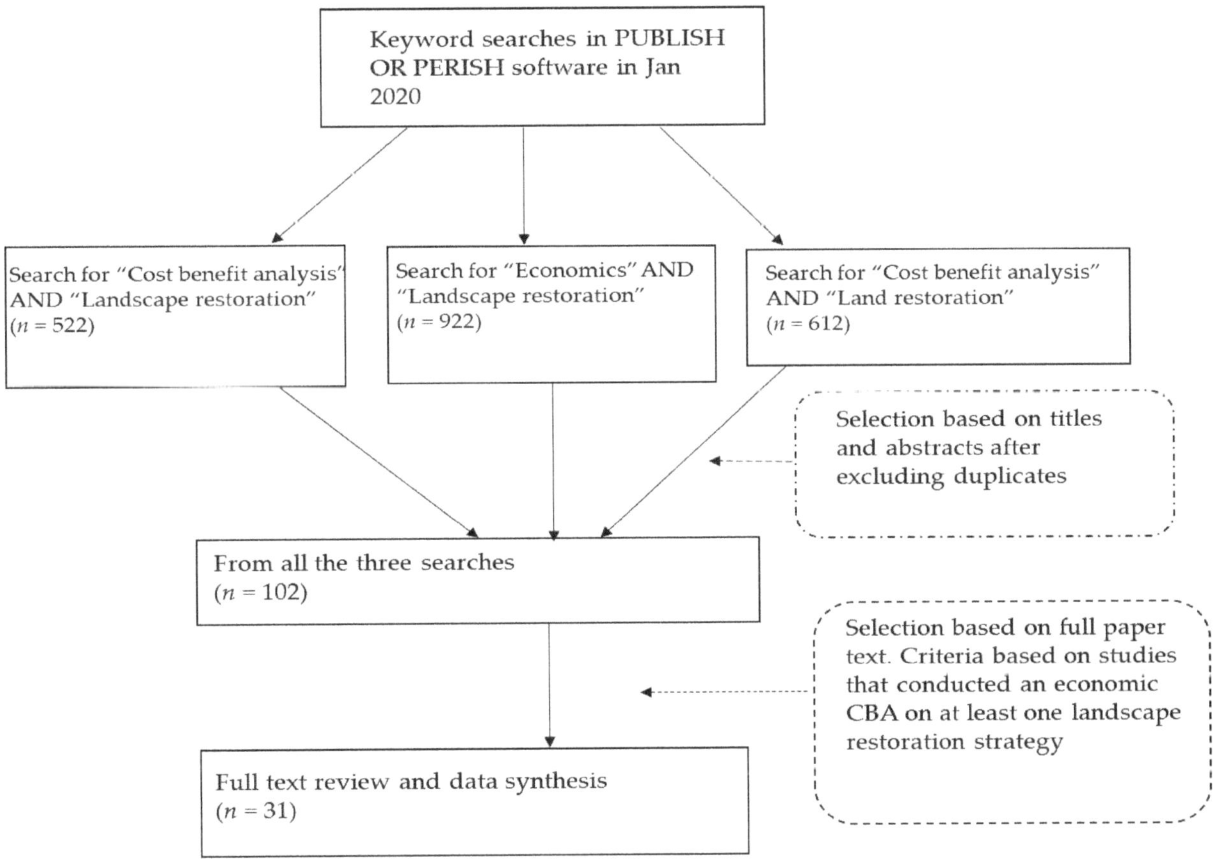

Figure 1. Review process of the studies on cost-benefit analysis (CBA) of landscape restoration.

4. Results

From the review process explained in the materials and method section, we found 31 publications of CBA on land restoration. In this chapter, we discuss these publications based on several attributes.

4.1. Country of Focus and Study Year

The 31 studies under review were conducted in about 20 countries distributed across five regions: Africa (10), Europe (2), N. America (2), S. America (3), and Asia & Middle East (3). Almost all the 31 studies reviewed, were conducted in a single country and only five focused on multiple countries. Of the five, three studies focused on a global perspective [18,27,28], while one focused on multiple African countries[8] [29], and one focused on several countries in Latin America[9] [7]. As shown in Figure 2, countries that have had the most CBA studies conducted there over the years had, on average, three studies and include South Africa, Brazil, Tanzania. Ethiopia, Kenya, and Vietnam, where each have had two CBA studies conducted.

8 Angola, Benin, Botswana, Burkina Faso, Burundi, Cameroon, Central African Republic, Chad, Congo, Côte D'Ivoire, Djibouti, DR Congo, Egypt, Eritrea, Ethiopia, Gabon, Ghana, Guinea, Kenya, Lesotho, Liberia, Madagascar, Malawi, Mali, Mauritania, Morocco, Mozambique, Namibia, Niger, Nigeria, Rwanda, Senegal, Sierra Leone, South Africa, Sudan, Swaziland, Togo, Tunisia, Uganda, UR of Tanzania, Zambia, Zimbabwe.

9 The focus Latin America countries include: Mexico, Argentina, and Chile.

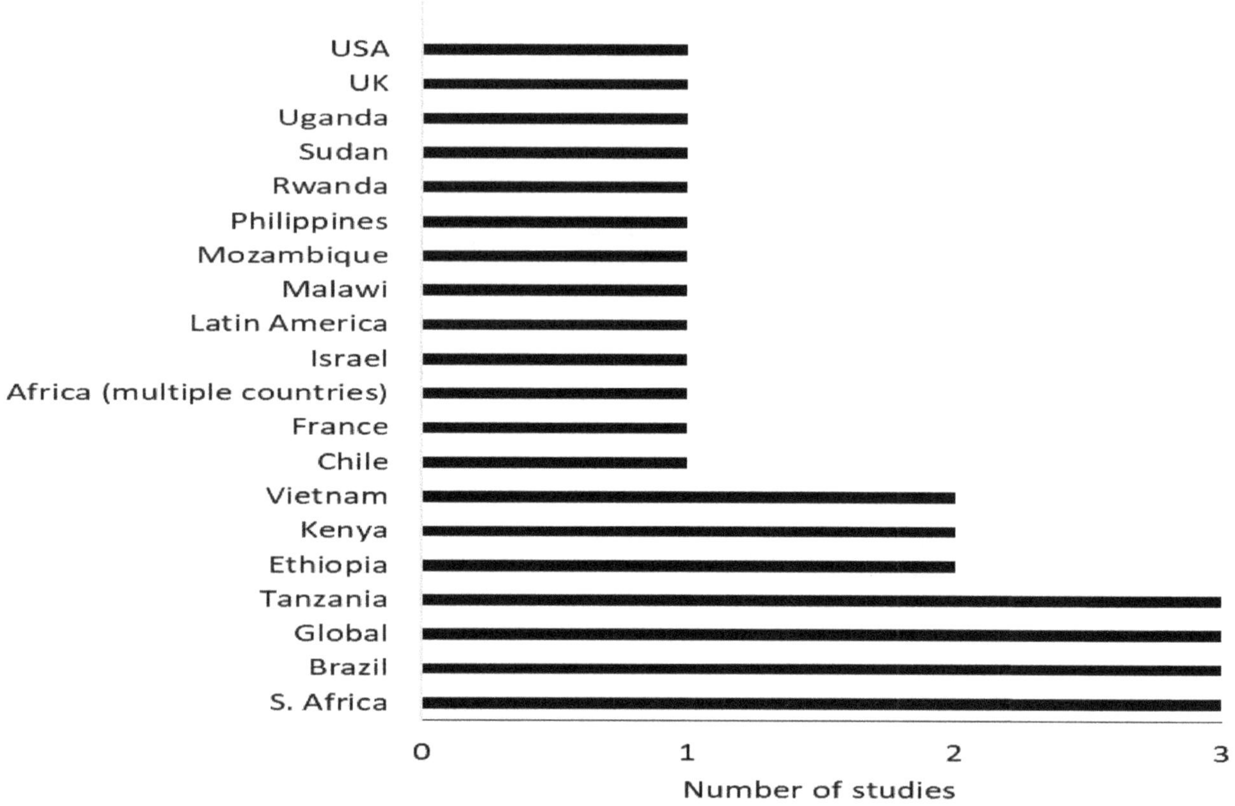

Figure 2. Distribution of CBA studies across various countries.

Figure 3 presents a trend of the year studies on restoration under review were conducted. The oldest study we reviewed was conducted in 1997 in France and focused on landscape restoration using hedgerows [30]. The was no study under review conducted between 1997 and 2005. The most recent studies were conducted in Kenya by Reference [31] and in Brazil by Reference [32].

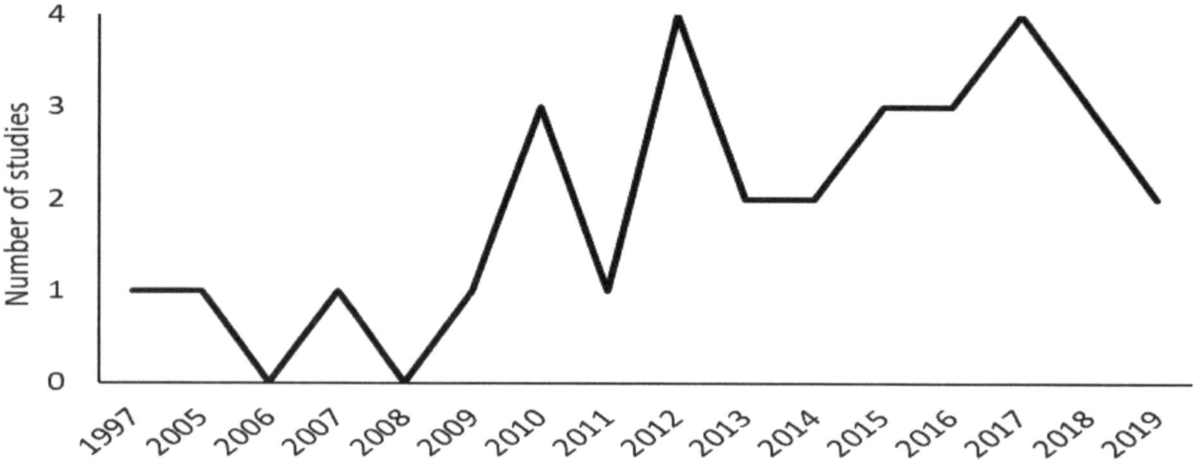

Figure 3. Year the CBA studies were conducted.

4.2. *Landscape Restoration Options*

As shown in Table 2, the CBA studies we reviewed were conducted for different types of restoration options/strategies, including: reforestation & afforestation, agroforestry, biofuel agroforestry, participatory forest management, woodlots establishment, sustainable land management practices, natural regeneration, assisted natural regeneration, mangrove restoration, clearing of invasive alien species, urban area restoration, and buffer areas restoration. The specific studies are presented in the Appendix A in Table A1. Some studies focused on only one type of restoration, while others conducted a comparison of different restoration options depending on the land use type. One of the most comprehensive CBA study we reviewed was conducted in Kenya by Reference [31] and compared returns for several landscape restoration strategies, including: afforestation or reforestation of degraded natural forests, rehabilitation of degraded natural forests, agroforestry in cropland, commercial tree and bamboo growing on potentially marginal cropland and un-stocked forest plantation forests, tree-based buffer zones along water bodies and wetlands, tree-based buffer zones along roads, and restoration of degraded rangelands. Overall, the most popular landscape restoration options for which CBA studies were conducted include: reforestation and afforestation (8), agroforestry (7), farmer-managed natural regeneration/Assisted natural regeneration (5), soil and water conservation practices (5), and establishment of woodlots (4). Further still, Reference [20] assessed the net present value of the Bonn Challenge, which is a global effort to restore 350 million hectares of degraded forest landscape.

Table 2. Landscape restoration options considered in the CBA studies reviewed.

Landscape Restoration Options	Number of Studies	Countries of Focus
Reforestation and Afforestation	8	Brazil, Chile, Ethiopia, Kenya, Uganda, USA, Chile, Tanzania
Agroforestry	7	Brazil, Ethiopia, Kenya, Sudan, Tanzania, Uganda, Malawi
Participatory forest management/FMNR/ANR	5	Brazil, Ethiopia, Malawi, Vietnam, Tanzania
Soil and water conservation measures and SLM practices, e.g., bunds, terracing, zero tillage,	5	Ethiopia, Kenya, Malawi, Vietnam, multiple countries in Africa
Establishment of woodlots	4	Ethiopia, Malawi, Uganda, Tanzania, Rwanda
Mangrove restoration (protective and planting)	3	Mozambique, Philippines, Vietnam
Natural regeneration	2	Uganda, Rwanda
River restoration/habitat restoration for river catchment	2	Israel, UK
Dryland forest restoration	2	Latin America, Chile
Alien vegetation clearing for water yield and tourism	2	South Africa (2)
Biofuels agroforestry or biofuel for energy, thus reducing deforestation	1	Tanzania,
Landscape restoration using hedgerows	1	France
Urban area restoration (green and blue infrastructure in urban areas)	1	Multiple countries
Tree based buffer zones along water bodies and wetlands	1	Kenya
Tree-based buffer zones along roads and riparian land	1	Kenya
Restoration of degraded rangelands	1	Kenya
Subtropical thicket restoration	1	South Africa

Figure 4 below shows reported NPV (positive or negative) by the various CBA studies for the different landscape restoration options. For some of the restoration options, all the studies conducted reported positive NPV; agroforestry (8 studies), soil and water conservation (5), mangrove restoration (3), and alien vegetation clearing (3). However, for some of the restoration strategies, some of the studies reported negative NPV; for reforestation and afforestation, the number of studies that reported positive NPV (4) was equal to those that reported negative NPV (4). For other restoration options—FMNR/ANR and woodlot establishment, the number of studies that reported positive NPV was higher than those which reported negative NPV. None of the restoration strategies had more studies that reported negative NPV compared to those that reported positive NPV.

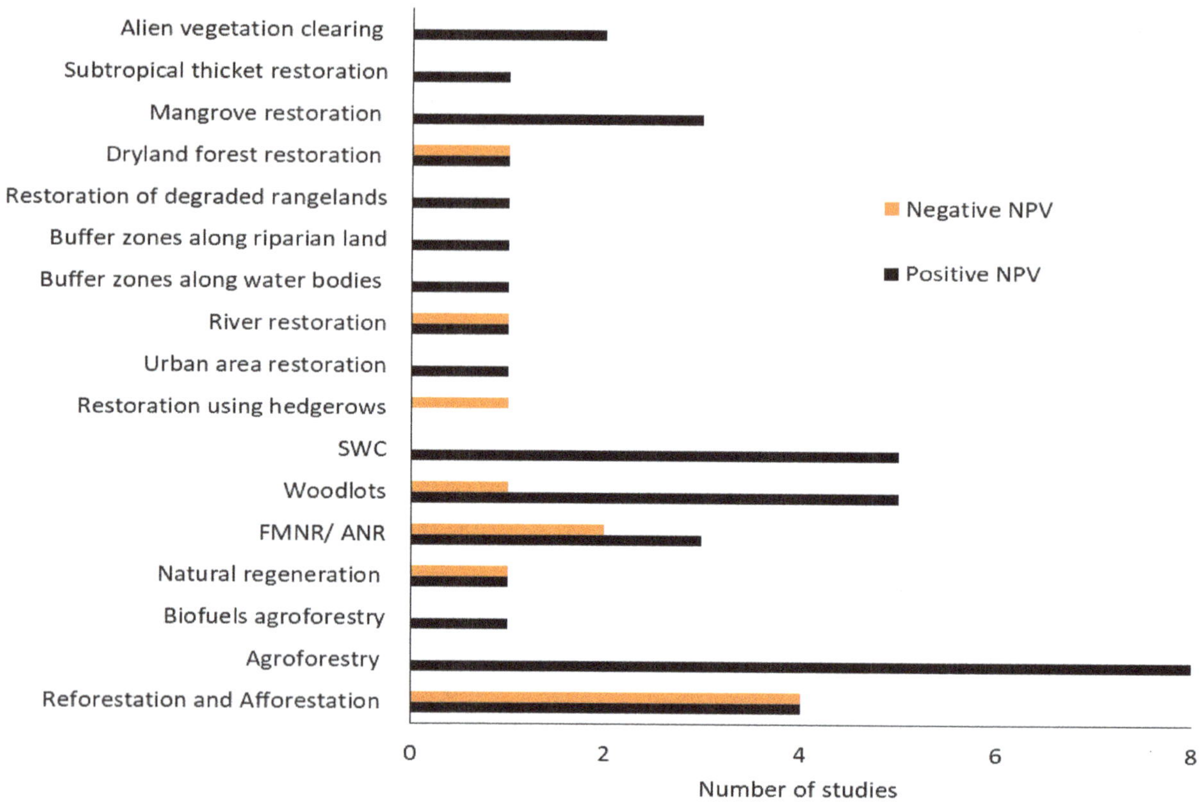

Figure 4. Reported Net Present Value (NPV) for the various landscape restoration type by the reviewed studies.

4.3. Age of Restoration in the CBA Studies

Figure 5 shows the time in years for which different studies under review considered. A higher number of CBA studies, (8) covered restoration benefits and costs for 16–20 years. Another six studies covered between 21–25 years. The maximum duration that was considered was 100 years by two studies, Reference [33] in Mozambique and [30] in France. A further three CBA studies, Reference [34–36], covered restoration benefits and costs for 50 years.

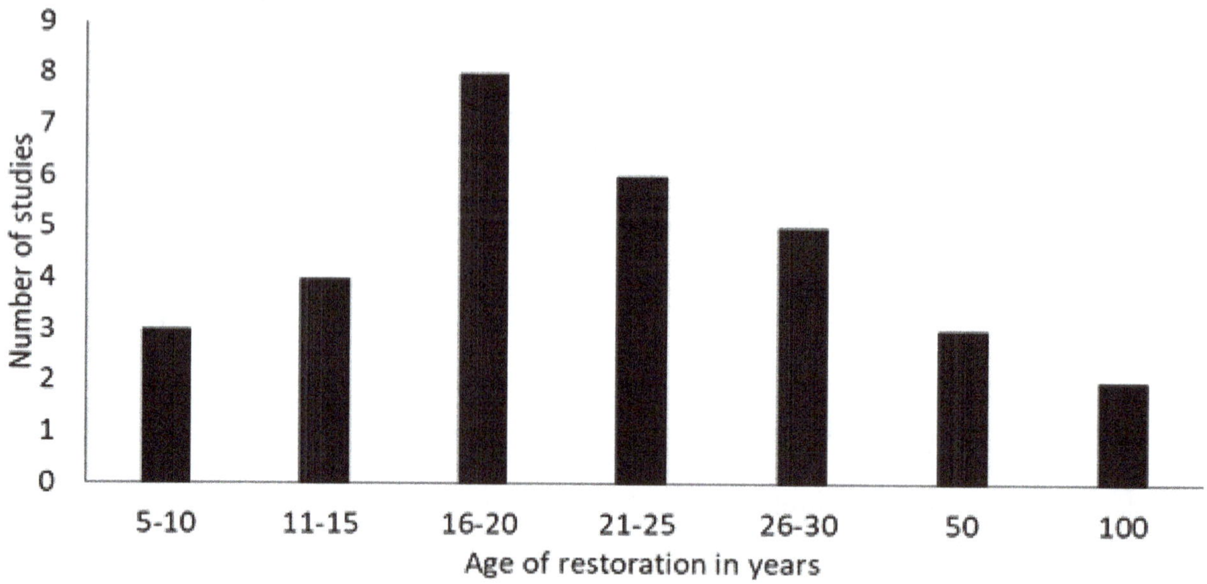

Figure 5. Restoration project timeline considered in the CBA studies.

4.4. Data Source

The CBA studies we reviewed sourced data from different sources; 12 studies used primary data, 11 applied secondary data, and the remaining 8 used both primary and secondary data. For primary data, various studies applied different data collection methods and techniques, as shown in Figure 6. Approximately seven studies used expert discussions, key informants' interviews (KIIs) or focus group discussions (FGDs). Others employed other data collection techniques, including: surveys, reviewing budgets, spatial analysis, and field observations.

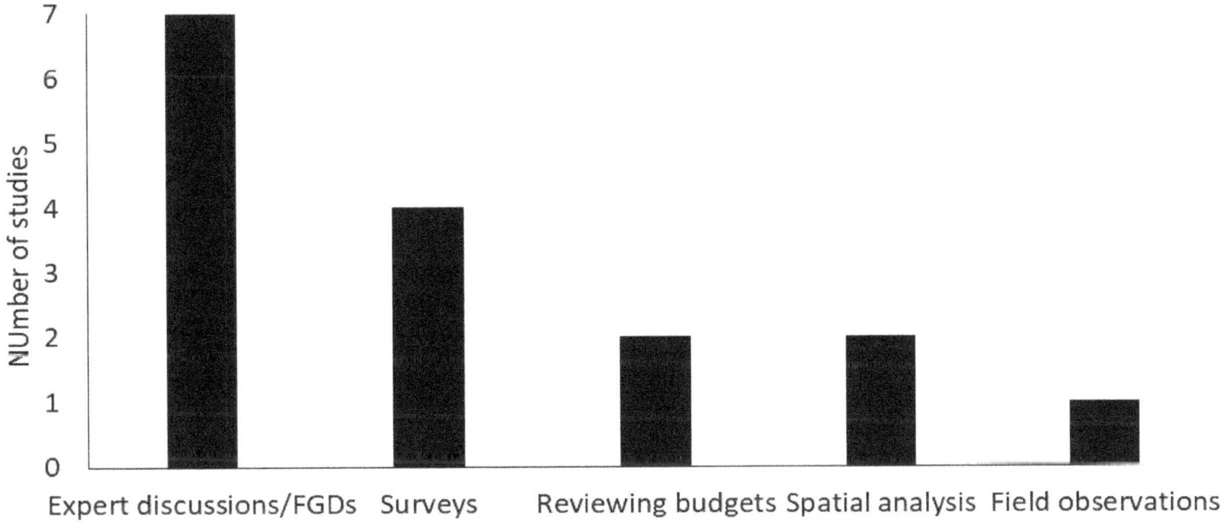

Figure 6. Primary data collection methods employed in the restoration CBA studies.

For those studies that considered indirect use and/or non-use benefits in their analysis, different approaches were applied in valuing these benefits since they mostly do not have a market value. Three studies applied the contingent valuation method, and at least one used either hedonic analysis, travel cost method, or benefit transfer method.

4.5. Benefits and Cost Components in the CBA Studies

Figure 7 shows the proportion of studies that considered different benefits and cost categories. Most of the studies accounted for the use-values only (either direct use, indirect use, or both) and only around 16% accounted for the total economic value of the project (both use and non-use values). About half the studies accounted for both direct and indirect use-values. Of these, the indirect use benefit that was mostly considered in these restoration studies was carbon sequestration. Other indirect use benefits that were accounted for include: erosion control, stormwater control, air regulation, temperature regulation, recreational value, avoided nutrient loss, nitrogen fixation, soil fertility improvement, and aquifer recharge. Non-use values that the reviewed studies accounted for include: aesthetic value, bequest (inheritance) values, and existence value.

The studies applied different valuation methods to value the respective benefits, as shown in Table 3. A more detailed explanation of these benefits by the specific study is provided in Appendix A, Table A1. The provisioning services were considered in almost all the studies. They were commonly assessed by collecting primary/secondary data on the quantities of these goods and valuing them at the current market price. Carbon sequestration was valued by either spatial analysis (e.g., Reference [7,37]) or sourcing the quantities from secondary data and valuing them at the market price of carbon. Contingent valuation method was applied by several studies in valuing benefits, including: air pollution & regulation [38], public benefits [30], cultural, aesthetic & recreation values [35], and inheritance and bequest values [39].

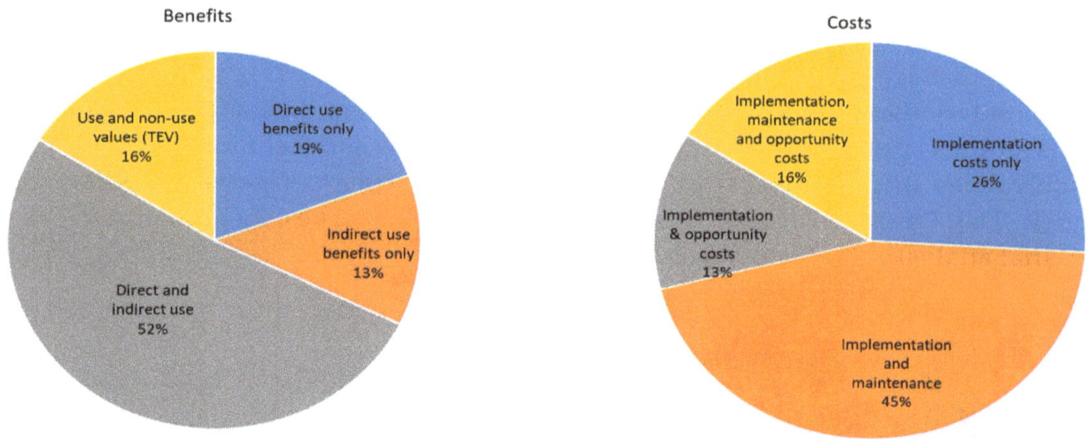

Figure 7. Benefits and cost categories considered in the CBA studies reviewed (n = 31).

Table 3. Valuation methods applied across various benefits.

Benefits	Valuation Methods
Timber, wood fuel, non-timber forest products (NTFPs), livestock, yield benefits	- Primary data valued at existing market price - Expert opinion
Tourism	- Value of tickets paid at the entrance - Travel cost method - Spatial analysis
Carbon sequestration	- Secondary data valued at the market price of carbon
Public benefits	- Contingent valuation methods, i.e., assessing willingness to pay for these benefits - Hedonic method
Air pollution and air regulation	- Contingent valuation method - Benefit transfer
Stormwater control/reduction	- Hedonic methods - Benefit transfer
Increased soil fertility	- Replacement cost method - Benefit transfer
Soil erosion control	- Avoided cost - Benefit transfer
Temperature regulation	- Benefit transfer
Cultural, aesthetic and recreation	- Contingent valuation method; assessing willingness to pay
Positive health effects	- Benefit transfer
Biodiversity recovery	- Secondary data with simulations - Benefit transfer
Inheritance and bequest values	- Contingent valuation methods
Total economic value for global studies	- Secondary data and benefit transfer particularly from the TEEB valuation database

Benefit transfer method was frequently applied in valuing many indirect use benefits, such as air pollution & regulation, biodiversity improvement, soil fertility, soil erosion control, temperature control, positive health effects, recreation, and stormwater control. A comprehensive example of the use of benefit transfer is in Reference [27], where they used data collected from 94 studies, from which they created a database of benefits and costs for conducting CBA in seven biomes. In addition, Reference [18] used secondary data and the comprehensive TEEB database due to Reference [40] in valuing benefits while assessing the CBA of forest landscape restoration within the Bonn Challenge for six different biomes. In the absence of site-specific valuation information, benefit transfer is an alternative to estimating non-existing values. It adapts existing valuation information to a new context (location or time), and it is principally useful when there are budget and time constraints with the collection of primary data [24]. However, there is a need to ensure that the ecological conditions are the same; otherwise, the value may be overstated or understated.

Hedonic pricing method was the least frequently applied valuation method; only one study used the approach in valuing stormwater control and air pollution control [34]. The study employed the hedonic models through controlling for the price of housing in different locations. The reason the hedonic method is rarely applied is that the data required can be quite intensive. Similarly, this method works well if markets can pick up quality differentials, which may not be the case for agricultural and forest land, due to the non-observability of some attributes [23]. Further still, Reference [31] applied the replacement cost approach to value the soil fertility and the avoided loss approach to value soil erosion control.

Three cost categories were considered in the studies we reviewed, as shown in Figure 8: implementation costs, maintenance costs (mostly annual costs of maintaining the restoration infrastructure), and opportunity costs (cost of foregone opportunities). All the CBA studies we reviewed considered implementation cost since this is the most direct cost in restoration and is easily captured. Approximately 25% considered only the implementation costs, and a further 45% covered both implementation and maintenance costs—only 16% of the reviewed studies included all the three cost categories. Implementation and maintenance costs were mostly sourced from the costs incurred in the projects. Implementation costs included the investment costs, such as seedlings, materials, and other inputs labor, and training costs, among others. Maintenance cost includes monitoring and transaction costs. For opportunity cost, on the other hand, the studies had to conduct baseline assessments to ascertain the value of foregone opportunities. For example, Reference [41] assessed the baseline situation of agriculture and grazing, which was translated into the opportunity cost of the land.

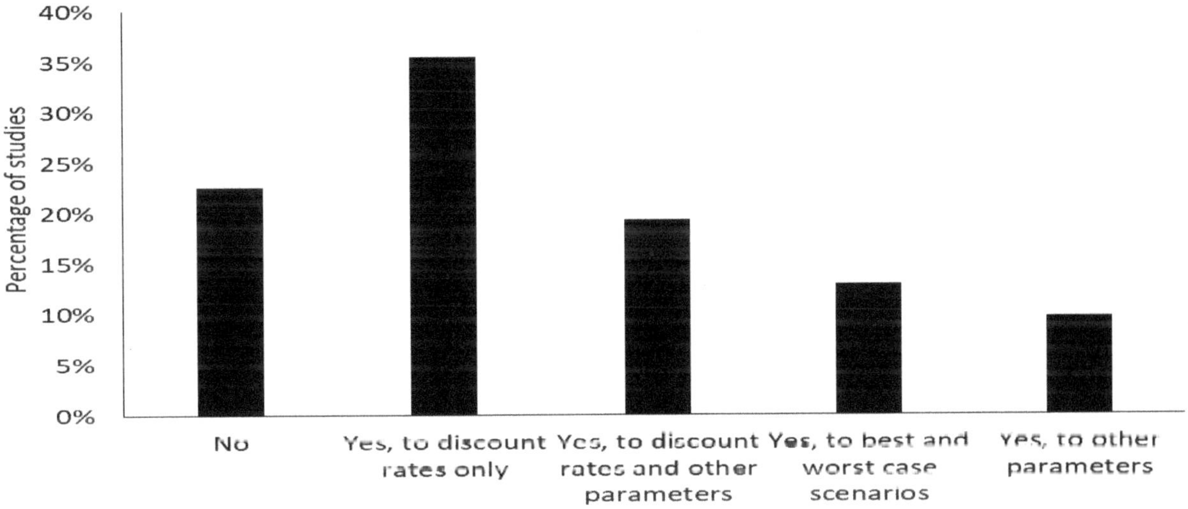

Figure 8. Sensitivity analysis across the CBA studies under review ($n = 31$).

4.6. *Sensitivity Analysis*

The uncertainties in the entire CBA process, such as fluctuating prices, discount rates, and unseen events within the restoration lifetime, can affect the estimated results especially when the CBA is conducted ex-ante. Hence, after conducting a CBA, it is necessary to conduct a sensitivity analysis by altering various parameters of the estimation, such as the discount rate or prices. Alternatively, one may conduct a Monte Carlo simulation. A Monte Carlo simulation is similar to sensitivity analysis in that it demonstrates how a project's profitability varies. However, instead of altering one input variable and analyzing how that changes in that variable affects the project viability, a Monte Carlo simulation attempts to model uncertainty across multiple inputs assumptions. The model is run thousands of times to understand different possible outcomes and the likelihood of them occurring [19]. Hence, it is more rigorous and robust compared to regular sensitivity analysis.

Figure 8 shows the proportion of studies under review that conducted a sensitivity analysis to test the validity of their results. Of the 31 studies we reviewed, none of the studies applied the rigorous Monte Carlo simulation. Approximately 23% of the studies did not conduct any form of sensitivity analysis. The majority of the studies (about 35%) conducted a sensitivity analysis by varying the discount rates only. Additionally, a further 19% varied the discount rates and other parameters, such as carbon prices, products prices, maintenance costs, and so on. Approximately 13% tested the validity of the CBA results by varying the best- and worst-case scenarios by assuming a very optimistic scenario where most of the assumptions hold and a very pessimistic scenario where most of the assumptions do not hold. Still, 10% of the reviewed studies only varied other parameters without varying the discount rate.

5. Discussion

We present a discussion of the results with respect to the quantity and quality of CBA for landscape restoration and the implications for moving forward with restoration in terms of planning, policies, and investments. There is a need to grow the quantity of evidence in specific areas, as well as economic evidence to make restoration attractive to investors. A number of areas for improvement emerge from our analysis. These include capturing all costs categories; going beyond direct use values; capturing public benefits; sensitivity analysis; and the need for standardization and or guidance. We briefly discuss each below.

5.1. Capturing All Costs Categories

Most of the studies do not account for all the costs associated with restoration, thus overstating the profitability of restoration. All the reviewed studies accounted for implementation costs, but relatively few covered all the cost categories (16%). The least accounted for cost category is the opportunity cost; probably because it is often difficult to estimate this cost since it is not a direct cost. Estimating the opportunity cost requires a baseline assessment to identify the foregone uses of the land. For example, Reference [41] assessed the baseline situation of agriculture and grazing, which was translated into the opportunity cost of the land. Most of the studies do not conduct baseline assessments since this involves committing more resources and time, making it challenging to provide an estimate of the opportunity cost. In addition, for some land uses, the opportunity cost may be negligible, especially if the land is highly degraded. However, degraded lands with high surface runoff can be used for water harvesting, providing a positive externality that may get lost if infiltration and vegetation water use increase as a result of landscape restoration [42].

In addition, there is a need to include maintenance and monitoring costs in accounting for the total economic costs. Most restoration projects fail to account for maintenance and monitoring costs since they view restoration as a one-time cost activity as opposed to a continuous activity—for example, tree planting as opposed to tree growing [43]. Tree planting is a one-time cost activity where only the implementation cost will be significant. On the other hand, tree-growing is a continuous activity, implying that maintenance and monitoring costs are significant and accountable, as well. Hence, all the three cost categories ought to be accounted for, in order for a cost-benefit analysis to reflect the actual economic viability of a restoration project.

5.2. Going Beyond Direct Use Values

A major challenge in conducting an economic CBA is that it is difficult and controversial to monetize social and environmental benefits. Environmental benefits are valued differently by different stakeholders, which disputes the findings of the analysis. For instance, greenhouse gases (GHG) mitigation, which is a global ecosystem service, maybe prioritized over reducing soil erosion which is a local benefit. Additionally, the value of some environmental benefits, such as supporting biodiversity, are often excluded because monetizing them is challenging [19]. However, it is still possible to include non-monetized values, such as biodiversity, in the decision-making.

Based on the stocktake, a significant proportion of existing CBA studies do not account for indirect use benefits, and an even more substantial proportion do not account for non-use benefits (84%). This is mostly because most of these benefits are "invisible" and do not have a market value, thus making valuing them quite challenging. Some of the studies that included the indirect and non-use values applied various methods for evaluating them, including: benefit transfer, contingent valuation method, travel cost method, hedonic pricing models, replacement cost, avoided loss, and spatial analysis. Benefit transfer and contingent valuation methods were frequently applied compared to the other methods. This is probably because, these methods can be applied in the valuation of almost all the benefits, such as air pollution control, biodiversity, inheritance and bequest value, cultural and aesthetic values, and so on [23,24]. Similarly, benefit transfer is generally cost-effective and is commonly applied when there is limited budgetary allocation.

On the other hand, some of the other methods (e.g., hedonic models, avoided loss, replacement costs) are data-intensive and, cannot be universally applied for most of the benefits. For example, only one study, Reference [34], employed the hedonic models in valuing stormwater control and air pollution control through controlling for the price of housing in different locations. This method works well if markets can pick up quality differentials, which may not be the case for agricultural and forest land, due to the non-observability of some attributes [23]. Hence, there is a need for understanding the methods that can be used to reasonably value the specific benefits arising from the project depending on the nature of the benefit/ecosystem services data availability, time and cost constraint, and so on. Conducting a comprehensive economic CBA for restoration is costly and time-consuming; thus, there is a need for restoration projects to budget for this.

5.3. Capturing Public Benefits and Implications for Government Investments in Restoration

In addition,, some of these benefits are public benefits attributable to other stakeholders beyond those directly targeted by the restoration projects. A comprehensive economic CBA ought to account for all these benefits. Otherwise, the estimated NPVs for these restoration projects are undervalued. Hence, to present a true picture of the profitability of restoration projects, future CBA studies should aim to capture all the benefits arising from the restoration projects—use and non-use benefits, as well as private and public benefits. This particularly useful for large-scale restoration projects where the benefits accrue to the broader public beyond the targeted stakeholders. Reference [20] also found that when the value of public goods and services are accounted for in the cost-benefit analysis, the benefits of large-scale restoration outweigh the costs and targets like the Bonn Challenge can be met efficiently.

There is also a need for information on public benefits to drive government investments in restoration. Investors require good information on costs and benefits for investment proofing and decision-making. To this end, there is a need for a cost-benefit analysis (CBA) database compiling existing data on landscape restoration costs and benefits more so information on indirect and public benefits [18].

5.4. Improving Sensitivity Analysis

CBA of restoration attempts to model or estimate the future; therefore, a certain degree of uncertainty is involved. For example, unforeseen events and climate change may affect productivity in ways that are difficult to predict, and they should be considered. Similarly, applying a discount rate is also an inherently subjective decision, but it is important for prioritizing near-term benefits versus long-term benefits [19]. Thus, CBA is based on certain assumptions that vary in their degree and level of confidence. The assumptions made during CBA include political and/or social assumptions that may not necessarily hold.

This calls for the need for sensitivity analysis in the CBA process to provide a robustness check for the results. While conducting sensitivity analysis, almost all the existing studies we reviewed conducted a direct sensitivity analysis by varying only one or just a few variables, mostly the discount rate, e.g., Reference [7,27,32] or carbon prices, e.g., Reference [35]. None of the studies applied a more rigorous sensitivity test, such as the Monte Carlo simulation approach. For more robust and comparable results, future CBA on restoration should consider more rigorous approaches for sensitivity analysis, such as the Monte Carlo simulation.

5.5. The Need for Standardization and Guidelines

During CBA of land restoration, it is difficult to monetize social or political considerations. Restoration options selected during CBA should produce maximum benefits for all, but this is usually not the case. Due to political reasons, benefits to one group may be valued more than the benefits of another group and such is not usually included in the CBA [19]. Data collection to be used for CBA of restoration can be time-consuming and expensive since the impacts of restoration transitions are felt over long-time periods. One requirement for a comprehensive CBA is to quantify all the impacts for each land use (degraded and restored) for the relevant time horizon of the project [21]. Predictions about the levels of inputs (i.e., costs) and the production of ecosystem services must be made for each year and each land use in a restoration transition. This can be the most challenging aspect of CBA because there is not always a complete scientific understanding of how complex natural systems work, especially when significant changes to their structure are made.

Closely related, lack of reliable data owing to poor data-keeping during the restoration period also affects the CBA results. It takes time to realize the actual profitability of these restoration investments since returns to landscape restoration projects are not immediate. For example, in the studies we reviewed, the restoration age considered was even up to 100 years for some projects, with the minimum being seven years. This requires data over several years, and most projects do not keep a record of this data. Hence, even for ex-post CBA evaluations, a lot of predictions and assumptions are involved in data generation. Thus, there is need to adopt standardized methods of data prediction if the results are to be comparable across different restoration projects in deciding the allocation of funds. Similarly, the Food and Agriculture Organization (FAO)"The Economics of Ecosystem Restoration" (TEER), points to need for a comprehensive tool on costs and benefits of Ecosystem Restoration (and FLR) interventions[10]. In an on-going project, TEER aims to "offer a reference point for the estimation of costs and benefits of future ecosystem restoration projects in all major biomes, based on information from comparable projects on which data are collected through a standardized framework". Of course, the question of whether a standardized approach is feasible remains, considering the diversity in landscapes and land-use practices. But such an initiative is a good starting point for providing the missing database on all costs and benefits categories, particularly the indirect use and non-use benefits, as well as maintenance and opportunity costs categories, which are rarely captured in CBA. It can also provide guidelines to be used broadly by supporting donors, investors, and a wider range of stakeholders10.

5.6. Economic Attractiveness of Restoration and Implications for Private and Impact Investments

From the analysis, proportionately, more studies reported positive NPVs for most of the restoration strategies. In fact, for some restoration options, all the studies conducted reported positive NPV: agroforestry, soil and water conservation, mangrove restoration, and alien vegetation clearing. None of the restoration strategies had more studies that reported negative NPV compared to those that reported positive NPV. Positive NPV and economic viability, as confirmed in these studies, is a good starting point for promoting investments and financing. Investors will only be attracted to landscape restoration if their risks are covered, or at least mitigated to an acceptable level [18]. CBA is a first step in documenting the economic viability of landscape restoration and providing empirical evidence that in the long-term, benefits accruing from restoration outweigh the vast investment costs associated with landscape restoration.

---

[10] https://www.vi-med.forestweek.org/sites/default/files/presentations/docs/c5-teer-garavaglia.pdf

5.7. *Insufficient CBA Evidence on Landscape Restoration*

Compared to the relatively large number of restoration projects and studies, few have conducted comprehensive CBA. These studies were skewed towards some regions and some restoration options. For example, almost half of the existing studies were conducted in Africa. Similarly, substantially more studies focused on reforestation & afforestation and agroforestry; this is probably because these are among the common landscape restoration options. However, there remains a gap in CBA studies for other popular restoration options, including soil and water conservation practices and establishment of woodlots.

Nonetheless, some of the studies took a global focus, and some assessed and compared CBA results over many restoration strategies. For example, one of the most comprehensive CBA study we reviewed was conducted in Kenya by Reference [31] and compared returns for several landscape restoration strategies in different landscapes. In addition, Reference [20] took a global focus by assessing the net present value of the Bonn Challenge. Such studies form a good starting point for building a comprehensive CBA database upon which resource allocation in restoration can be based.

Overall, a major reason for the relatively few CBA studies is because most restoration projects do not budget for a CBA study. This may be due to an assumption that such projects always yield positive gains which may not necessarily hold. However, owing to scarce resources and the growing global demand for restoration, CBA studies can provide empirical evidence of restoration options with good returns on investment under different landscapes. Conducting a comprehensive economic CBA for landscape restoration is costly, data-intensive, and time-consuming; thus, there is a need for restoration projects to budget for this adequately.

6. Conclusions

This study set out to understand the breadth and depth of current CBA applications in landscape restoration, in a bid to find ways of improving its usefulness in planning, investments, and policies related to land restoration. Thirty-one out of 2056 studies were found to meet the CBA study selection criteria, i.e., they had conducted an economic CBA on at least one landscape restoration strategy. Three of these studies were of global character, while more than half covered African countries, with about two each covering Europe, Asia, Latin America, and the Middle East, respectively. Agroforestry, afforestation, reforestation, and assisted natural regeneration seem to be the most studied with at least five studies each. Other forms of land restoration are lagging. Most studies show a positive NPV for at least one restoration option, pointing to and confirming that restoration can be a viable private investment. Because most studies do not capture public benefits, evidence for public investments remains thin and could potentially hamper prioritization of government investments where resources are scarce. The study also identifies a number of areas for improvement in CBA from the stocktake. These include capturing all costs categories, including opportunity costs and maintenance and monitoring costs; going beyond direct use values; capturing public benefits; conducting thorough sensitivity analysis; the need for standardization and or guidance; and the insufficient CBA evidence on landscape restoration. Overall, the limited extent and depth in landscape restoration CBA studies suggest a great need to improve both quantity and quality in order to better inform planning, policies and investments in landscape restoration.

Author Contributions: Conceptualization, P.W., P.A.M., and L.D.; methodology, P.W. and E.G.; validation, P.W., P.A.M., and E.G.; formal analysis, P.W. and E.G.; investigation, E.G.; resources, P.W.; data curation E.G. and P.W.; writing—original draft preparation, P.W.; writing—review and editing, P.W., P.A.M, E.G., and L.D.: supervision, P.A.M.; project administration, P.M.; funding acquisition, P.A.M. and L.D. All authors have read and agreed to the published version of the manuscript.

Appendix A

Table A1. A summary of studies conducted to assess CBA of Landscape restoration following a systematic review.

No.	Paper	Country	Type of Restoration	Data Source	Years	Benefits	Costs	Net Present Value	Sensitivity Analysis	Private or Communal
1	Baig et al. (2016) [44]	Philippines	Ecosystem-based adaptation using mangrove protection and planting	Secondary data	20	Total economic values (direct use, non-direct use values and non-use values	Implementation costs,	Positive	Yes, to discount rate	Communal
2	Becker et al. (2018) [39]	Israel	Full and partial river restoration	Primary data through CVM and travel cost methods	varies	Total economic values; Use values (direct and optional use values) and non-use values (inheritance (bequest) and existence values)	Restoration costs (fixed value and the yearly value of maintenance)	Positive	No	Communal
3	Birch et al. (2010) [7]	Latin America	Dryland forest restoration	Primary data through spatial analysis	20	Carbon sequestration, NTFPs, timber, tourism and livestock products (benefit between restoration and BAU scenario)	Implementation costs (fencing and fire suppression), opportunity costs (cost foregone from livestock production from forest expansion	Positive	Yes, to discount rates and market price of carbon	Private and communal
4	Bonnieux and Le Goffe (1997) [30]	France	Landscape restoration using hedgerows	WTP (to assess public benefits	100	Firewood, timber and public benefits	Planting and regenerating costs, maintenance costs	Negative	No	Communal
5	Chadourne et al. (2012) [34]	USA (Tennessee)	Forest landscape restoration	Primary data-hedonic models were used to	50	Indirect use values (air pollution mitigation and stormwater control)	Explicit costs (land acquisition, labor, seedlings, materials) and amenity value	Positive	No	Private and communal
6	De Groot et al. (2013) [27]	Global	Restoration of 9 different biomes	Secondary data Reviews of 94 studies	20	Total Economic value of all services	Implementation costs, Maintenance costs	Positive for the various restoration types considered	Yes, to discount rate and to worst-case and best-case scenarios	Communal
7	ELD Initiative and UNEP (2015) [29]	Africa (FDjibouti,	Sustainable land management against soil erosion	Secondary data mostly from FAO and world bank data	15	Avoided crop damages from erosion control	SLM establishment cost and SLM maintenance costs	Positive	Yes, by varying discount rates, prices of cereals, capital and maintenance costs.	Private and communal

Table A1. *Cont.*

No.	Paper	Country	Type of Restoration	Data Source	Years	Benefits	Costs	Net Present Value	Sensitivity Analysis	Private or Communal
8	Elmqvist et al. (2015) [28]	Global	Urban areas restoration (Green and blue infrastructure in urban areas)	Secondary data (review)-Benefit transfer	20	Ecosystem services (pollution and air regulation, carbon sequestration, stormwater reduction, temperature regulation, recreation, positive health effects)	Costs for planning, preparation, modest soil restoration, plant propagation, and planting both for grasslands and woodlands	Positive	Yes, to discount rates and max/min benefits and costs	Communal
9	Verdone and Seidl (2017) [20]	Global	CBA of FLR within the Bonn Challenge within six different biomes	Primary and secondary data from TEEB	varies	Total Economic value both direct and indirect benefits	Bonn challenge cost of restoration	Positive	No	Private and communal
10	Gasparinetti et al. (2019) [32]	Brazil (South Amazon)	FLR through agroforestry with cocoa, coffee and Guarana	Primary data	30	Direct outputs and ecosystem services	Maintenance, fencing, labor costs, machine costs	Positive	Yes, to discount rates	Private
11	Hofer et al. (2010) [45]	Brazil (Amazon)	Reforestation (land use from pastures to forest for carbon sequestration (carbon for credits)	Secondary data	20	Carbon sequestration	Opportunity, implementation, and transaction costs	Negative	Yes, to different carbon prices	Communal
12	Narayan et al. (2017) [33]	Mozambique	Mangrove restoration to shelter against storms and flooding	Secondary data from an adaptation project conducted in 2013	100	Reduction in storm damages to houses, fish production, aquaculture, apiculture, carbon sequestered by growing mangroves	Costs of buying the seedlings; labor for planting, maintenance, and support staff; and hydrological restoration	Positive	Yes, to different carbon prices and discount rates	Private and communal
13	Newton et al. (2012) [35]	UK	Habitat restoration for river catchment	Primary data	10 and 50 years	Marginal value of benefits- carbon, timber, crops, livestock and recreational, aesthetic and cultural values	Initial capital investment and annual maintenance costs	Negative	Yes, to different carbon prices and discount rates	Communal
14	Pistorius et al. (2017) [37]	Ethiopia	FLR-(1) Afforestation/ reforestation (2) participatory forest management (3) sustainable woodland management (4) restoration of afro-alpine or sub-afro-alpine (5) establishment of woodlots	Primary data-spatial analysis and expert opinion	20	Provisioning services (timber & NTFPs), Carbon sequestration	Investment costs and labor costs	Positive except for the afro-alpine slope restoration	No	Communal

Table A1. *Cont.*

No.	Paper	Country	Type of Restoration	Data Source	Years	Benefits	Costs	Net Present Value	Sensitivity Analysis	Private or Communal
15	Mills et al. (2007) [36]	South Africa	Restoration of natural capital through Subtropical Thicket Restoration	Secondary data with simulations	50	livestock and game production, harvesting plant products (assuming natural recovery of biodiversity), and carbon sequestration	Transaction costs (including costs of verification of carbon stocks), labor costs, opportunity costs	Positive	Yes, to various parameters including biomass growth rate,	Communal
16	Holmes et al. (2007) [46]	South Africa	Restoring Natural Capital Following Alien Plant Invasions in Fynbos Ecosystems	Projections from secondary data	30	Direct and indirect use benefits	Clearing costs, installation costs	Positive	Yes, to discount rates	Communal
17	Rizzetti et al. (2018) [47]	Vietnam	FLR through ANR, extended acacia rotation, native species rotation, SWC	Projections from secondary data	23, 30, 2	Crop income, income from timber	Labor costs, seedling cost	Positive	No	Private and communal
18	Schiappacasse et al. (2012) [38]	Chile	Dryland forest restoration thru reforestation using native trees	Primary data using contingent valuation method	25	WTP for forest restoration for the entire pollution of the city	Implementation costs, operating costs,	Negative	Yes, to discount rate	Communal
19	Currie et al. (2009) [48]	South Africa	Alien vegetation clearing for water yield and tourism	Primary data and projections	15	water and tourism benefits (tourism benefits involved revenue from the sale of tickets)	Costs of alien invasive plant removal, gully-erosion repair and reseeding with indigenous plants	Positive	Yes, to discount rates and with realistic and pessimistic scenarios	Communal
20	Silva and Nunes (2017) [49]	Brazil Amazon	Forest restoration through sustainable forest management (legal logging) and agroforestry	Secondary	11	Timber from logging, financial benefits of AFS (timber and NTFPs)	Implementation costs, transaction costs, opportunity costs (loss from agriculture and livestock)	Negative	Yes, to discount rates and different scenarios	Communal
21	Tuan and Tinh (2013) [50]	Vietnam	Mangrove restoration	Secondary and primary using CVM to value non-use values and market methods to value use values	22	Direct use values, indirect use values and non-use values. WTP for non-use values,	Maintenance and protection costs, mangrove restoration	Positive	Yes, to discount rates	Private and communal
22	Monela (2005) [51]	Tanzania	FLR through agroforestry and silviculture (Ngitili)	Primary data through expert evaluation and literature review	20	Direct use values (timber and other NTFPs), time saved in collecting firewood and water,	Total project cost for restoration	Positive	Yes, to discount rates	Private and communal

Table A1. *Cont.*

No.	Paper	Country	Type of Restoration	Data Source	Years	Benefits	Costs	Net Present Value	Sensitivity Analysis	Private or Communal
23	Cheboiwo et al. (2019) [31]	Kenya	Afforestation or reforestation of degraded natural forests, Rehabilitation of degraded natural forests, Agroforestry in cropland, Commercial tree and bamboo growing on potentially marginal cropland and un-stocked forest plantation forests, Tree-based buffer zones along water bodies and wetlands, Tree-based buffer zones along roads and restoration of degraded rangelands)	Expert discussions, activity restoration budgets and extensive review of various land use literature. Benefits and opportunity costs were valued using market prices, avoided cost/replacement cost and benefit transfer approaches	30	Direct (crop harvests, timbers and NTFPs) and indirect use values (carbon sequestration, soil erosion control and increased soil fertility)	Implementation costs, opportunity costs, monitoring and maintenance costs	Positive for the various restoration types considered	Yes, to discount rates	Private and communal
24	Ministry of Natural resources, energy and mining-Malawi (2017) [3]	Malawi	Conservation agriculture, agroforestry. FMNR, Community plantations and private woodlots, Natural forest management	Primary data	20	Direct and indirect use benefits	Implementation costs and opportunity costs	Positive for the various restoration types considered	Yes, to discount rates	Private and communal
25	Ministry of Water & Env-Uganda (2016) [52]	Uganda	Reforestation and afforestation, woodlots, Agroforestry, Natural regeneration	Budgets, expert discussions and secondary sources	30	Direct and indirect use benefits	Implementation costs	Positive for the various restoration types	Yes, to discount rates	Private and communal
26	FAO and UNHCR (2018) [53]	Tanzania	Wood-energy rehabilitation (Afforestation and reforestation), Agroforestry, Rehabilitation of degraded native forests	Primary data (field observations, KIIs, FGDs) secondary data	10	Direct benefits (wood fuel) and indirect use benefits (carbon sequestration)	Implementation, operational and opportunity costs	Positive for wood energy plantations and agroforestry but negative for the rest	Yes, to discount rates and wood prices	Private and communal
27	Aymeric et al. (2014) [54]	Sudan	SLM through *A. senegal* Agroforestry	Secondary data	25	Direct use benefits (fuelwood and Gum arabica) and indirect use benefits (N fixation, avoided nutrient loss, aquifer recharge, carbon sequestration)	Implementation and maintenance costs	Positive	Yes, to discount rates	Private

Table A1. *Cont.*

No.	Paper	Country	Type of Restoration	Data Source	Years	Benefits	Costs	Net Present Value	Sensitivity Analysis	Private or Communal
28	Ministry of Natural resources-Rwanda (2014) [55]	Rwanda	Agroforestry, Well managed woodlots, Natural forest regeneration, protective forests	Primary and secondary data through simulations and predictions	20–30	Direct use benefits (crops, wood) and indirect use benefits (carbon sequestration and erosion control	Implementation, operational and monitoring costs		Yes, to discount rates	Private and communal
29	Tesfaye et al. (2016) [56]	Ethiopia	Soil conservation measures (soil bunds, stone bunds, Fanya juu bunds)	Primary data	27	Yield increment from implementation of bunds	Investment and maintenance costs	Positive	Yes, to investment and maintenance costs and the market price of yield	Private
30	Wiskerke et al. (2010) [41]	Tanzania	A small-scale forestation project for carbon sequestration, a short rotation woodlot and a Jatropha plantation	Primary data (expert opinions and field survey) and secondary data	7	Direct benefits (wood fuel, electricity from jatropha, etc.) and indirect benefits (avoided deforestation, improved health, indirect economic benefits)	Production costs and opportunity costs	Positive for woodlots and jatropha for electrification and soap production, negative for forestation for C credits	No	Private and communal
31	Onduru and Muchena (2011) [57]	Kenya	SWC practices, such as mulching, zero tillage, stone lines, contour ridges, micro catchments with bananas, terracing, and others	Primary data	15	Incremental yield benefits from the adoption of SWC practices	Investment and maintenance costs	Positive	Yes, to discount rates	Private

References

1. Intergovernmental Platform on Biodiversity and Ecosystem Services (IPBES). *Summary for Policymakers of the Global Assessment Report on Biodiversity and Ecosystem Services of the Intergovernmental Science-Policy Platform on Biodiversity and Ecosystem Services*; IPBES Secretariat: Bonn, Germany, 2019.
2. Ding, H.; Altamirano, J.C.; Anchondo, A.; Faruqi, S.; Verdone, M.; Wu, A.; Ortega, A.A.; Zamora Cristales, R.; Chazdon, R.; Vergara, W. *Roots of Prosperity: The Economics and Finance of Restoring Land*; World Resources Institute: Washington, DC, USA, 2017.
3. Ministry of Natural Resources; Energy and Mining—Malawi. *Forest Landscape Restoration Opportunities Assessment for Malawi*; NFLRA: Lilongwe, Malawi; IUCN: Gland, Switzerland; WRI: Washington, DC, USA, 2017.
4. Crookes, D.J.; Blignaut, J.N.; De Wit, M.P.; Esler, K.J.; Le Maitre, D.C.; Milton, S.J.; Mitchell, S.A.; Cloete, J.; De Abreu, P.; Fourie, H.; et al. System dynamic modelling to assess economic viability and risk trade-offs for ecological restoration in South Africa. *J. Environ. Manag.* **2013**, *120*, 138–147. [CrossRef]
5. Gichuki, L.; Brouwer, R.; Davies, J.; Vidal, A.; Kuzee, M.; Magero, C.; Walter, S.; Lara, P.; Oragbade, C.; Gilbey, B. *Reviving Land and Restoring Landscapes: Policy Convergence between Forest Landscape Restoration and Land Degradation Neutrality*; IUCN: Gland, Switzerland, 2019.
6. Yirdaw, E.; Tigabu, M.; Monge, A. Rehabilitation of degraded dryland ecosystems—Review. *Silva Fenn.* **2017**, *51*, 1–32. [CrossRef]
7. Birch, J.C.; Newton, A.C.; Aquino, C.A.; Cantarello, E.; Echeverría, C.; Kitzberger, T.; Schiappacasse, I.; Garavito, N.T. Cost-effectiveness of dryland forest restoration evaluated by spatial analysis of ecosystem services. *Proc. Natl. Acad. Sci. USA* **2010**, *107*, 21925–21930. [CrossRef]
8. Society for Ecological Restoration International Science & Policy Working Group. *The SER International Primer on Ecological Restoration*; Society for Ecological Restoration International: Washington, DC, USA, 2004; Volume 2, pp. 206–207.
9. Ciccarese, L.; Mattsson, A.F.; Pettenella, D. Ecosystem services from forest restoration: Thinking ahead. *New For.* **2012**, *43*, 543–560. [CrossRef]
10. Wortley, L.; Hero, J.-M.; Howes, M.J. Evaluating Ecological Restoration Success: A Review of the Literature. *Restor. Ecol.* **2013**, *21*, 537–543. [CrossRef]
11. Chazdon, R.L.; Brancalion, P.H.S.; Lamb, D.; Laestadius, L.; Calmon, M.; Kumar, C. A Policy-Driven Knowledge Agenda for Global Forest and Landscape Restoration. *Conserv. Lett.* **2016**, *10*, 125–132. [CrossRef]
12. Menz, M.H.M.; Dixon, K.W.; Hobbs, R.J. Hurdles and Opportunities for Landscape-Scale Restoration. *Science* **2013**, *339*, 526–527. [CrossRef] [PubMed]
13. Shono, K.; Cadaweng, E.A.; Durst, P.B. Application of Assisted Natural Regeneration to Restore Degraded Tropical Forestlands. *Restor. Ecol.* **2007**, *15*, 620–626. [CrossRef]
14. Holl, K.; Aide, T. When and where to actively restore ecosystems? *For. Ecol. Manag.* **2011**, *261*, 1558–1563. [CrossRef]
15. Suding, K.N.; Higgs, E.; Palmer, M.; Callicott, J.B.; Anderson, C.B.; Baker, M.; Gutrich, J.J.; Hondula, K.L.; LaFevor, M.C.; Larson, B.M.H.; et al. Committing to ecological restoration. *Science* **2015**, *348*, 638–640. [CrossRef]
16. Dellasala, D.A.; Martin, A.; Spivak, R.; Schulke, T.; Bird, B.; Criley, M.; Van Daalen, C.; Kreilick, J.; Brown, R.; Aplet, G. A Citizen's Call for Ecological Forest Restoration: Forest Restoration Principles and Criteria. *Ecol. Restor.* **2003**, *21*, 14–23. [CrossRef]
17. Van Noordwijk, M.; Gitz, V.; Minang, P.A.; Dewi, S.; Leimona, B.; Duguma, L.; Pingault, N.; Meybeck, A. People-Centric Nature-Based Land Restoration Through Agroforestry: A Typology. *Land* **2020**, *9*, 251. [CrossRef]
18. FAO; Global Mechanism of the UNCCD. *Sustainable Financing for Forest and Landscape Restoration*; Discussion paper; Food and Agriculture Organization: Rome, Italy, 2015.
19. Gromko, D.; Pistorius, T.; Seebauer, M.; Braun, A.; Meier, E. *Economics of Forest Landscape Restoration. Estimating impacts, costs and benefits from ecosystem services*; Unique Forestry and Land Use: Freiburg, Germany, 2019.
20. Verdone, M.; Seidl, A. Time, space, place, and the Bonn Challenge global forest restoration target. *Restor. Ecol.* **2017**, *25*, 903–911. [CrossRef]

21. Verdone, M. *A Cost-Benefit Framework for Analyzing Forest Landscape Restoration Decisions*; IUCN: Gland, Switzerland, 2015.
22. Pascual, U.; Muradian, R.; Brander, L.; Gómez-Baggethun, E.; Martín-López, B.; Verma, M.; Armsworth, P.; Christie, M.; Cornelissen, H.; Eppink, F.; et al. The economics of valuing ecosystem services and biodiversity. In *The Economics of Ecosystems and Biodiversity: Ecological and Economic Foundations*; UNEP: Hertfordshire, UK, 2010; pp. 183–256.
23. Gundimeda, H.; Markandya, A.; Bassi, A.M. TEEBAgriFood methodology: An overview of evaluation and valuation methods and tools. In *TEEB for Agriculture & Food: Scientific and Economic Foundations*; UN Environment: Geneva, Switzerland, 2018.
24. Chiputwa, B.; Ihli, H.J.; Wainaina, P.; Gassner, A. Accounting for the invisible value of trees on farms through valuation of ecosystem services. In *The Role of Ecosystem Services in Sustainable Food Systems*; Elsevier BV: Amsterdam, The Netherlands, 2020; pp. 229–261.
25. Ring, I.; Hansjürgens, B.; Elmqvist, T.; Wittmer, H.; Sukhdev, P. Challenges in framing the economics of ecosystems and biodiversity: The TEEB initiative. *Curr. Opin. Environ. Sustain.* **2010**, *2*, 15–26. [CrossRef]
26. Malkamäki, A.; D'Amato, D.; Hogarth, N.J.; Kanninen, M.; Pirard, R.; Toppinen, A.; Zhou, W. A systematic review of the socio-economic impacts of large-scale tree plantations, worldwide. *Glob. Environ. Chang.* **2018**, *53*, 90–103. [CrossRef]
27. De Groot, R.S.; Blignaut, J.N.; Van Der Ploeg, S.; Aronson, J.; Elmqvist, T.; Farley, J. Benefits of Investing in Ecosystem Restoration. *Conserv. Biol.* **2013**, *27*, 1286–1293. [CrossRef]
28. Elmqvist, T.; Setälä, H.; Handel, S.; Van Der Ploeg, S.; Aronson, J.K.; Blignaut, J.N.; Gomez-Baggethun, E.; Nowak, D.J.; Kronenberg, J.; De Groot, R. Benefits of restoring ecosystem services in urban areas. *Curr. Opin. Environ. Sustain.* **2015**, *14*, 101–108. [CrossRef]
29. ELD Initiative; UNEP. The Economics of Land Degradation in Africa: Benefits of Action Outweigh the Costs. A Complementary Report to the ELD Initiative. 2015. Available online: www.eld-initiative.org (accessed on 20 January 2020).
30. Bonnieux, F.; Le Goff, P. Valuing the Benefits of Landscape Restoration: A Case Study of the Cotentin in Lower-Normandy, France. *J. Environ. Manag.* **1997**, *50*, 321–333. [CrossRef]
31. Cheboiwo, J.; Langat, D.; Muga, M.; Kiprop, J. *Economic Analysis of Forest Land Restoration Options in Kenya*; Ministry of Environment and Forestry: Nairobi, Kenya, 2019.
32. Gasparinetti, P.; Brandão, D.O.; Araujo, V.; Araujo, N. *Economic Feasibility Study for Forest Landscape Restoration Banking Models: Cases from Southern Amazonas State*; CSF-Brazil and WWF-Brazil: Brasilia, Brazil, 2019.
33. Narayan, T.; Foley, L.; Haskell, J.; Cooley, D.; Hyman, E. *Cost-Benefit Analysis of Mangrove Restoration for Coastal Protection and an Earthen Dike Alternative in Mozambique*; USAID: Washington, DC, USA, 2017.
34. Chadourne, M.H.; Cho, S.-H.; Roberts, R.K. Identifying Priority Areas for Forest Landscape Restoration to Protect Ridgelines and Hillsides: A Cost-Benefit Analysis. *Can. J. Agric. Econ. Can. D'agroeconomie* **2012**, *60*, 275–294. [CrossRef]
35. Newton, A.C.; Hodder, K.; Cantarello, E.; Perrella, L.; Birch, J.C.; Robins, J.M.; Douglas, S.J.; Moody, C.E.; Cordingley, J. Cost-benefit analysis of ecological networks assessed through spatial analysis of ecosystem services. *J. Appl. Ecol.* **2012**, *49*, 571–580. [CrossRef]
36. Mills, A.J.; Turpie, J.K.; Cowling, R.M.; Marais, C.; Kerley, G.I.; Lechmere-Oertel, R.G.; Sigwela, A.M.; Powell, M. Assessing costs, benefits, and feasibility of restoring natural capital in subtropical thicket in South Africa. In *Restoring Natural Capital, Science, Business, and Practice*; Island Press: Washington, DC, USA, 2007; pp. 179–187.
37. Pistorius, T.; Carodenuto, S.; Wathum, G. Implementing Forest Landscape Restoration in Ethiopia. *Forests* **2017**, *8*, 61. [CrossRef]
38. Schiappacasse, I.; Nahuelhual, L.; Vásquez, F.; Echeverría, C. Assessing the benefits and costs of dryland forest restoration in central Chile. *J. Environ. Manag.* **2012**, *97*, 38–45. [CrossRef] [PubMed]
39. Becker, N.; Greenfeld, A.; Shamir, S.Z. Cost–benefit analysis of full and partial river restoration: The Kishon River in Israel. *Int. J. Water Resour. Dev.* **2018**, *35*, 871–890. [CrossRef]
40. Van der Ploeg, S.; de Groot, R.S. *The TEEB Valuation Database–A Searchable Database of 1310 Estimates of Monetary Values of Ecosystem Services*; Foundation for Sustainable Development: Wageningen, The Netherlands, 2010.
41. Wiskerke, W.; Dornburg, V.; Rubanza, C.; Malimbwi, R.; Faaij, A.P. Cost/benefit analysis of biomass energy supply options for rural smallholders in the semi-arid eastern part of Shinyanga Region in Tanzania. *Renew. Sustain. Energy Rev.* **2010** *14*, 148–165. [CrossRef]

42. Singh, R.; van Noordwijk, M.; Chaturvedi, O.; Garg, K.K.; Dev, I.; Wani, S.P.; Rizvi, J. Public co-investment in groundwater recharge in Bundelkhand, Uttar Pradesh, India. In *Sustainable Development through Trees on Farms: Agroforestry in Its Fifth Decade*; van Noordwijk, M., Ed.; World Agroforestry (ICRAF) Southeast Asia Regional Program: Bogor, Indonesia, 2019.
43. Duguma, L.; Centre, W.A.; Minang, P.; Aynekulu, E.; Carsan, S.; Nzyoka, J.; Bah, A.; Jamnadass, R. *From Tree Planting to Tree Growing: Rethinking Ecosystem Restoration Through Trees*; ICRAF: Nairobi, Kenya, 2020.
44. Baig, S.P.; Rizvi, A.; Josella, M.; Palanca-Tan, R. *Cost and Benefits of Ecosystem Based Adaptation: The Case of the Philippines*; IUCN: Gland, Switzerland, 2015.
45. Hofer, C.T. *Carbon Finance & Cattle Externalities in the Brazilian Amazon: Pricing Reforestation in Terms of Restoration Ecology*; University of Connecticut: Storrs, CT, USA, 2010.
46. Holmes, P.M.; Richardson, D.M.; Marais, C. Cost and benefits of restoring natural capital following alien plant invasions in Fynbos ecosystems in South Africa. In *Restoring Natural Capital, Science, Business, and Practice*; Island press: Washington, DC, USA, 2007; pp. 188–197.
47. Rizzeti, D.; Swaans, K.; Holden, J.; Brunner, J.; Le, T.; Nguyen, T. *Assessing Opportunities in Forest Landscape Restoration in Quang Tri, Vietnam*; IUCN: Gland, Switzerland, 2018; Volume 46.
48. Currie, B.; Milton, S.J.; Steenkamp, J. Cost–benefit analysis of alien vegetation clearing for water yield and tourism in a mountain catchment in the Western Cape of South Africa. *Ecol. Econ.* **2009**, *68*, 2574–2579. [CrossRef]
49. Silva, D.; Nunes, S. *Evaluation and Economic Modelling of Forest Restoration in the State of Pará, Eastern Brazilian Amazon*; Imazon: Pará, Brasil, 2017.
50. Tuan, T.H.; Tinh, B.D. *Cost-Benefit Analysis of Mangrove Restoration in Thi Nai Lagoon, Quy Nhon City, Vietnam*; IIED: London, UK, 2013.
51. Monela, G.C.; Chamshama, S.A.O.; Mwaipopo, R.; Gamassa, D.M. *A Study on the Social, Economic and Environmental Impacts of Forest Landscape Restoration in Shinyanga Region, Tanzania*; Final Report to the Ministry of Natural Resources and Tourism and IUCN; IUCN: Gland, Switzerland, 2005.
52. Ministry of Water and Environment—Uganda. *Forest Landscape Restoration Opportunity Assessment Report for Uganda*; IUCN: Gland, Switzerland, 2016.
53. Gianvenuti, A.; Vyamana., V.G. *Cost–Benefit Analysis of Forestry Interventions for Supplying Woodfuel in a Refugee Situation in the United Republic of Tanzania*; Food and Agriculture Organization of the United Nations (FAO): Rome, Italy; United Nations High Commissioner for Refugees (UNHCR): Geneva, Switzerland, 2018.
54. Aymeric, R.; Myint., M.M.; Westerberg, V. An economic valuation of sustainable land management through agroforestry in eastern Sudan. In *Report for the Economics of Land Degradation Initiative by the International Union for Conservation of Nature*; ELD: Nairobi, Kenya, 2014; Available online: www.eld-initiative.org (accessed on 17 January 2020).
55. Ministry of Natural Resources—Rwanda. *Forest Landscape Restoration Opportunity Assessment for Rwanda. MINIRENA: Kigali Kigali, Rwanda*; IUCN: Gland, Switzerland; WRI: Washington, DC, USA, 2014.
56. Tesfaye, A.; Brouwer, R.; Van Der Zaag, P.; Negatu, W. Assessing the costs and benefits of improved land management practices in three watershed areas in Ethiopia. *Int. Soil Water Conserv. Res.* **2016**, *4*, 20–29. [CrossRef]
57. Onduru, D.D.; Muchena, F.N. *Cost-Benefit Analysis of Land Management Options in the Upper Tana, Kenya*; Green water credits Reports 15; ISRIC—World Soil Information: Wageningen, The Netherlands, 2011.

10

Carbon Storage Potential of Silvopastoral Systems of Colombia

Ermias Aynekulu [1,*], Marta Suber [1], Meine van Noordwijk [1], Jacobo Arango [2], James M. Roshetko [1] and Todd S. Rosenstock [1]

[1] World Agroforestry (ICRAF), United Nations Avenue, P.O. Box 30677, Nairobi 00100, Kenya; m.suber@cgiar.org (M.S.); m.vannoordwijk@cgiar.org (M.v.N.); j.roshetko@cgiar.org (J.M.R.); t.rosenstock@cgiar.org (T.S.R.)

[2] International Centre for Tropical Agriculture (CIAT), Km 17, Recta Cali–Palmira CP, Apartado Aéreo 6713, Cali 763537, Colombia; j.arango@cgiar.org

* Correspondence: e.betemariam@cgiar.org

Abstract: Nine Latin American countries plan to use silvopastoral practices—incorporating trees into grazing lands—to mitigate climate change. However, the cumulative potential of scaling up silvopastoral systems at national levels is not well quantified. Here, we combined previously published tree cover data based on 250 m resolution MODIS satellite remote sensing imagery for 2000–2017 with ecofloristic zone carbon stock estimates to calculate historical and potential future tree biomass carbon storage in Colombian grasslands. Between 2000 and 2017, tree cover across all Colombian grasslands increased from 15% to 18%, with total biomass carbon (TBC) stocks increasing from 0.41 to 0.48 Pg. The range in 2017 carbon stock values in grasslands based on ecofloristic zones (5 to 122 Mg ha^{-1}) suggests a potential for further increase. Increasing all carbon stocks to the current median and 75th percentile levels for the respective eco-floristic zone would increase TBC stocks by about 0.06 and 0.15 Pg, respectively. Incorporated into national C accounting, such Tier 2 estimates can set realistic targets for silvopastoral systems in nationally determined contributions (NDCs) and nationally appropriate mitigation actions (NAMAs) implementation plans in Colombia and other Latin American countries with similar contexts.

Keywords: agroforestry; carbon sequestration; climate change mitigation; grazing management; land restoration; nationally determined contribution; silvopastoral; tree cover

---

1. Introduction

The ability to limit global warming below 2 °C, the target set by the Paris Climate Agreement, is predicated on significant greenhouse gas (GHG) emission reductions in the agriculture and land use sector (AFOLU) [1,2]. While land use accounts for nearly 25% of net annual anthropogenic GHG emissions [3], it also offers many options to mitigate climate change. Estimates suggest that land use interventions could generate up to 30% of the emission reductions and carbon sequestration needed to meet the Paris Agreement's ambition [4,5]. Most agriculture and land use mitigation interventions target the reduction of emissions from livestock systems, rice systems, deforestation, and nitrogen fertilizer. However, the integration of trees in crop and pasture lands, also known as agroforestry, is another potentially significant intervention option [6].

Nine Latin American countries have identified silvopastoral systems (SPS), agroforestry systems where trees are managed or planted in pasturelands, as a priority action to mitigate climate change within their nationally determined contributions (NDCs) to the United Nations Framework Convention on Climate Change (UNFCCC) [7,8] and the Nationally Appropriate Mitigation Actions (NAMAs), two documents that prescribe national climate response priorities. The focus on livestock in Latin American countries is due to the economic importance of the sector and its leverage on national emissions. Indeed, the livestock sector accounts for 58–70% of the overall agricultural emissions in Latin America [9], making greening milk and meat production a critical aspect of meeting national goals.

Silvopastoral systems sequester carbon in biomass increasing the amount of carbon in the landscape. Silvopastoral systems accumulate more than 2 Mg C per ha per year above and below ground biomass and additional 0.5 Mg C per ha per year in the soil [10]. Grazing grasslands occupy more than 550 million hectares of land in Latin America. Therefore, scaling up silvopastoral systems would seemingly have the potential to increase carbon in the landscape despite the modest change per unit of land. Estimates of the climate change mitigation benefits of scaling up silvopastoral systems at the national level, however, are largely unavailable. Zomer et al. [11,12] provided a first estimate of t C stocks in tree biomass in crop lands at global scale but did not specifically consider trees in grasslands (i.e., silvopastoral systems). Griscom et al. [5], on the other hand, estimate silvopastoral C but mix these estimates together with other land management options to estimate a total mitigation potential of land restoration. However, countries need estimates of silvopastoral potential to set targets for climate action [13].

This study estimated the climate change mitigation benefits of expanding silvopastoral systems in Colombia to inform realistic inclusion of silvopastoral systems in NDC and NAMA planning options for Colombia and other Latin American countries with similar contexts. Expanding on the earlier analysis for agricultural lands [11,12], we explored historical C stocks for Columbian grassing lands in 2000 and 2017, before considering options to increase C stock density with the silvopastoral systems in the local context. Specific questions were: (i) What changes in tree cover in Columbian pastureland are evident from satellite imagery? (ii) What changes in C stock do such changes in tree cover imply, using ecofloristic-zone-specific C density estimates? (iii) Evaluating the existing statistical distribution, which further changes may be feasible?

Colombia was selected because of its active promotion of silvopastoral systems. For example, the Colombian government and National Livestock Federation (FEDEGAN) have a strategy to reduce GHG emissions from the agricultural sector by 13.46 Mt CO_2e yr^{-1} by 2030. The conversion of current grasslands to SPS is among the priority mitigation activities [14]. Importantly, Tapasco et al. [15] also identified SPS as the most promising policy option for achieving this goal, and Lerner et al. [16] notes that the country is in an ideal position to create integrated plans for sustainable cattle intensification, including SPS, conservation, and restoration initiatives.

2. Methods

2.1. Conceptual Approach

For this analysis, a methodology earlier applied to agricultural lands [8] was adapted to spatially quantify carbon sequestration from tree cover in grasslands and estimate potential carbon stocks under widespread implementation of SPS. The steps were to define the extent of grasslands in the target area; derive the current level of tree cover; estimate the carbon stocks in above and belowground biomass; estimate the potential increase with shifts in tree cover from the current to the 25th percentile, median, and 75th percentile of 2017 biomass carbon stocks; and quantify the difference between current and potential tree cover and carbon sequestration ("silvopastoral carbon gap"). A detailed description of the methodology used for this analysis is found in Zomer et al. [11,12].

2.2. Study Area

Given that both the Colombian government and independent researchers have identified SPS as a priority in achieving Colombia's agricultural emission reduction goals [15], we focused on Colombia as a case study. As of 2015, Colombia had more than 12 million ha of grassland (13% of the national land area) scattered throughout tropical moist deciduous forests (77% of grassland), tropical rainforest

(12%), and other ecofloristic zones (Figure 1). The largest grazing areas are found in the sparsely populated Llanos Orientales and Orinoquia regions of the Orinoco River basin. In other regions of the country, smaller grassland areas are interspersed with various land uses, including croplands and settlements.

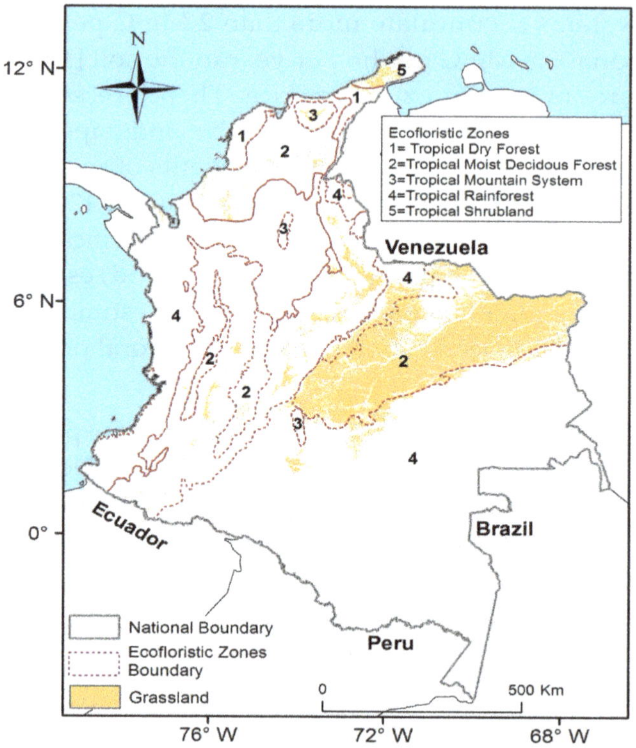

Figure 1. Grassland cover in 2015 (source: European Space Agency, 2017a) and ecofloristic zones of Colombia.

2.3. Data Collection and Analysis

Grassland area was derived at 300 m spatial resolution from the European Space Agency's (ESA) annual global land cover data. Per ESA recommendations, the ESA land use classes grass land, mosaic herbaceous cover > 50% and tree and shrub < 50% were combined to produce the grassland designation used in this analysis [17]. Percent tree cover was derived from previously published tree cover data based on the 250 m resolution Moderate-Resolution Imaging Spectroradiometer (MODIS) 44B Version 6 Vegetation Continuous Field remote sensing product [18]. This product is a continuous, quantitative representation of global land surface cover at 250 m spatial resolution. Ground cover gradations are based on percent tree cover, percent non-tree cover, and percent non-vegetated (bare) [18]. A detailed description of the methodology used for this analysis is found in Zomer et al. [11,12].

We used the default Intergovernmental Panel on Climate Change (IPCC) Tier 1 total carbon stock values for each ecofloristic zone as the minimum potential carbon stock values for 0% tree cover [19] (Table 1). We then estimated the total carbon sequestration potential of SPS by ecofloristic zone using the methodology recommended by IPCC Tier 1 and detailed in Zomer et al. [12]. The biomass carbon value of the equivalent (or most similar) Global Land Cover (GLC) 2000 Mixed Forest class [19] was used as a surrogate aboveground biomass carbon value for each ecofloristic zone to simulate full tree cover (100%). Belowground carbon sequestration potential was calculated using aboveground sequestration potential values and the root/shoot ratio for each of the ecofloristic zones as per IPCC Tier 1 values (Table 1) [19]. Below and aboveground C sequestration potential values were then summed to determine total potential carbon stock values for 100% tree cover. A carbon fraction value of 0.47 was uniformly applied to determine the carbon content of dry woody biomass. We then assumed a linear increase in biomass carbon from 0% to 100% tree cover for each ecofloristic zone. We used the Tier 1 value when tree cover was 0% and the maximum value for Mixed Forest when tree cover was 100%.

Table 1. IPCC Tier 1 values of above ground carbon stocks in grassland under different ecofloristic zones.

Ecofloristic Zone	Area		Above Ground Carbon Stocks Mg ha^{-1}		Root to Shoot Ratio
	ha	%	Minimum	Maximum	
Tropical dry forest	67,758	0.6	4	126	0.28
Tropical Moist deciduous forest	9,487,070	77	8	128	0.24
Tropical mountain system	1,076,967	8.8	6	87	0.27
Tropical rainforest	1,469,597	12	8	193	0.37
Tropical shrubland	206,495	1.7	4	126	0.28
Total	12,307,887	100			

We used the 2015 grassland area as a basis for 2000, 2008, and 2017 tree cover and biomass assessments to control for any change in grassland area over time. We used the 2017 25th, median, and 75th percentile carbon stock values in each ecofloristic zone to calculate carbon gaps and the current carbon sequestration potential. We identified areas with carbon stock values below the 25th percentile in each ecofloristic zone as those with the greatest potential to increase carbon stock. Similarly, we considered areas with carbon stock values between the 25th percentiles and medium values as having medium potential and areas between the median and 75th percentile values as having low potential to increase carbon stocks. Areas with greater than the 75th percentile tree cover values were considered as not suitable for increasing carbon stocks. This approach is particularly useful for spatially targeting SPS-based mitigation actions [20].

3. Results

3.1. Estimates of Tree Cover on Grassland in 2000, 2008, and 2017

Tree cover increased between 2000 and 2017. For presentation purposes, we classified these results into tree cover classes (Table 2). Land area with tree cover greater than 10% increased from 55% in 2000 to 74% in 2008 and 73% in 2017. The percentage of grasslands with tree cover > 30% increased with time, from 10% in 2000 to 12% in 2008 and 13% in 2017. Total tree cover across all grasslands increased by about 20% from 2000 to 2017.

Table 2. Grassland (ha) tree cover (%) in 2000, 2008, and 2017.

Tree Cover (%)	2000		2008		2017	
	ha	(%)	ha	%	ha	%
≤10	5,507,090	45	3,210,585	26	3,319,775	27
11–20	3,883,571	32	4,861,344	39	4,700,027	38
21–30	1,715,737	14	2,708,461	22	2,709,611	22
>30	1,201,488	10	1,527,496	12	1,578,474	13
Total	12,307,887	100	12,307,887	100	12,307,887	100

Tree cover increased on grassing lands between 2000 and 2017, but tree cover dynamics and the relative increase or decrease depended on the location. Colombian grassing lands show between 0% and 84% tree cover (Figure 2). The relative change on any given parcel increased up to 76% over this period while others decreased by 80%.

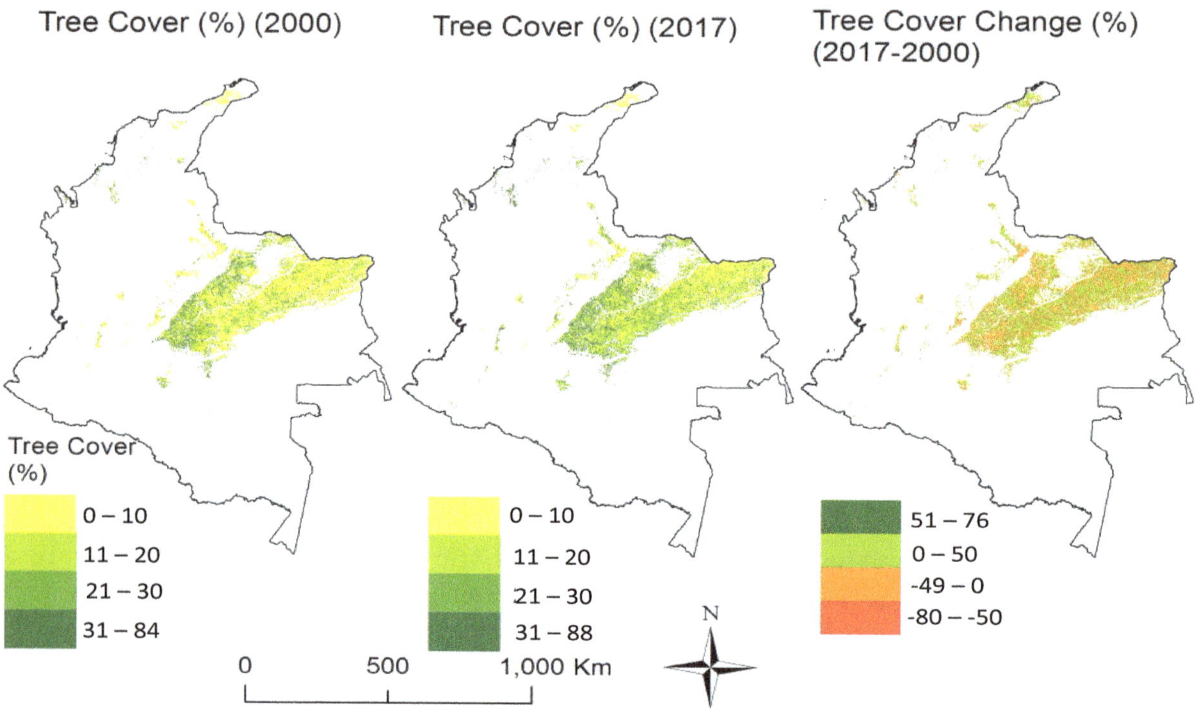

Figure 2. Tree cover change from 2000 to 2017 in grasslands.

3.2. *Biomass Carbon Stocks*

The TBC stock increased by 17% from 2000 to 2017 (Table 3). TBC stocks in Colombian grassland were 0.41 and 0.48 petagrams (Pg) C in 2000 and 2017, respectively (Table 3).

Table 3. Average and total biomass carbon stocks on Colombian grassland in 2000 and 2017.

Biomass Carbon	Average (SD) (Mg ha^{-1})			Total (Pg C)		
	2000	2017	Change	2000	2017	Change
Above ground	27 (14)	31 (14)	4	0.33	0.38	0.05
Total biomass	34 (18)	39 (18)	5	0.41	0.48	0.07

Land area with a C stock < 10 Mg ha^{-1} decreased from 2000 to 2017, and land area with C stocks of 26–100 Mg ha^{-1} increased (Table 4).

Table 4. Land area carbon stocks in 2000 and 2017.

Total Carbon Stocks (Mg ha^{-1})	2000 ha	%	2008 ha	%	2017 ha	%
≤10	489,333	4	610,508	5	355,760	3
11–25	4,967,851	40	2,715,698	22	3,063,775	25
26–50	4,937,931	40	6,199,358	50	6,207,735	50
51–75	1,185,878	10	1,968,009	16	1,796,833	15
76–100	514,348	4	575,421	5	643,622	5
>100	212,545	2	238,892	2	240,161	2
Total	12,307,887	100	12,307,887	100	12,307,887	100

The grassland biomass carbon stock maps are indicated in Figure 3.

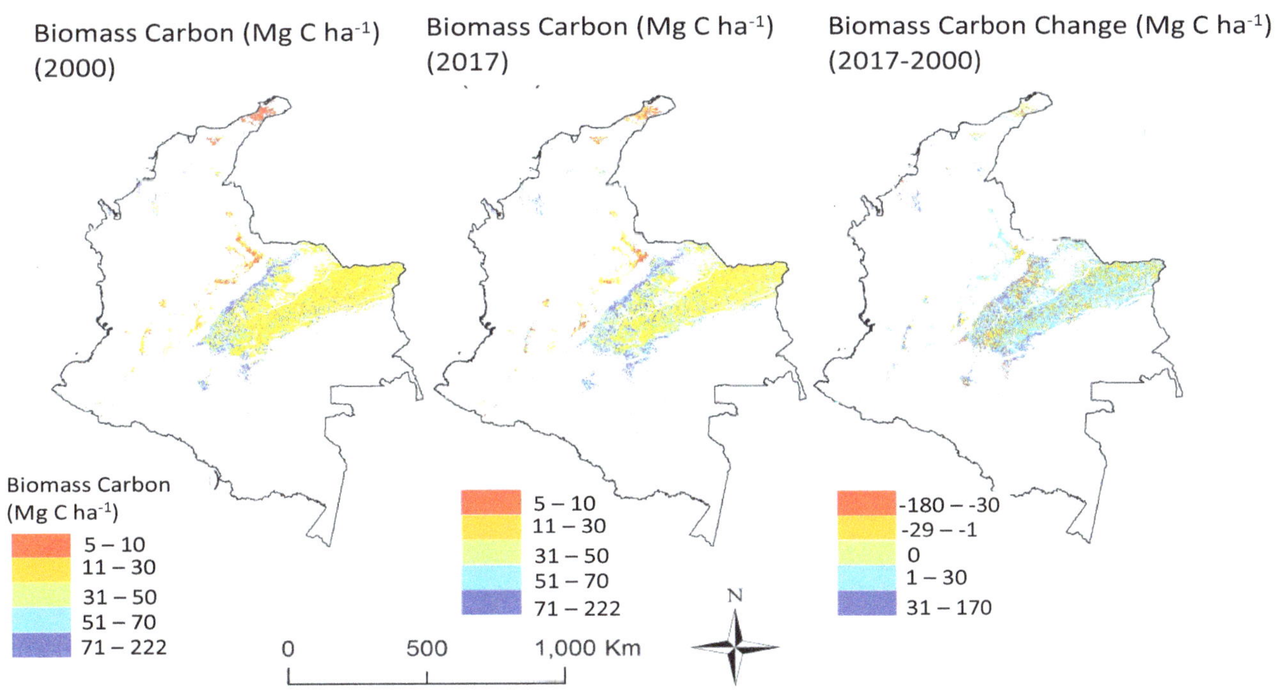

Figure 3. Biomass carbon stocks change (Mg ha^{-1}) from 2000 to 2017.

3.3. Carbon Stock Gaps by Ecofloristic Zones

Average 2017 carbon stock values ranged from 5 and 122 Mg ha^{-1}. The highest average stock per hectare was recorded in the tropical rainforest ecofloristic zone, which accounts for 12% of total grassland. The highest carbon stock density and variation occurred in tropical rainforest, while the lowest carbon stock per hectare and the lowest variation were found in shrubland ecofloristic zones (Figure 4). About 73% of the total grassland had carbon stocks less than the < 75th percentile. Increasing these stocks to the current 75th percentile level would increase total stocks by approximately 0.15 Pg C (0.57 Pg CO_2e). By the same token, about 46% of grasslands have carbon stocks lower than the median (Figure 4). Increasing these carbon stocks to the median value would increase total stocks by 0.06 Pg C (0.2 Pg CO_2e).

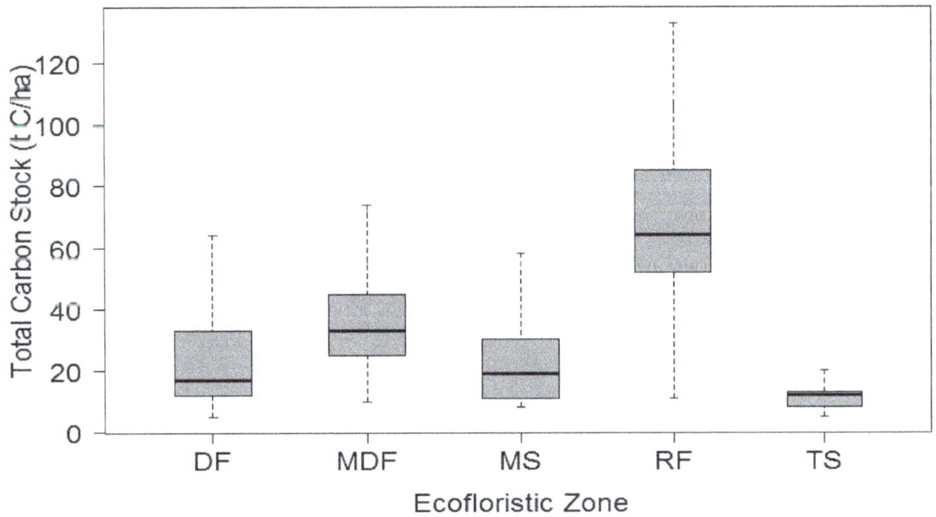

Figure 4. Variation in percentile carbon stocks in tropical dry forest (DF), moist deciduous forest (MDF), mountain system (MS), rain forest (RF), and tropical shrubland (TS).

Figure 5 indicates the spatial distribution of grazing lands with different levels of biomass carbon gaps. This helps identify priority areas for silvopastoral system interventions in Colombia.

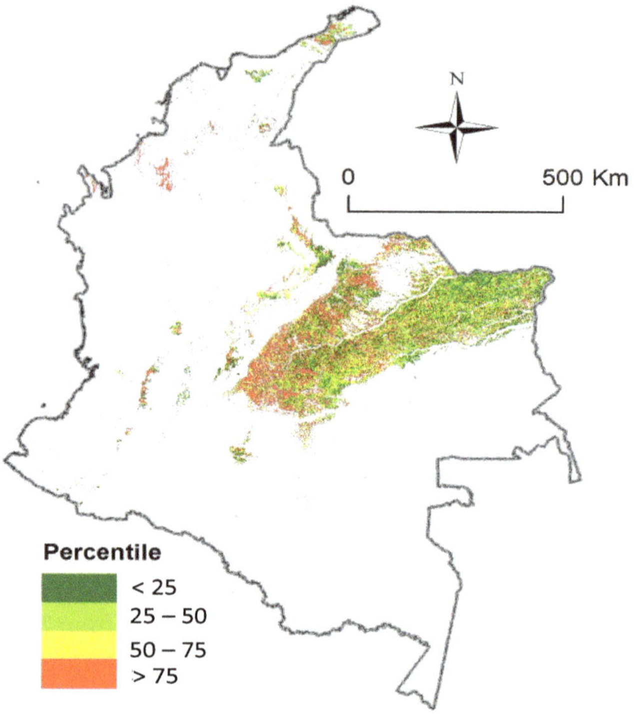

Figure 5. Priority areas for increasing Colombian grassland carbon stocks.

4. Discussion

Tree cover across all grasslands showed an increasing trend between 2000 and 2017. Trends in tree cover in grasslands found here support the same conclusions of more general studies on tree cover change, e.g., Sánchez-Cuervo et al. [21], who found that woody vegetation showed an increasing trend from 2001 to 2010 at the national scale in Colombia. This was mainly attributed to woody regrowth of trees following land abandonment resulting from armed conflicts, economic development, and increase in rainfall [21,22]. However, Sánchez-Cuervo et al. [21] reported that woody cover decreased by about 5% between 2000 and 2010, which was mainly due to intensive management of grasslands. The difference between these earlier results and ours may simply be a result of our study analyzing a longer time period, and the inconsistency in trends may be a function of changing management dynamics.

Zomer et al. [11,12] reported that the contribution of trees on global arable land was over 4 times higher than when estimated with IPCC default values [19]. Chapman et al. [23] also reported that trees in the global crop and pasture lands store about 3.07 and 3.86 Pg carbon in their aboveground biomass, respectively. Similarly, this study found that carbon stocks in the grassland of Colombia has likely also been underestimated. We found mean values of 34 Mg C ha^{-1} in 2000 and 39 Mg C ha^{-1} in 2017, which is more than four times larger than the IPCC Tier 1 global estimate of 8 Mg C ha^{-1} [19]. On the contrary, the global study by Liu et al. [24] reported that the TBC in grasslands and croplands did not show significant change during 1993–2010. Liu et al. [24] used harmonized Vegetation Optical Depth (VOD) data for 1993 onwards derived from a series of passive microwave satellite sensors to estimate above ground canopy in forest and non-forest land use types [25].

Zomer et al. [11,12] reported an average biomass carbon stock of 53 Mg ha^{-1} in Colombian arable land in 2000, demonstrating that trees in productive systems can significantly increase carbon stocks in Colombia. In a meta-analysis, Feliciano et al. [26] found that SPS that use controlled grazing practices and appropriate pasture species can increase average aboveground carbon sequestration by 2.29–6.54 Mg ha^{-1} yr^{-1}. López-Santiago et al. [27] reported that *Leucaena leucocephala* (shrub legume) and *Panicum maximum* grass based SPS contain higher aboveground (19.6 ± 1.6 Mg ha^{-1}) and belowground (7.7 ± 0.90 Mg ha^{-1}) biomass compared with deciduous tropical forest and grass monoculture systems in Mexico. This led to a rapid expansion of SPS in Mexico. Our study showed that average Colombian grasslands contained 34 Mg ha^{-1} in the same year. While 36% less than that of arable land, these significant levels of carbon in silvopastoral systems suggest a large opportunity for climate change mitigation, especially when considering the existing and planned areal extent of silvopastoral systems in Colombia. About 13% and 12% of the total area of Colombia was grazing and cultivated lands, respectively, in 2015 [28].

The potential carbon stock of SPS is notable for its ability to accumulate carbon over relatively short time frames. Increasing soil carbon in Colombian grasslands could take decades [29]. In contrast, carbon stocks in tree biomass can be accumulated in less than a decade. Trees have the additional benefit in that they help maintain and increase soil carbon stocks. Given the swift action necessary to limit global warming below the <2° climate change goal (by 2030 according to the IPCC), integrating trees into multi-use landscapes like SPS is an important pathway to meeting mitigation goals in a timely manner. Importantly, trees also regulate soil conditions, including organic matter content, fertility, structure, erosion resistance, and moisture content, during extreme events such as heavy precipitation and drought. Maintaining a hospitable soil environment has a dramatic effect on the presence of N-fixing bacteria, which are crucial in cropping systems with little or no N fertilizer inputs [12]. In turn, all of these mechanisms extend and improve productivity, both in terms of quality and quantity.

Silvopastoral system grasslands go beyond carbon stock potential to offer myriad co-benefits. SPS have direct positive impacts on the livelihoods of producers and environmental quality [30]. Trees create micro-climates that help protect crops and livestock from sun, wind, and extreme temperatures [31,32]. SPS have also been shown to reduce plant and animal production seasonality, reduce ruminal methane production, and increase biodiversity and natural pest control mechanisms as compared to conventional systems [7]. It is important to note that there may be significant trade-offs associated with integrating trees into some production systems. Smallholders rely on these systems as their source of livelihood. As such, increasing tree cover in SPS must be tested locally as one of the suites of potential climate-smart agricultural solutions with the aim of developing a project portfolio that minimizes tradeoffs and maximizes co-benefits.

The management of silvopastoral system trees can minimize trade-offs through complementary uses that augment climate resiliency and diversify household income. Complementary uses may include, timber plantations, living fences, tree alleys, windbreaks, fuelwood, perennial crops, and fodder banks [30,33]. For example, *Albizia saman* (syn. *Samanea saman*) trees improved forage production under their canopy [34], and their palatable pods are suitable as a dry season feed supplement [35]. Smallholder agroforestry systems can continue to increase their carbon stocks while also producing timber through select harvesting of high economic value tree species [36]. Colombian pastoral systems face serious complications because of climate change. The country is well poised to create national initiatives around SPS and other mitigation initiatives [16]. Indeed, the Colombian livestock sector will need to increase carbon efficiency in order to remain competitive on the international market. Our study reveals significant potential for addressing climate change mitigation and an array of co-benefits through implementation of SPS in Colombian grasslands. This suggests that SPS can significantly contribute to the NDCs and NAMAs of Colombia and other Latin American countries, as well as addressing other national commitments such as the United Nations 2030 Sustainable Development Goals and national regulatory targets.

However, scaling up SPS will not be straightforward. Calle et al. [37] demonstrated the importance of adequate technical assistance to farmer adoption of SPS in Colombia. Similarly, it is equally important to consider the quality and source of genetic material when promoting tree planting programs [38]. Market incentives including factors related to lower costs and/or higher benefits are important for silvopastoral system technology adoption [39]. In spite of the growing international demand for carbon-efficient and environmentally friendly animal products [16], very few producers will plant trees in grasslands just for the sake of climate change mitigation [40]. However, awareness of the implications of land use decisions and the increased productivity and resiliency of SPS over conventional systems, in conjunction with adequate technical and policy support [15], has been shown to incentivize adoption [41].

The MODIS data are low-resolution and may overestimate carbon stock estimates in grasslands by including different land uses like wooded areas and forest patches into grazing land pixels [23]. However, our study showed the potential of SPS to mitigate climate by sequestering more carbon in woody biomass. We assumed a linear increase in biomass carbon stocks with increasing tree cover, which may not be accurate in all cases. For example, species functional traits play a role in carbon dynamics by influencing species productivity [42].

We considered grasslands as proxy for silvopastoral systems as maps detailing silvopastoral systems were not identified. It is important to considered trade-offs across expected ecosystem services from increasing tree cover in grassland biomes. For example, a study on the effects of increased woody vegetation following fire suppression in Brazilian Cerrado reported a decline in plant and ant species by 27% and 35%, respectively [43]. Bond et al. [44] and Parr et al. [45] also reported that a large-scale increase in tree cover in grasslands could impact African grassland biomes. Several similar comments [46,47] were generated by the optimistic analysis of the global tree restoration potential by Bastin et al. [48].

As discussed in Lusiana et al. [49], however, common measures of pixel-level uncertainty in both classification and assigned C stock do not stand in the way of fairly narrow confidence intervals for aggregated data as used in national GHG accounting. The specific reduction of uncertainty with aggregation depends on the spatial structure of land cover change. For an Indonesian forest margin landscape, the study concluded that for the 1 km^2 scale of aggregation error was reduced below a defined tolerance. Further studies of this type on SPS in Columbia would be relevant.

The study is not without limitations. Data used in this study were generated on a global scale and thus may have some inaccuracies when down-scaled to a national analysis. Tree cover was measured via remote sensing and interpreted from MODIS Vegetation Continuous Fields (VCF) product; as such, it is an estimate of percentage crown cover, not tree density or tree biomass per se, and is likely to underestimate or overestimate tree cover [11]. Remotely sensed data should be validated with ground-level measurements [10], but this was not possible for logistical reasons during this study. We assumed a linear increase in biomass carbon stocks with increasing tree cover, which provides an approximation of biomass. We considered grasslands as proxy for silvopastoral systems as silvopastoral systems are a type of grazing system. Our estimates can inform the potential for further increase in C storage in SPS, but do not indicate yet what types of policy change and farmer innovation will be needed to achieve this.

5. Conclusions

We estimated carbon stocks in Colombian grasslands using remotely sensed tree cover data from MODIS and IPCC Tier 1 values. The results help clarify current and potential grassland carbon stocks at a national scale. Tree cover and carbon stocks increased significantly from 2000 to 2017. There remains high spatial variability, and regions with low tree cover have significant potential for increasing carbon stocks. This approach, along with ground-level data validation, could be useful in creating targets for planning NDCs and NAMAs in Latin American countries. Co-benefits include improved productivity and socioeconomic outcomes, climate resilience, and environmental conservation. Moving forward, it will be essential to create an enabling environment for the implementation of SPS in terms of policy, land governance, investment, capacity development, and international cooperation.

Author Contributions: Conceptualization, E.A., M.S., J.A. and T.S.R.; Formal analysis, E.A.; Funding acquisition, T.S.R.; Investigation, E.A. and T.S.R.; Methodology, E.A., M.S. and M.v.N.; Project administration, T.S.R.; Resources, M.S., J.A., T.S.R.; Writing—original draft, E.A., M.v.N. and T.S.R.; Writing—review and editing, E.A., M.S., M.v.N., J.A., J.M.R. and T.S.R. All authors have read and agreed to the published version of the manuscript.

References

1. Wollenberg, E.; Richards, M.; Smith, P.; Havlík, P.; Obersteiner, M.; Tubiello, F.N.; Herold, M.; Gerber, P.; Carter, S.; Reisinger, A.; et al. Reducing emissions from agriculture to meet the 2 °C target. *Glob. Chang. Biol.* **2016**, *22*, 3859–3864. [CrossRef] [PubMed]
2. Smith, P. Soil carbon sequestration and biochar as negative emission technologies. *Glob. Chang. Biol.* **2016**, *22*, 1315–1324. [CrossRef]
3. Roe, S.; Streck, C.; Obersteiner, M.; Frank, S.; Griscom, B.; Drouet, L.; Fricko, O.; Gusti, M.; Harris, N.; Hasegawa, T.; et al. Contribution of the land sector to a 1.5 °C world. *Nat. Clim. Chang.* **2019**, *9*, 817–828. [CrossRef]
4. Witkowski, K.; Medina, D. *Agriculture in the New Climate Action Plans of Latin America (Intended Nationally Determined Contributions)*; Inter-American Institute for Cooperation on Agriculture: San Jose, Costa Rica, 2016.
5. Griscom, B.W.; Adams, J.; Ellis, P.W.; Houghton, R.A.; Lomax, G.; Miteva, D.A.; Schlesinger, W.H.; Shoch, D.; Siikamäki, J.V.; Smith, P.; et al. Natural climate solutions. *Proc. Natl. Acad. Sci. USA* **2017**, *114*, 11645–11650. [CrossRef] [PubMed]
6. IPCC. *Climate Change and Land: An IPCC Special Report on Climate Change, Desertification, Land Degradation, Sustainable Land Management, Food Security, and Greenhouse Gas Fluxes in Terrestrial Ecosystems*; Shukla, P.R., Skea, J., Buendia, E.C., Masson-Delmotte, V., Pörtner, H.-O., Roberts, D.C., Zhai, P., Slade, R., Connors, S., van Diemen, R., et al., Eds.; IPCC: Geneva, Switzerland, 2019.
7. Cardona, C.A.C.; Naranjo Ramiírez, J.F.; Tarazona Morales, A.M.; Murgueitio Restrepo, E.; Charaá Orozco, J.D.; Ku Vera, J.; Solorio Saánchez, F.J.; Flores Estrada, M.X.; Solorio Saánchez, B.; Barahona Rosales, R. Contribution of intensive silvopastoral systems to animal performance and to adaptation and mitigation of climate change. *Rev. Colomb. Cienc. Pecu.* **2014**, *27*, 76–94.
8. Calle, Z.; Murgueitio, E.; Chará, J.; Molina, C.H.; Zuluaga, A.F.; Calle, A. A Strategy for Scaling-Up Intensive Silvopastoral Systems in Colombia. *J. Sustain. For.* **2013**, *32*, 677–693.
9. Van Dijk, S.; Tennigkeit, T.; Wilkes, A. *Climate-Smart Livestock Sector Development: The State of Play in NAMA Development*; CGIAR Research Program on Climate Change, Agriculture and Food Security (CCAFS): Copenhagen, Denmark, 2015.
10. Cardinael, R.; Umulisa, V.; Toudert, A.; Olivier, A.; Bockel, L.; Bernoux, M. Revisiting IPCC Tier 1 coefficients for soil organic and biomass carbon storage in agroforestry systems. *Environ. Res. Lett.* **2018**, *13*, 12. [CrossRef]
11. Zomer, R.; Trabucco, A.; Coe, R.; Place, F.; van Noordwijk, M.; Xu, J. *Trees on Farms: An Update and Reanalysis of Agroforestry's Global Extent and Socio-Ecological Characteristics*; World Agroforestry Centre (ICRAF): Bogor, Indonesia, 2014.
12. Zomer, R.J.; Neufeldt, H.; Xu, J.; Ahrends, A.; Bossio, D.; Trabucco, A.; Van Noordwijk, M.; Wang, M. Global Tree Cover and Biomass Carbon on Agricultural Land: The contribution of agroforestry to global and national carbon budgets. *Sci. Rep.* **2016**, *6*, 29987. [CrossRef]
13. Rosenstock, T.S.; Wilkes, A.; Jallo, C.; Namoi, N.; Bulusu, M.; Suber, M.; Mboi, D.; Mulia, R.; Simelton, E.; Richards, M.; et al. Making trees count: Measurement and reporting of agroforestry in UNFCCC national communications of non-Annex I countries. *Agric. Ecosyst. Environ.* **2019**, *284*, 106569. [CrossRef]
14. FEDEGAN. *Cifras de Referencia del Sector Ganadero Colombiano*; FEDEGAN Federacioún Colombiana de Ganaderos: Bogota, Colombia, 2018.
15. Tapasco, J.; LeCoq, J.F.; Ruden, A.; Rivas, J.S.; Ortiz, J. The Livestock Sector in Colombia: Toward a Program to Facilitate Large-Scale Adoption of Mitigation and Adaptation Practices. *Front. Sustain. Food Syst.* **2019**, *3*, 61. [CrossRef]
16. Lerner, A.M.; Zuluaga, A.F.; Chará, J.; Etter, A.; Searchinger, T. Sustainable Cattle Ranching in Practice: Moving from Theory to Planning in Colombia's Livestock Sector. *Environ. Manag.* **2017**, *60*, 176–184. [CrossRef] [PubMed]

17. ESA. *Land Cover CCI: Product User Guide Version 2.0*; UCL-Geomatics: London, UK, 2017.
18. Dimiceli, C.; Carroll, M.; Sohlberg, R.; Kim, D.H.; Kelly, M.; Townshend, J.R.G. *MOD44B MODIS/Terra Vegetation Continuous Fields Yearly L3 Global 250m SIN Grid V006*; LP DAAC: Sioux Falls, SD, USA, 2020.
19. Ruesch, A.; Gibbs, H.K. *New IPCC Tier-1 Global Biomass Carbon Map For the Year 2000*; Oak Ridge National Laboratory: Oak Ridge, YN, USA, 2008.
20. Heiskanen, J.; Liu, J.; Valbuena, R.; Aynekulu, E.; Packalen, P.; Pellikka, P. Remote sensing approach for spatial planning of land management interventions in West African savannas. *J. Arid Environ.* **2017**, *140*, 29–41. [CrossRef]
21. Sánchez-Cuervo, A.M.; Aide, T.M.; Clark, M.L.; Etter, A. Land Cover Change in Colombia: Surprising Forest Recovery Trends between 2001 and 2010. *PLoS ONE* **2012**, *7*, e43943. [CrossRef]
22. Fagua, J.C.; Baggio, J.A.; Ramsey, R.D. Drivers of forest cover changes in the Chocó-Darien Global Ecoregion of South America. *Ecosphere* **2019**, *10*, e02648. [CrossRef]
23. Chapman, M.; Walker, W.S.; Cook-Patton, S.C.; Ellis, P.W.; Farina, M.; Griscom, B.W.; Baccini, A. Large climate mitigation potential from adding trees to agricultural lands. *Glob. Chang. Biol.* **2020**, *26*, 4357–4365. [CrossRef]
24. Liu, Y.Y.; Van Dijk, A.I.J.M.; De Jeu, R.A.M.; Canadell, J.G.; McCabe, M.F.; Evans, J.P.; Wang, G. Recent reversal in loss of global terrestrial biomass. *Nat. Clim. Chang.* **2015**, *5*, 470–474. [CrossRef]
25. Liu, Y.Y.; De Jeu, R.A.M.; McCabe, M.F.; Evans, J.P.; Van Dijk, A.I.J.M. Global long-term passive microwave satellite-based retrievals of vegetation optical depth. *Geophys. Res. Lett.* **2011**, *38*. [CrossRef]
26. Feliciano, D.; Ledo, A.; Hillier, J.; Nayak, D.R. Which agroforestry options give the greatest soil and above ground carbon benefits in different world regions? *Agric. Ecosyst. Environ.* **2018**, *254*, 117–129. [CrossRef]
27. López-Santiago, J.G.; Casanova-Lugo, F.; Villanueva-López, G.; Díaz-Echeverría, V.F.; Solorio-Sánchez, F.J.; Martínez-Zurimendi, P.; Aryal, D.R.; Chay-Canul, A.J. Carbon storage in a silvopastoral system compared to that in a deciduous dry forest in Michoacán, Mexico. *Agrofor. Syst.* **2019**, *93*, 199–211. [CrossRef]
28. ESA. *300 m Annual Global Land Cover Time Series from 1992 to 2015*; ESA Climate Change Initiative: Oxford, UK, 2017.
29. Steiner, J.L.; Franzluebbers, A.J.; Neely, C.; Ellis, T.; Aynekulu, E. Enhancing Soil and Landscape Quality in Smallholder Grazing Systems. In *Enhancing Soil and Landscape Quality in Smallholder Grazing Systems*; Lal, R., Stewart, B.A., Eds.; CRC Press: Boca Raton, FL, USA, 2014; pp. 63–112.
30. Montagnini, F.; Ibrahim, M.; Murgueitio Restrepo, E. Silvopastoral Systems and Climate Change Mitigation in Latin America. *Bois For. Trop.* **2013**, *316*, 3–16. [CrossRef]
31. Dinesh, D.; Campbell, B.; Bonilla-Findji, O.; Richards, M. (Eds.) *10 Best Bet Innovations for Adaptation in Agriculture: A Supplement to the UNFCCC NAP Technical Guidelines*; CGIAR Research Program on Climate Change, Agriculture and Food Security (CCAFS): Wageningen, The Netherlands, 2017.
32. Eekhout, J.; de Vente, J. Assessing the effectiveness of Sustainable Land Management for large-scale climate change adaptation. *Sci. Total Environ.* **2018**, *654*, 85–93. [CrossRef] [PubMed]
33. Chara, J.; Reyes, E.; Peri, P.; Otte, J.; Arce, E.; Schneider, F. *Silvopastoral Systems and Their Contribution to Improved Resource Use and Sustainable Development Goals: Evidence from Latin America*; The Food and Agriculture Organization of the United Nations: Roma, Italy, 2018.
34. Durr, P.; Gangel, J. Enhanced forage production under Samanea saman in a subhumid tropical grassland. *Agrofor. Syst.* **2002**, *54*, 99–102. [CrossRef]
35. Durr, P. The biology, ecology and agroforestry potential of the raintree, Samanea saman (Jacq.) Merr. *Agrofor. Syst.* **2001**, *51*, 223–237. [CrossRef]
36. Roshetko, J.; Lasco, R.; Delos Angeles, M. Smallholder agroforestry systems for carbon storage. *Mitig. Adapt. Strateg. Glob. Chang.* **2007**, *12*, 219–242. [CrossRef]
37. Calle, A.; Montagnini, F.; Zuluaga, A.F. Farmer's perceptions of silvopastoral system promotion in Quindío, Colombia. *Bois For. Trop.* **2009**, *300*, 79–94. [CrossRef]
38. Roshetko, J.; Dawson, I.; Urquiola, J.; Lasco, R.; Leimona, B.; Weber, J.; Bozzano, M.; Lillesø, J.; Graudal, L.; Jamnadass, R. To what extent are genetic resources considered in environmental service provision? A case study based on trees and carbon sequestration. *Clim. Dev.* **2018**, *10*, 755–768. [CrossRef]
39. Iiyama, M.; Derero, A.; Kelemu, K.; Muthuri, C.; Kinuthia, R.; Ayenkulu, E.; Kiptot, E.; Hadgu, K.; Mowo, J.; Sinclair, F.L. Understanding patterns of tree adoption on farms in semi-arid and sub-humid Ethiopia. *Agrofor. Syst.* **2017**, *91*, 271–293. [CrossRef]

40. Mbow, C.; Van Noordwijk, M.; Luedeling, E.; Neufeldt, H.; Minang, P.A.; Kowero, G. Agroforestry solutions to address food security and climate change challenges in Africa. *Curr. Opin. Environ. Sustain.* **2014**, *6*, 61–67. [CrossRef]
41. Murgueitio, E.; Calle, Z.; Uribe, F.; Calle, A.; Solorio, B. Native trees and shrubs for the productive rehabilitation of tropical cattle ranching lands. *For. Ecol. Manag.* **2011**, *261*, 1654–1663. [CrossRef]
42. Lohbeck, M.; Winowiecki, L.; Aynekulu, E.; Okia, C.; Vågen, T.G. Trait-based approaches for guiding the restoration of degraded agricultural landscapes in East Africa. *J. Appl. Ecol.* **2018**, *55*, 59–68. [CrossRef]
43. Abreu, R.C.R.; Hoffmann, W.A.; Vasconcelos, H.L.; Pilon, N.A.; Rossatto, D.R.; Durigan, G. The biodiversity cost of carbon sequestration in tropical savanna. *Sci. Adv.* **2017**, *3*, e1701284. [CrossRef] [PubMed]
44. Bond, W.J.; Stevens, N.; Midgley, G.F.; Lehmann, C.E.R. The Trouble with Trees: Afforestation Plans for Africa. *Trends Ecol. Evol.* **2019**, *34*, 963–965. [CrossRef] [PubMed]
45. Parr, C.L.; Gray, E.F.; Bond, W.J. Cascading biodiversity and functional consequences of a global change-induced biome switch. *Divers. Distrib.* **2012**, *18*, 493–503. [CrossRef]
46. Friedlingstein, P.; Allen, M.; Canadell, J.G.; Peters, G.P.; Seneviratne, S.I. Comment on "The global tree restoration potential". *Science* **2019**, *366*, eaay8060. [CrossRef]
47. Skidmore, A.K.; Wang, W.; de Bie, K.; Pilesjo, P. Comment on "The global tree restoration potential". *Science* **2019**, *366*, eaaz0111. [CrossRef]
48. Bastin, J.F.; Finegold, Y.; Garcia, C.; Mollicone, D.; Rezende, M.; Routh, D.; Zohner, C.M.; Crowther, T.W. The global tree restoration potential. *Science* **2019**, *364*, 76–79. [CrossRef]
49. Lusiana, B.; van Noordwijk, M.; Johana, F.; Galudra, G.; Suyanto, S.; Cadisch, G. Implications of uncertainty and scale in carbon emission estimates on locally appropriate designs to reduce emissions from deforestation and degradation (REDD+). *Mitig. Adapt. Strateg. Glob. Chang.* **2014**, *19*, 757–772. [CrossRef]

11

Infiltration-Friendly Agroforestry Land uses on Volcanic Slopes in the Rejoso Watershed, East Java, Indonesia

Didik Suprayogo [1,*], Meine van Noordwijk [1,2,3], Kurniatun Hairiah [1], Nabilla Meilasari [1], Abdul Lathif Rabbani [1], Rizki Maulana Ishaq [1] and Widianto Widianto [1]

1. Soil Science Department, Faculty of Agriculture, Brawijaya University, Jl. Veteran no 1, Malang 65145, Indonesia; m.vannoordwijk@cgiar.org (M.v.N.); K.HAIRIAH@CGIAR.ORG (K.H.); nabilla.meilasari@gmail.com (N.M.); agroforestry@ub.ac.id (A.L.R.); rizky.maulana17@gmail.com (R.M.I.); widianto@gmail.com (W.W.)
2. World Agroforestry Centre, ICRAF, Indonesia Office, Bogor 16001, Indonesia
3. Plant Production Systems, Wageningen University, 6708 PB Wageningen, The Netherland

* Correspondence: Suprayogo@ub.ac.id or Suprayogo09@yahoo.com

Abstract: Forest conversion to agriculture can induce the loss of hydrologic functions linked to infiltration. Infiltration-friendly agroforestry land uses minimize this loss. Our assessment of forest-derived land uses in the Rejoso Watershed on the slopes of the Bromo volcano in East Java (Indonesia) focused on two zones, upstream (above 800 m a.s.l.; Andisols) and midstream (400–800 m a.s.l.; Inceptisols) of the Rejoso River, feeding aquifers that support lowland rice areas and drinking water supply to nearby cities. We quantified throughfall, infiltration, and erosion in three replications per land use category, with 6–13% of rainfall with intensities of 51–100 mm day^{-1}. Throughfall varied from 65 to 100%, with a zone-dependent intercept but common 3% increase in canopy retention per 10% increase in canopy cover. In the upstream watershed, a tree canopy cover > 55% was associated with the infiltration rates needed, as soil erosion per unit overland flow was high. Midstream, only a tree canopy cover of > 80% qualified as "infiltration-friendly" land use, due to higher rainfall in this zone, but erosion rates were relatively low for a tree canopy cover in the range of 20–80%. The tree canopy characteristics required for infiltration-friendly land use clearly vary over short distances with soil type and rainfall intensity.

Keywords: entrainment; erosion; forest conversion; overland flow; soil macroporosity; throughfall; water balance

---

1. Introduction

Water access for all, the Sustainable Development Goal 6 of the Agenda 2030 agreed by the United Nations [1], not only refers to drinking water and sanitation. It requires the protection of "infiltration-friendly" land uses in upland watersheds as a source of clean water [2]. Sufficient groundwater recharge is important to the sustainable management of groundwater resources to maintain streamflow throughout the year, as well as to feed springs [3,4]. While much of the public discourse is in terms of forest versus agriculture, thresholds for specific soil and climate regimes are needed within the intermediate agroforestry spectrum of land uses [5]. Thresholds to critical hydrological functions are likely dependent on local context but need to be understood to guide natural resource management in the challenging trade-offs between local and external priorities [6]. Hydrological functions, and their sensitivity to climate change, can be characterized by a number of metrics [7]. Much of the literature on forest hydrology is concerned with reductions in annual water yield due to increased canopy interception and/or tree water use by fast-growing forest stands [8], without a distinction between (fast) overland and (slower, infiltration-dependent) subsurface flow pathways. A recent review [9] found that the recovery of annual river flow with the age of planted forest is an exception rather than a rule. However, the recovery of infiltration with tree cover can increase dry season flows [10] without increasing annual water yield. Changes in streamflow regime will reflect both changes in evapotranspiration (ET) and in infiltration after the change in land use under given climatic conditions [11,12]. On the other hand, a high tree (*Acacia auriculiformis*) canopy, without an understory and permanent litter layer, was associated with high erosion rates due to high-impact drips from the leaves [13]. Empirical data and process-level understanding is needed of these diverse and partly contradictory effects of tree cover, especially in human-managed land uses.

Agroforestry systems with high canopy densities can, if a permanent litter layer is present, maintain high infiltration rates and can positively impact on hydrologic functions through: (1) a green canopy cover at the tree and understory level, (2) land surface roughness, (3) litter at the soil surface, and (4) water uptake by trees and other vegetation [14,15]. Five aspects that hydrologically differentiate natural forest from open-field agriculture, with intermediate functionality for managed forest, plantations, agroforestry, and trees outside forest [16,17], are: (1) the leaf area index (LAI) that allows photosynthesis when stomata are open and transpiring, and that, along with leaf morphology and rainfall intensity, determines canopy interception, retention, and subsequent evaporation, (2) the surface litter that prevents crusting and supports infiltration [18] while reducing soil evaporation and reduces the entrainment of soil particles if overland flow still occurs, (3) the soil macroporosity that governs infiltration and allows for the aeration of deeper soil layers between rainfall events while recovering at a decadal time scale after reforestation [19,20], (4) the root systems that govern water extraction from deeper soil layers, in conjunction with the phenology of the aboveground canopy [21], and (5) possible influences on rainfall events [22,23]. Each of these five aspects has its own dynamic (time constants) and dependency on the type of trees and their management, challenging the definition of hydrologically adequate land use choices. Rather than prescribing, independent of soil types and slope, the type and quantity of tree cover that is needed, as tends to happen in forest zoning, it may help if limits to infiltration-friendly land use (focused on the third function) can be operationalized in a local context. In terms of watershed hydrology, infiltration-friendly land uses can be interpreted as any land use that allows high rates of water infiltration so that surface runoff is a small (to be defined in local context) fraction of rainfall and the watershed functions of flow buffering and erosion control are secured (to specified standards). River flow in watersheds that provide perfect buffering might theoretically be constant every day, but in practice, a "flow persistence" metric of about 0.85 is hard to surpass [24]. Flow buffering is essential for climate resilience [25] and high flow persistence metrics are desirable, as they directly relate to peak flow transmission [26].

The discussion on forests and watershed functions in Indonesia became based on specific theories about underlying mechanisms and measurements in the 1930s [27–30], but at the policy level, a generic dichotomy between "forest" and "non-forest" conditions was maintained. The Indonesian spatial planning law prescribes that 30% forest cover is needed in all local government entities to secure hydrological forest functions [31]. As the 30% norm originated in studies of flooding risk in relation to spring snow melt in Switzerland [32], a more nuanced and process-based understanding is needed to underpin effective policies on desirable forest cover, especially in densely populated Java, where agroforests are common. In Southeast Asia, 8.5% of the global human population lives on 3.0% of the land area. With 7.9% of the global agricultural land base, the region has 14.7% and 28.9% of such land with at least 10% and 30% tree cover, respectively, and is the world's primary home of "agroforests" [33].

On densely populated Java, volcanic slopes are home to large numbers of farmers, while also serving as sources of water for lowland agriculture and the rapidly growing cities. The shrinking area of state-managed forests is no longer able to secure the required watershed functions, but at least part of the agroforestry managed by farmers can meet the required hydrological functions [34].

In the Rejoso Watershed (Pasuruan, East Java, Indonesia), numerous stakeholders depend on the watershed functions of densely populated mountain slopes to meet their water demand. These include local communities, farmers using water irrigation, the Regional Water Company, and bottled water industries. A major infrastructure is planned to bring water to Surabaya and the surrounding urban centers. However, the quantity and quality of the water at the source of the pipe have been decreasing over the past 10 years, putting the infrastructure investment at risk. Decreasing water resources are likely due to land use changes in the recharging area of the Rejoso Watershed on the northern slopes of the Bromo-Semeru volcanic mountain range, and/or decreased pressure on artesian wells across the land due to increased extraction for paddy rice fields. Among the hydrologic functions, infiltration is critical, as water travels through the subsoil to artesian wells at the foot of the volcano, in addition to surface rivers.

This research in the Rejoso Watershed using runoff and soil erosion plot-scale studies under natural rainfall [35] within the locally relevant range of land cover types was thus designed to assess which land uses can maintain infiltration rates under local peak rainfall intensity and restrict soil erosion to acceptable levels. The specific questions were:

1. Which existing land uses limit infiltration below the required rates at peak rainfall events?
2. Which factors that are directly observable, such as tree cover, litter layer thickness, or surface roughness, can be used to define thresholds for "infiltration-friendly land use"?
3. Do the answers to questions 1 and 2 need to be differentiated between the upper and middle watershed, with current vegetable production and agroforestry as dominant land uses, respectively?

2. Materials and Methods

2.1. Study Area

The Rejoso Watershed, is located on the northern slope of Mount Bromo, covering 16 sub-districts in Pasuruan District, East Java Province, Indonesia. The Rejoso Watershed is located between 7°37′13.35″ and 7°55′18.63″ South, and between 112°48′32.51″ to 113°55′55″ East (Figure 1).

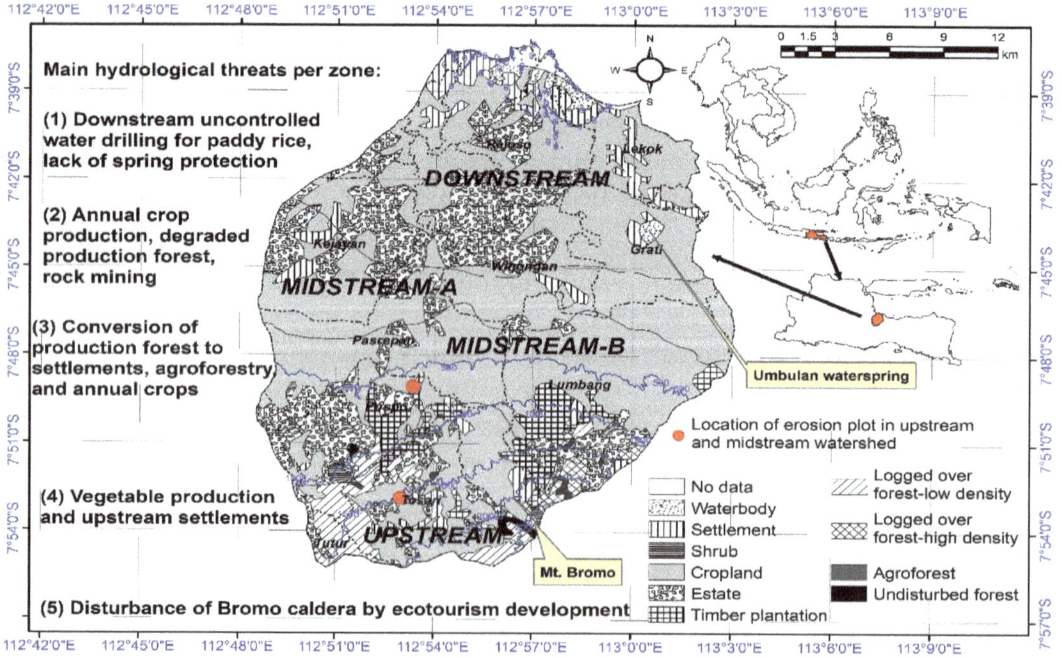

Figure 1. A. Position of the Rejoso Watershed in East Java as part of the Indonesian archipelago in Southeast Asia. B. The Rejoso Watershed from upstream (at the bottom) to sea level and land uses considered to be a hydrological threat; purple indicates open soil, green tree cover (base map in [36]).

The Rejoso Watershed covers an area of 634 km^2 with a hydrologic (watershed) length of the main channel of about 22 km. This study was conducted in two locations, namely in the upstream (above 800 m a.s.l.) and midstream (400–800 m a.s.l.) sections, with the dominant vegetation (land cover) selected for each location (Figure 1).

Climatic conditions that influence hydrology and erosion are largely determined by the influence of the northwest and southwest monsoons. The northwest monsoon, picking up large amounts of moisture over the Indian Ocean, brings in most of the annual precipitation in the area, and predominates during the period from November through April. Although there is considerable variation in the amount and distribution of rainfall from year to year, most places in the watershed receive about 91% of the rainfall during the November–May wet season (monthly rainfall > 100 mm) in the upper stream and about 91% of the rainfall during the November–April wet season in the midstream (Figure 2). Due to topographic influences, there is considerable spatial variation in annual precipitation as well, ranging from 1655 mm to 3675 mm, with extreme yearly rainfall in 2010, with an annual precipitation of 5298 mm over 24 years of rainfall measurements (1990–2013) in the upper stream compared with an annual precipitation ranging from 1020 mm to 2603 mm in the midstream. The May to October period is considered the dry season. Then the southeast monsoon predominates, bringing much smaller amounts of precipitation due to the lower atmospheric moisture caused by lower temperatures in the Southern Hemisphere at this time of the year. The annual precipitation in the upstream (average annual precipitation = 2488 mm) is higher than the in midstream (average annual precipitation = 1632 mm). Over 24 years of measurements, the maximum daily rainfall in the upper stream and midstream ranged from 80 mm day^{-1} to 200 mm day^{-1} and 60 mm day^{-1} to 320 mm day^{-1}, respectively. Based on Schmidt–Ferguson climate classification, the upper stream and midstream are considered rather wet (C) and average (D), respectively.

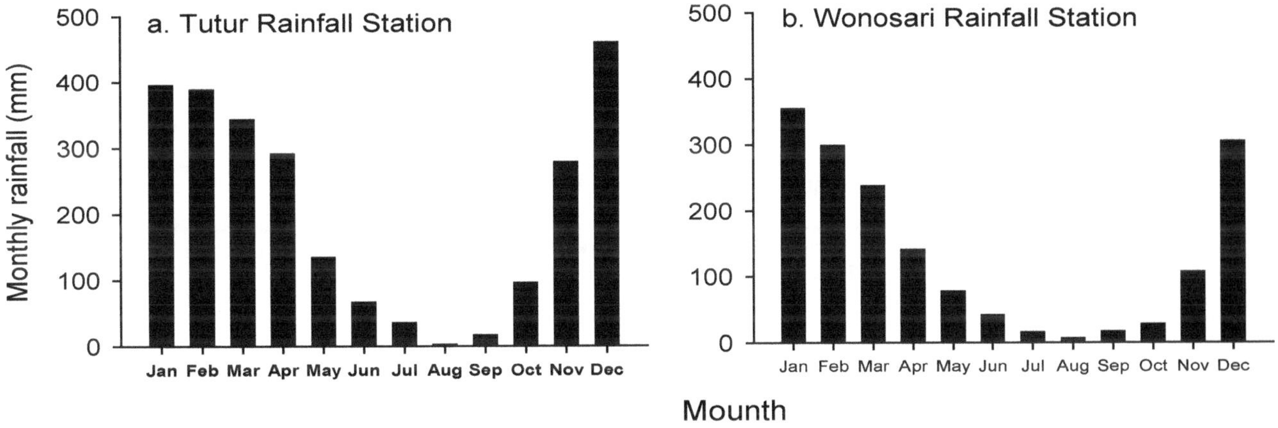

Figure 2. Monthly rainfall distribution in the Rejoso Watershed from the average of 24 years of rain events (1990–2013) in the (**a**) upstream (Tutur Rainfall Station) and (**b**) midstream (Wonosari Rainfall Station).

The Rejoso Watershed consists of four types of soil, namely: Andisols, Inceptisols, Alfisols, and Entisols. Andisols are mainly found on the upper slopes of the volcano. Andisols have a distinct black to very dark brown surface horizon rich in organic matter, which usually overlies a brown to dark yellowish-brown subsoil. The clay fraction is dominated by allophane. Andisols are highly permeable, porous with low bulk density, have high water-holding capacity, and a crumb structure. The most common texture is sandy loam. These soils have high inherent fertility and are highly erodible only when seriously disturbed. The middle and some lower volcanic slopes, consisting of easily weatherable permeable tuffs and ash deposits, give rise to deep soils—Inceptisols and Alfisols. Inceptisols have only limited horizon differentiation. Their texture ranges from deep friable clays to clay loams. Alfisols are soils which have an accumulation of clay in the subsoil. Their texture ranges from loam to clay loam in the topsoil and clay loam to clay in the subsoil. Both soils have moderate to high inherent fertility but are highly susceptible to erosion. The fourth group, Entisols, are soils that lack horizon development and are found on volcanic sands, ashes, and tuffs. Entisols occur on recent and sub-recent lahars of the Bromo volcano. Entisols with a coarse texture are extremely erodible and have very low water-holding capacities. Permanent vegetative cover, and especially diversified tree crops and agroforestry or forestry, are the most suitable land utilization types to prevent erosion.

2.2. Land Cover Types Compared

In important research traditions associated with the Universal Soil Loss Equation (USLE) [37], the quantification of erosion requires a "bare soil" reference, expressing the degree of protection provided by vegetation relative to this "control". Fortunately, bare soil is rare in this landscape, and it would be considered an extreme, rather than a standard agricultural point of reference. Artificially clearing land to allow such treatment to be measured would give results that are hard to be interpreted, as soil changes after clearing would lead to a time-dependence of the results, rather than being an unambiguous point of reference. By referring to the more process-based Rose equation [38], separating overland flow as a transport medium and "entrainment" as a soil characteristic relative to the energy-dependent transport capacity of such flow, we do not depend on the USLE framework (that infers that soil loss is universal, but does not account for its counterpart process, sedimentation [39]) but can focus on existing land covers and associated land uses in the landscape.

In both the upstream and midstream parts of the catchment, four dominant land use systems were assessed (Table 1), spatially replicated in three separate measurement plots. Upstream land uses included old and young pine plantations (production forest) and highland vegetable crops with variations in tree canopy cover in the landscape on steep (30–60%) to very steep (>60%) land with imperfect ridge terraces. Midstream land uses included production forest, multistrata coffee-based agroforestry, clove-based agroforestry, and several mixed agroforestry types with variations in tree canopy cover in the landscape on moderately steep (15–30%) and steep (30–60%) land with bench terraces sloping outward.

Table 1. Land use, vegetation, soil conservation measure, and slope of measurement plots.

Code	Land Use	Vegetation (the Average Height of Trees)	Terracing	Slope (Plot Level, %)
		Upstream Rejoso Watershed		
UT1	Old production forest	Pine (*Pinus merkusii*) (34 m) + grass	None	35–40
UT2	Young production forest	Pine (11 m) + grass	None	50–60
UT3	Agroforestry	Strip cemara (*Casuarina junghuniana*) (13 m) + cabbage	None	40–50
UT4	Arable land	Banana, maize, carrot	None	40–50
		Midstream Rejoso Watershed		
MT1	Old production forest	Mixed pine (28 m) or mahogany (*Swietenia macrophylla*) (12 m), banana, salak (*Salacca zalacca*), taro (*Colocasia esculenta*), elephant grass (*Miscanthus giganteus*).	Bench terrace sloping outward	3–8
MT2	Agroforestry	Coffee-based (2 m) mix with durian (*Durio zibethinus*) (10 m), mahogany (9 m), *Leucaena leucocephala* (8 m), *Paraserianthes falcataria* (11 m), *Albizia saman* (11 m), dadap (*Erythrina variegata*) (11 m), banana	Bench terrace sloping outward	3–8
MT3	Agroforestry	Clove (*Syzygium aromaticum*) (8 m), banana	Bench terrace sloping outward	3–8
MT4	Agroforestry	Mango (*Mangifera indica*) (10 m), durian (10 m), *Randu kapuk* (*Ceiba pentandra*) (11 m), maize, cassava, groundnut	Bench terrace sloping outward	3–8

2.3. *Quantifying Terms of Water and Soil Balance*

2.3.1. Overview

As forest and tree cover can influence various steps in the chain from rainfall to streamflow and erosion, we aimed to quantify (1) the direct effect of tree canopies on the retention of part of the rainfall (followed by direct evaporation), versus the fraction reaching the soil surface by throughfall and stemflow, (2) the partitioning of the latter into infiltration and overland flow, (3) the entrainment of soil particles into this overland flow. Within the time and resources available we did not assess (4) the seepage of groundwater beyond the root zone and access by vegetation, (5) the pathways and release of groundwater into streams, (6) the routing of overland flow into streams, or (7) the in-field (or riparian filter zone) sedimentation of entrained soil particles, beyond the scale of the measurement plots. However, we did characterize the vegetation and soil characteristics that influence the various processes.

2.3.2. Rainfall and Throughfall

Rain gauges, outside of the direct influence of tree canopies, were installed in four observation locations (with adjacent erosion plots) upstream and four observation locations midstream of the Rejoso Watershed. In each runoff plot, throughfall was measured with five replications. The throughfall gauges below the tree canopy had a horizontally placed 30 cm diameter funnel 120 cm above the soil surface and a collector bottle with a volume of 1.5 dm^3 placed with bamboo as a support. Throughfall and rainfall were collected every day for two months of the rainy season, from March to May 2017. Attempts were made to also quantify stemflow, but due to technical problems with the method used, no reliable data were obtained, and the results are not shown here.

2.3.3. Water Infiltration and Soil Erosion Measurement

Water infiltration was quantified in each land cover type via its complement, surface runoff, and expressed in the runoff/throughfall ratio. As the throughfall for quantified infiltration was measured below the tree canopy, the amounts were the net of canopy retention and possible stemflow. Surface runoff was measured in 6 m × 2 m plots protected from surface run-on, with two drums at the lower end to collect surface runoff and sediment concentrations for soil erosion measurements (Figure 3).

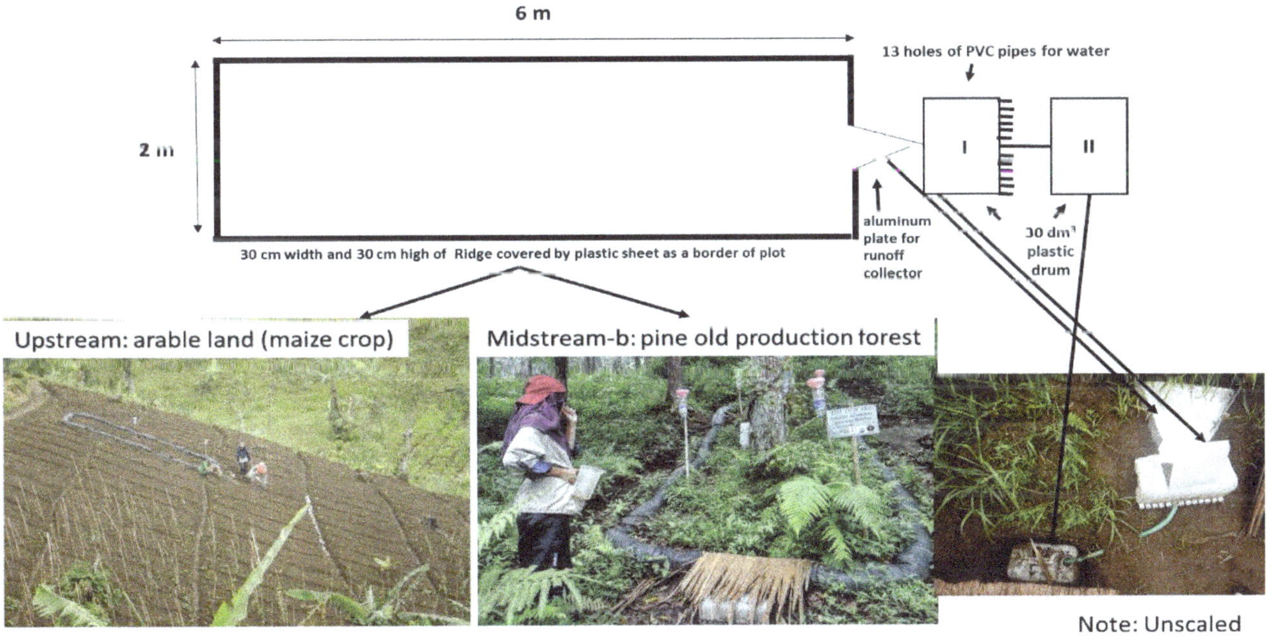

Figure 3. Runoff and soil erosion plot design.

In each plot, the water flow was collected into two collection drums with a capacity of 30 dm^3. The first drum had a divider system channeling into 13 channels (PVC pipes) with equal diameters and level positions, with one connected with a second drum. The volume of water flowing from each pipe was measured to calibrate the water volume proportion entering into the second drum. The potential capacity of the runoff collector thus was (30 dm^3 * 13) + 30 dm^3 = 420 dm^3 for 12 m^2 or 35 mm. We did nor encounter situations where the second tank overflowed. Runoff samples at each plot were collected every day and the rain that occurred during the measurement period was measured by measuring the water depth in each drum. The amount of runoff in each rain event was calculated using Equations (1) and (2):

$$R_t = V_{d-I} + (13 * V_{d-II}) \tag{1}$$

$$V_d = 1000 * (D * L * W) \tag{2}$$

where R_t is total runoff (dm^3), V_d is the water volume in drums I and II (dm^3), L = length and W = width of drum (cm), and D is the water depth in each drum (cm). The total runoff was then divided by the area of the plot (2 m × 6 m) to convert to mm. Data could be compared to a classification developed elsewhere [40] that indicates a runoff coefficient of 0.14 as adequate for Andisols, and 0.20 for Inceptisols.

Soil erosion in each rain event was determined by collecting 1 dm^3 of runoff sediment in each drum. The sample was filtered with "newsprint" and dried in an oven with a temperature of 105 °C to get the weight of the sediment (S). In earlier studies, we found that effluent from this readily available filter material had a negligible sediment concentration [41]. Erosion (E) in each rain event was calculated using Equation (3):

$$E = ((V_{d-1} * S) + (13 * (V_{d-2} * S))) * \left(\frac{10^{-2}}{A}\right) \tag{3}$$

where E is soil erosion (Mg ha^{-1}), S is sediment (g dm^{-3}), and A is the area of the plot (m^2).

2.4. Determination of Soil Properties

Three bulk mineral soil samples were collected from each layer of soil of 0–10 cm, 10–20 cm, 20–30 cm, 30–40, and 40–50 cm for soil texture analysis and each layer of soil of 0–10 cm, 10–20 cm, and 20–30 cm for soil bulk density, particle density, total soil porosity, and soil organic matter content. Particle size distribution (particles < 2 mm) was determined with the Bouyoucos densimeter method [42] after H_2O_2 pre-treatment and after samples had been dispersed in 5% sodium hexametaphosphate and 5% dispersing solution. Bulk density (oven dry weight per unit volume) was measured for a block-sized sample (20 cm × 20 cm × 10 cm = 4000 cm^3) collected in field moisture conditions (modified from [43]). Particle density was measured by the pycnometer method. Total soil porosity (∅), the percentage of the total soil volume that is not filled by solid (soil) particles [44], was calculated from bulk density data and particle density using Equation (4):

$$\varnothing = \left(1 - \frac{\rho_b}{\rho_p}\right) \times 100\% \tag{4}$$

where ∅ is porosity (%), ρ_b is bulk density (g cm^{-3}), and ρ_p is particle density (g cm^{-3}).

Soil organic carbon (SOC) was determined by dichromate oxidation [45]. Soil infiltration was measured by the standard double-ring infiltrometer test [46]. The double-ring infiltrometer as often constructed from a thin-walled steel pipe with inner and outer cylinder diameters of 20 and 30 cm, respectively.

The soil macro-porosity was measured using the methylene blue method, by looking at the blue distribution pattern of the methylene blue solution in the soil profile. The methylene blue solution (70 g methylene blue per 200 L of water) was gradually poured into the ground, which had been bound by a metal frame measuring 100 cm × 50 cm × 30 cm (Figure 3) and left for 36 h until the methylene blue solution soaked into the soil. Methylene blue will pass through soil macropores but be absorbed by micropores and soil surfaces. After all the methylene blue solution had disappeared from the soil surface, the top 5 cm of soil was removed from a 100 cm × 100 cm sample area and infiltration patterns were recorded, before a further 10 cm of soil was removed for a second map (at 15 cm below the soil surface), and a further 10 cm of soil for a third map (25 cm below soil surface). For mapping blue patches in each horizontal plane, a transparent sheet of plastic was placed on the surface and all visible blue patches were mapped with marker pens. Blue distribution patterns, redrawn on tracing paper, were photocopied for analysis of the black-and-white pattern of the fraction of soil involved in macropore flow with the IDRISI computer program.

2.5. Other Plot Characteristics

2.5.1. Canopy Cover

The canopy cover can be defined as the percentage of tree canopy area occupied by the vertical projection of tree crowns [47]. The percentage of canopy cover is measured by scathing the shadow of sunshine at ground level using 10 m × 10 m sheets of white paper. The canopy projection when the sun was overhead was drawn to scale on white paper in each of the four quadrants of the 20 m × 20 m plots, after which the shaded areas were cut out and weighed separately. Canopy cover was calculated according to Equation (5):

$$\%\text{Canopy Cover} = \frac{\text{W Canopy}}{\text{W Total}} \times 100 \tag{5}$$

where %Canopy Cover is the percentage of tree canopy cover, W Canopy is the paper weight representing canopy cover and W Total is the paper weight representing the total area of observation, respectively.

2.5.2. Understory and Litter

Understory vegetation and litter were measured according to the rapid carbon stock appraisal protocol [48], using 50 cm × 50 cm samples for fresh weight, with subsamples dried for dry weight determination.

2.5.3. Land Surface Roughness

Surface roughness was measured in each plot as the standard deviation of elevation measured every 30 cm along a thread (thin rope) installed 30 cm from the surface vertically, horizontally, and diagonally over the erosion plot [49]. The measurement of the difference in elevation was set to a pixel size of 30 cm × 30 cm. Each plot was divided into six pixels for a 2 m plot width and 20 pixels for a 6 m plot length, so there were 120 pixels (N). Pixels were made on a flat plane 30 cm from the ground point of reference with a thin rope. In each center, the pixel was measured vertically parallel to the thin rope towards the surface of the ground with a ruler. The results of the measurements of height differences in each pixel were used to calculate Ra with the equation:

$$\text{Ra} = \frac{1}{\text{N}} \sum_{n=1}^{\text{N}} |\, \text{h}_\text{n} \,| \tag{6}$$

where N = Number of pixels in the patch and h_n = difference of elevation between the nth pixel in the patch and the mean value.

2.6. Data Analysis

To answer the first research question, the null hypothesis was that within the forest to open field agriculture continuum of any observed difference in soil hydrological functions could be due to random variation. To see if that null hypothesis could be rejected, we examined differences in soil infiltration, runoff coefficient, and soil erosion between the dominant land uses in the upstream and midstream with Fisher's Least Significant Differences (LSD) test. Fisher's LSD test, which establishes differences between groups defined for independent samples, was used for hypothesis testing, given that the data met the requirements for normality and the homogeneity of variances. A probability level of 0.05 was set for rejecting the null hypothesis of no difference in tests of statistical significance. We used the GenStat 15th edition software for Fisher's LSD tests. The soil infiltration, runoff coefficient, and soil erosion were then compared with the soil infiltration category [50], existing infiltration adequacy standards, and acceptable soil erosion rates. An acceptability threshold, below which soil erosion is less than an "agriculturally permissible" rate (E_{apr}, Mg ha^{-1} year^{-1}), was derived as:

$$E_{apr} = \left(\frac{\text{Depth of soil} * \text{Factor of soil depth}}{\text{Time horizon}}\right) * \text{Soil bulk density} \quad (7)$$

Both Andisols and Inceptisols are deep (beyond 120 cm soil depth) and have a soil depth factor of 1.0. We chose 400 years as a time horizon. Given the average soil bulk density of Andisols (0.83 g cm^{-3}) and Inceptisols (0.99 g cm^{-3}), we obtained E_{apr} Andisol = 24.9 Mg ha^{-1} year^{-1} and E_{apr} Inceptisol 29.7 Mg ha^{-1} year^{-1}.

For the second research question, we tested a number of plot scale characteristics as possible indicators of "infiltration-friendly" plot characteristics: tree canopy cover, understory vegetation, litter necromass, and land surface roughness. Linear regression relationships between the surface runoff/rainfall ratio or soil erosion and the amount of rainfall, tree canopy cover, understory, litter, and land surface roughness were determined using SigmaPlot version 10.0. While a search for "explanatory" factors might have explored multiple regression, our focus was single indicators that could be used as proxies for follow-up discussions with farmers about adjusting land use.

The third research question required the analysis of data for the first two research questions, with the expectation that any thresholds for acceptable hydrological disturbance could be zone-specific, given variations in rainfall, soil type, and the specific characteristics of land use and vegetation.

3. Results

3.1. Rainfall and Throughfall

Within the measurement period, 31 rainy days were recorded (Figure 4). Rainfall variation between the upstream and midstream observation plots was relatively high, with an average of 520 mm (range 476–556 mm among 12 rain gauge measurements), and an average of 666 mm (range 541–840 mm among 12 rain gauge measurements), respectively. In the upstream and midstream areas, 71% and 57% of the rainy days had < 20 mm day^{-1} ("light rain"), 24% and 31% had "moderate" rainfall (21–50 mm day^{-1}) and 6% and 13% "heavy" rain (51–100 mm day^{-1}), respectively; none had "very heavy rain" (>100 mm day^{-1}). Such rain conditions indicate that the rain erosivity in the midstream is higher than that of the upstream.

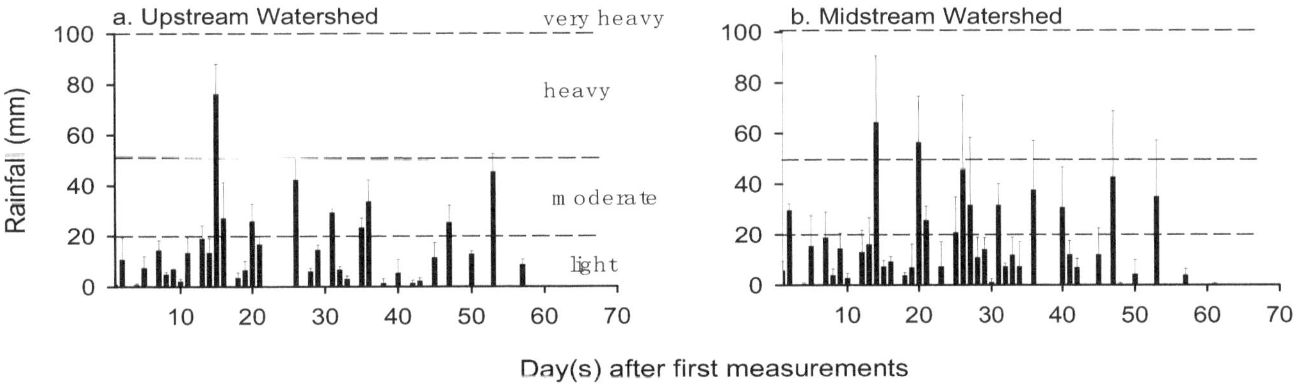

Figure 4. Distribution of rainfall during observation starting on March 03, 2017 in the Rejoso Watershed.

In the upstream area, the old production forest obtained a throughfall/rainfall ratio of 0.73 (standard deviation (SD) = 0.05), while for open-field agriculture it was 0.94 (SD = 0.5) (Figure 5a). For young production forests of *Casuarina junghuniana*-based agroforestry, the throughfall/rainfall ratio was 0.83 (SD = 0.05). In the midstream, the throughfall/rainfall ratio in agroforestry systems with tree canopies of 87%, 75%, and 52% were 0.81 (SD = 0.07) 0.81 (SD = 0.07), and 0.1 (SD = 06), respectively (Figure 5b). For agroforestry with low cover (26%), the throughfall/rainfall ratio was 0.96 (SD = 0.01).

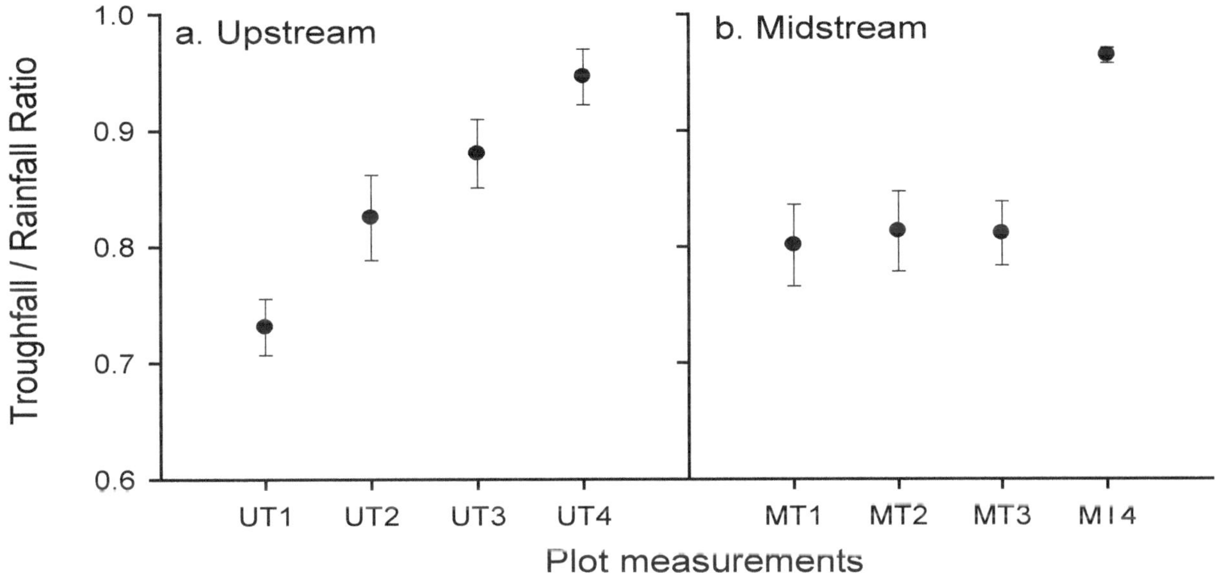

Figure 5. The throughfall/rainfall ratio variability in measured runoff plots (**a**) upstream, (**b**) midstream.

3.2. Soil Properties

The Andisols in the upstream area had a 40–60% silt fraction in all soil measured layers; the Inceptisols had a higher clay fraction (Appendix A). The upstream area had a lower bulk density and higher soil porosity, with a lower clay content than the midstream area (Table 2). The soil organic carbon content varied from 0.65 to 2.12%.

Table 2. Bulk density, particle density, soil porosity, macro-porosity, and organic C of runoff plots.

Location Code at Soil Depth (cm)	Bulk Density ($g\ cm^{-3}$) *			Particle Density ($g\ cm^{-3}$) *			Soil Porosity (%) *			Soil Macro- Porosity (%)			C_{org} (%) *		
	0–10	10–20	20–30	0–10	10–20	20–30	0–10	10–20	20–30	0–10	10–20	20–30	0–10	10–20	20–30
Upstream Rejoso Watershed: Andisols															
UT1	0.87a	0.81a	0.83a	2.16a	2.23a	2.31a	60a	63a	64c	8.0b	5.2b	0.9a	2.05bc	1.61c	1.79b
UT2	0.85a	0.86a	0.82a	2.27a	2.30a	2.33a	63a	63a	65c	5.1ab	1.5a	0.3a	2.46c	1.56bc	1.78b
UT3	0.81a	0.84a	0.85a	2.14a	2.12a	2.28a	62a	60a	63b	4.7ab	2.1ab	1.4a	1.17a	0.58a	0.71a
UT4	0.84a	0.88a	0.84a	2.28a	2.29a	2.08a	63a	62a	60a	3.0a	0.3a	0.1a	1.35ab	1.06ab	0.92a
LSD	0.07	0.13	0.12	0.17	0.21	0.38	4	5	1	3.52	3.4	1.8	0.85	0.50	0.50
Midstream Rejoso Watershed: Inceptisols															
MT1	0.83a	0.85a	0.83a	2.20a	2.28a	2.20a	62c	63a	62b	13.6ab	7.0bc	2.5c	1.73a	1.87a	1.65b
MT2	0.96b	0.91a	0.91a	2.42b	2.38a	2.21a	60bc	62a	59ab	16.1b	8.3c	1.8bc	2.22a	1.59a	1.84b
MT3	1.03bc	0.96a	0.94ab	2.38b	2.36a	2.40a	57ab	59a	61b	11.7a	3.4ab	0.9ab	2.19a	1.61a	1.01a
MT4	1.09c	1.04a	1.04b	2.38b	2.33a	2.33a	54a	55a	55a	11.4a	0.8a	0a	1.71a	1.36a	1.12a
LSD	0.10	0.24	0.11	0.15	0.17	0.22	4	10	4	4.0	3.9	1.0	0.84	0.54	0.41

* The same letter indicates no statistically significant differences between locations with Fisher's Least Significant Differences (LSD) test ($p < 0.05$).

3.3. Land Characteristics Related to Runoff and Soil Erosion

Production forests in the upstream area had a lower tree canopy cover than those midstream but higher than those in agroforestry systems (Table 3). Agroforestry in the upstream area had a very low tree canopy cover because trees were planted only along field edges. Midstream agroforestry gardens ranged from high (75%) to low (26%) canopy cover. Understory vegetation was more prominent upstream than midstream. Litter layer necromass and land surface roughness were generally aligned with tree canopy cover.

Table 3. Canopy cover, understory vegetation, litter necromass, and soil roughness of the sample plots.

Code	Land Cover	Tree Canopy Cover (%) *	Understory Vegetation (Mg ha^{-1}) *	Litter (Mg ha^{-1}) *	Soil Roughness (%) *
Upstream Rejoso Watershed					
UT1	Old production forest	55b	10.1b	9.2b	8.5a
UT2	Young production forest	40b	10.5b	2.0a	7.0a
UT3	Agroforestry	4a	10.1b	2.1a	9.5a
UT4	Arable land	0a	3.7a	0.3a	7.7a
LSD		15	5.6	3.7	4.6
Midstream Rejoso Watershed					
MT1	Old Production Forest	87c	2.5a	9.8b	7.6b
MT2	Agroforestry	75c	2.5a	4.8a	5.4ab
MT3	Agroforestry	52b	2.1a	5.2a	2.8a
MT4	Agroforestry	26a	1.3a	3.5a	2.0a
LSD		14	2.6	2.4	4.5

* The same letter indicates no statistically significant differences between locations with Fisher's LSD test ($p < 0.05$).

3.4. Runoff and Soil Erosion

Decreasing tree canopy cover in agroforestry systems significantly increased the surface runoff/rainfall ratio or the surface runoff/throughfall ratio (Table 4). In these results, the relationship between surface runoff with rainfall or throughfall was in line, and the ratio of surface runoff/rainfall was further used. The ratio of surface runoff/rainfall is also known as the runoff coefficient.

Table 4. Rainfall, runoff, ratio runoff/rainfall, and soil erosion in the runoff plots in each land cover type.

Code	Land Cover	Rainfall (mm)	Runoff (mm) *	Runoff/ Rainfall Ratio *	Runoff/ Throughfall Ratio *	Soil Erosion (Mg ha^{-1}) *
Upstream Rejoso Watershed						
UT1	Old production forest	555	14.3a	0.03a	0.04a	5.86a
UT2	Young production forest	492	13.2a	0.03a	0.03a	1.47a
UT3	Agroforestry	476	203.3b	0.43b	0.56c	120.98b
UT4	Arable land	556	225.7b	0.41b	0.43b	163.22b
LSD			46.3	0.09	0.11	87
Midstream Rejoso Watershed						
MT1	Old Production Forest	616	80.2a	0.13a	0.16a	3.07a
MT2	Agroforestry	841	316.3c	0.38b	0.48b	2.88a
MT3	Agroforestry	616	228.8b	0.37b	0.46b	6.63ab
MT4	Agroforestry	541	344.9c	0.64c	0.66c	10.33b
LSD			86.6	0.12	0.12	4.22

* The same letter indicates no statistically significant differences between locations with Fisher's LSD test ($p < 0.05$).

Infiltration rates in the andisols of the upper watershed were all above 45 mm hour^{-1} (Figure 6a). In the midstream area, forest plots had a high infiltration rate, but in the agroforestry systems infiltration rates were low and the apparent declined with decreasing tree cover was not statistically significant (Figure 6b).

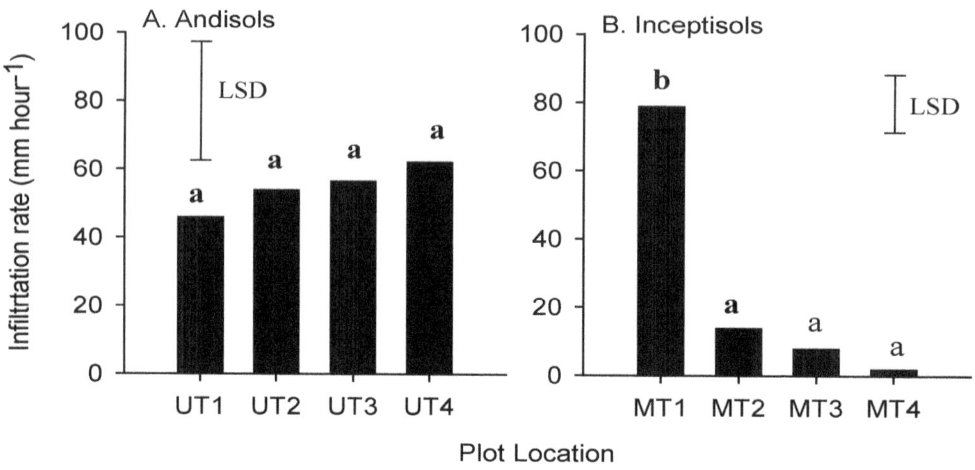

In the upstream area, with decreasing tree canopy cover, the surface runoff/rainfall ratio increased 16-fold compared to production forest (Figure 7a). In the midstream area, agroforestry systems with a tree canopy cover > 80% were still able to support low surface runoff (Figure 7b). With a tree canopy cover of < 80%, surface runoff increased rapidly on days with moderate rainfall (20–50 mm day^{-1}) (Figure 7b).

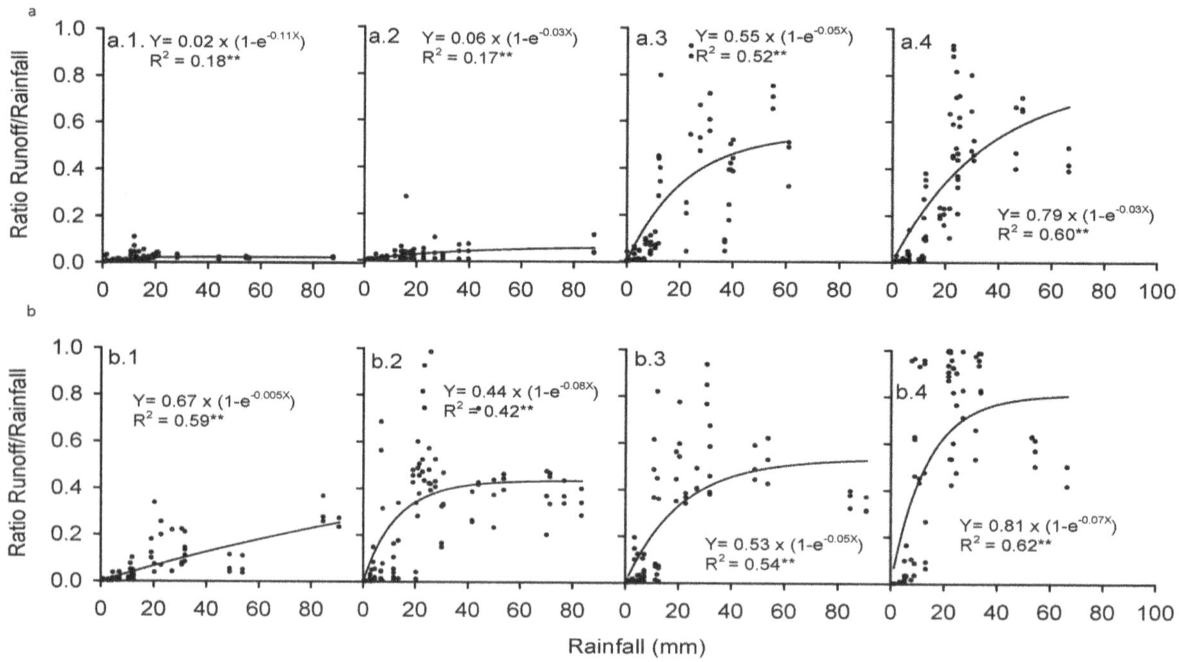

Figure 7. The relationship between surface runoff/rainfall ratio and the amount of rainfall in production forest and agroforestry systems in (**a**) the upstream Rejoso Watershed, under (a.1) 55% canopy cover of pine-based old production forest, (a.2) 40% canopy cover of pine-based young production forest, (a.3) 5% canopy cover of *Casuarina*-based agroforestry with cabbage crop, (a.4) 0% tree canopy cover of arable land (maize crop); (**b**) the midstream Rejoso Watershed under (b.1) 87 % canopy cover of pine/mahogany-based old production forest, (b.2) 75% canopy cover of coffee-based agroforestry, (b.3) 52% canopy cover of clove-based agroforestry, (b.4) 26% canopy cover of mixed tree and crop-based agroforestry.

In production forests with a closed tree canopy cover, soil erosion rates were low (Table 4 and Figure 8a.1,a.2,b.1). These production forests still had a protective understory vegetation that contributed to litter necromass and surface roughness (Table 3), controlling splash erosion. Upstream, with a reduction in tree cover, canopy soil erosion increased dramatically from 20 to 110 times the rates measured in forested plots (Table 4). Erosion rates in all plots increased with the amount of rainfall (Figure 8a.3,a.4). Midstream agroforestry systems had erosion rates ranging from 2.8 to 10.3 Mg ha^{-1} in the measurement period (Table 4). As annual rainfall is approximately three times what was recorded in the measurement period, with similar rainfall intensities, these erosion rates are to be multiplied by a factor of three, leading to 9–31 Mg ha^{-1} year^{-1}.

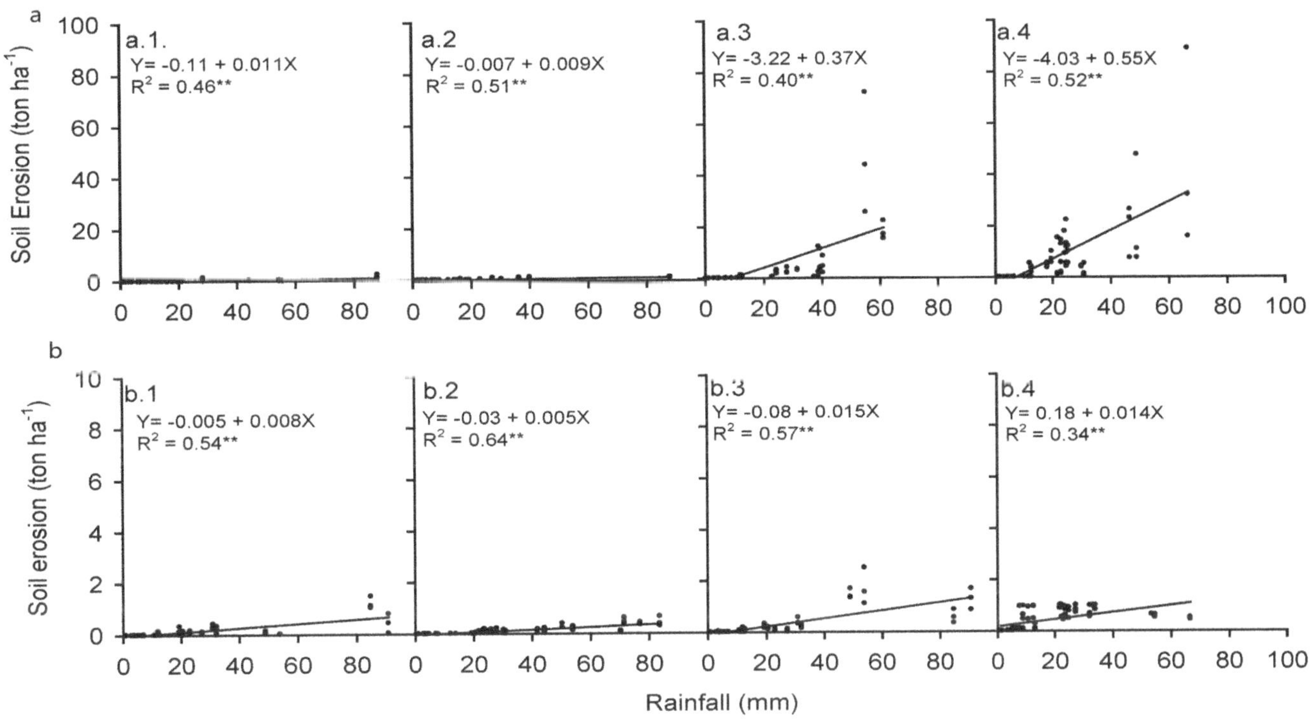

Figure 8. Soil erosion in relation to daily rainfall rates in production forest and agroforestry in (**a**) the upstream Rejoso Watershed, under (a.1) 55% canopy cover of Pine-based old production forest, (a.2) 40% canopy cover of pine-based young production forest, (a.3) 5% canopy cover of *Casuarina*-based agroforestry with cabbage crop, (a.4) 0% tree canopy cover of arable land (maize crop); (**b**) the midstream Rejoso Watershed under (b.1) 87% canopy cover of pine/mahogany-based old production forest, (b.2) 75% canopy cover of coffee-based agroforestry, (b.3) 52% canopy cover of clove-based agroforestry, (b.4) 26% canopy cover of mixed tree and crop-based agroforestry.

3.5. *Thresholds for Infiltration-Friendly Land Use*

Increasing tree canopy cover, while maintaining understory vegetation and litter necromass, is a strong indicator of watershed health and the main driver of low surface runoff (or high soil infiltration) and low soil erosion in production and agroforestry forest systems in the Rejoso Watershed (Figures 8a.1, 9a.1, 10b.1 and 11b.1).

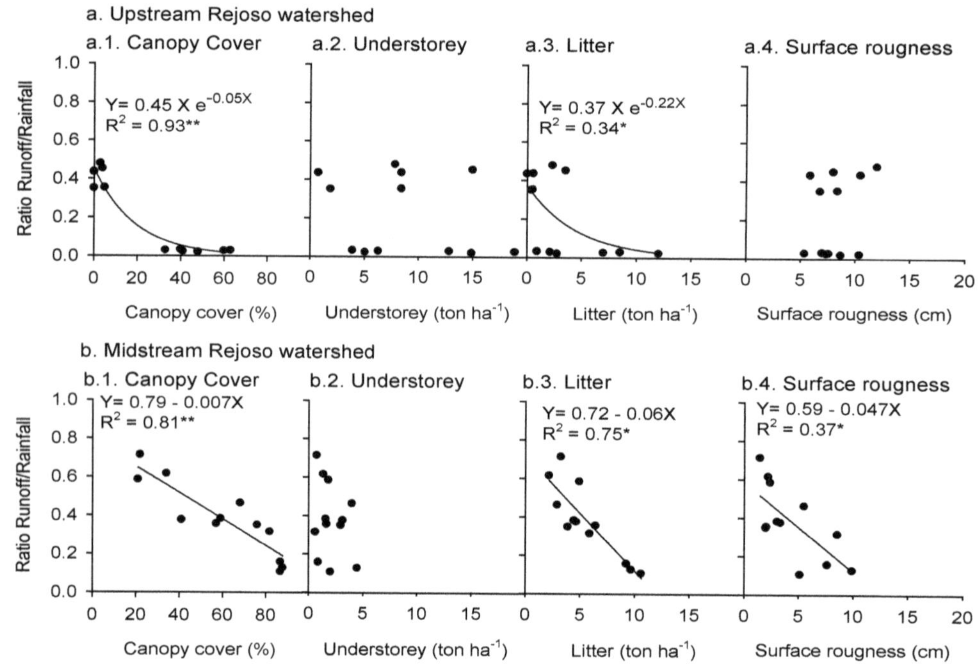

Figure 9. The runoff/rainfall ratio as a function of tree canopy cover, understory vegetation, litter necromass, and land surface roughness in the (**a**): upstream, (**b**): midstream.

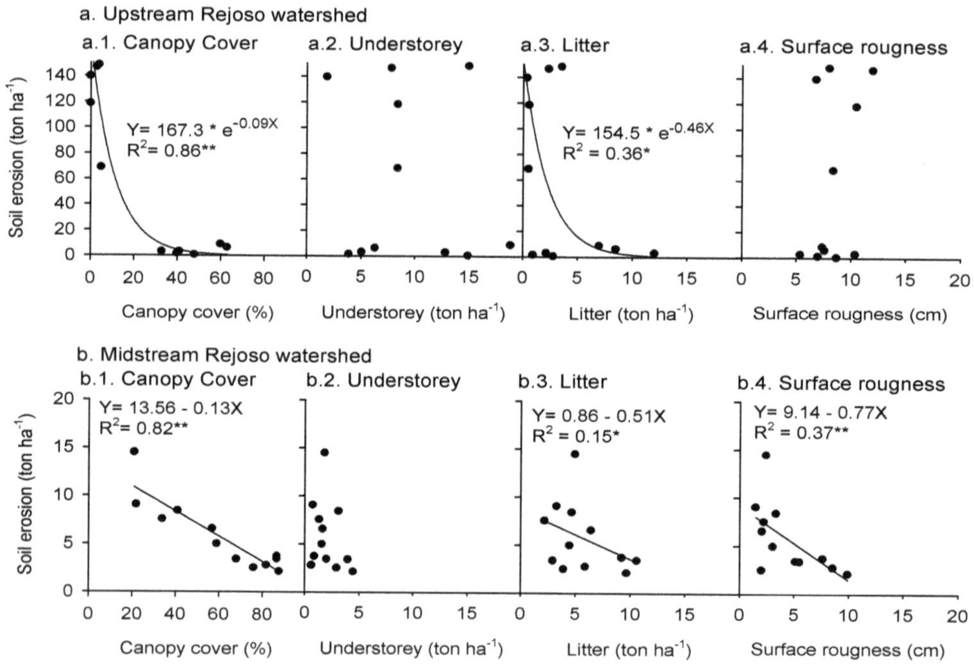

Figure 10. Soil erosion in relation to tree canopy cover, understory vegetation, litter necromass, and land surface roughness in the (**a**): upstream, (**b**): midstream.

Understory vegetation theoretically can reduce splash impacts on the soil and supports infiltration, as does the litter necromass present. However, the result of this study indicated that understory had no statistically significant relationships with runoff coefficient and soil erosion (Figures 9a.3,b.3 and 10a.3,b.3). Land surface roughness, in contrast to litter necromass, had no consistent relationship with runoff or erosion (Figures 9a.4 and 10a.4).

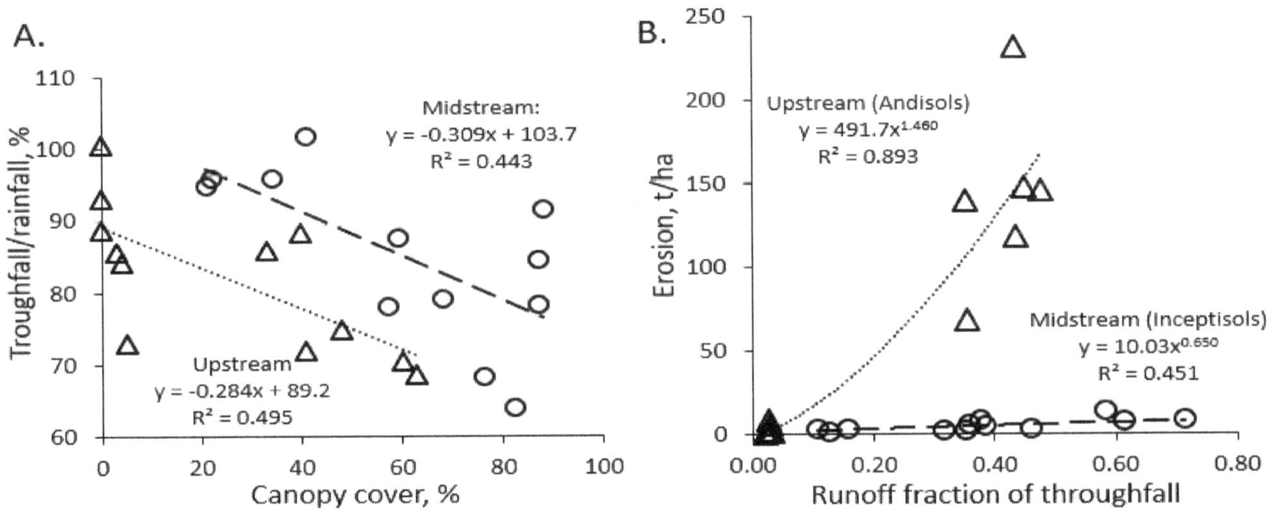

Figure 11. Comparison of canopy effects on throughfall in the two zones (**A**) and the relationship between erosion and surface runoff for the two soil types (**B**).

Summarizing (Figure 11), we found a similar slope (3% more canopy water retention per 10% canopy cover) but a 10% higher canopy retention overall in the upstream area (with lower rainfall intensity and more small events) and a strong difference between the two soil types in erosion per unit surface runoff, offsetting the higher infiltration in Andisols.

4. Discussion

Our study only covered two months' worth of data, rather than the recommended 3–5 years for such studies. Measurements included one-fourth of the mean annual rainfall, and the validity of the result may be primarily limited by the assumption that it represented a fair sample of the rainfall intensities that can be expected in the landscape. With a disproportional fraction of erosion normally associated with extreme events (compare the curvature of responses in Figure 11B), scaling up our comparison among land cover types to a multi-year basis may underestimate the relevance of controlling overland flows.

The first research question tested the hypothesis that, along the forest to open field agriculture continuum, there is a significant decrease in soil hydrological functions. The results of the present study confirmed that the conversion of high-density forest to land uses with a lower tree canopy significantly decreased soil infiltration rates (Table 2). The results of this study align with previous studies that showed that decreases in ground cover resulted in decreases in soil infiltration rates [51]. Forests and coffee agroforestry have been shown to reduce surface runoff and erosion compared with coffee monoculture [52]. Soil infiltration into Andisols both under deciduous and pine forest was higher than that in cropland in a study on the Canary Islands [53]. A study in China [54] found that the soil infiltration rate of forest was greater than that of agroforestry. A meta-analysis [55] concluded that converting any land use type with permanent vegetation cover (grassland, shrub, or forest) to seasonal cropland leads to a decline in the soil infiltration rate, harming soil and water conservation, while agroforestry improved the soil infiltration capacity compared to cropland and plantations.

The degradation of the soil hydrological functions of forest could be attributed to the decrease in soils' macroporosity, organic matter content, and increased soil bulk density (Table 2), which had relevance to the decreasing infiltration rates (Figure 8). Among various land use patterns, plant root activities are important factors affecting soil infiltration [56]. The reason why cropland has a lower infiltration rate than the land use types with a high density of trees compared with those with a low density of trees in forest may be verified by the fact that soils beneath the canopies of woody plants had a more extensive distribution of plant roots and a greater number of macropores, which are biologically produced pores [57,58], which created a positive feedback on infiltration [59,60]. The soil macroporosity, needed for effective infiltration, is the result of a continuous process of compaction and the filling in of macropores with fine soil particles, and the creation of biogenic channels (formed by old tree roots, earthworms, and other soil engineers) or abiotic processes (cracks). As no heavy machinery is used in any of these land use systems, compaction is restricted to human feet and motorbikes on specific tracks. The formation of old tree root channels can cause long time lags between land cover change and soil macroporosity [61,62], obscuring relations between current tree cover and soil hydrologic functions. "Fallows" were found to be intermediate between forests and grasslands in terms of infiltration in Madagascar [63]. Recovery of infiltration after the reforestation of grasslands in the Philippines was found to be a matter of decades rather than years [64]. In studies elsewhere in Indonesia, forest soils had more macropores and higher surface infiltration rates than monoculture coffee plantations [52]. Land use changes, especially from forest to cropland, have caused remarkable changes in soil properties, including the loss of organic matter and increases in bulk density [65], which lead to decreased infiltration rates [66]. Some researchers suggested a positive relationship between soil organic matter and infiltration rate [67,68].

Our study results can be compared to earlier studies in the volcanic uplands of West and East Java. Sediment delivery to streams increased after a clear-felling and replant operation in the Citarum Basin (W. Java), which involved the delayed flushing of material trapped during forest clearance and the incipient gullying of trails created by farmers involved in the replanting program, rather than by in-field erosion [69]. A study [70] of the Kali Konto catchment in East Java similarly concluded that "despite their relatively small areal extent (5% in the study area), rural roads, trails and settlements are significant producers of runoff and sediment at the catchment scale and should be included in watershed management programs designed to reduce catchment sediment yields and reservoir siltation." Soil conservation practices that transform slopes to relatively flat terrace beds and steep terrace risers are, in the absence of vegetation, still subject to erosion, with splash impacts on the terrace risers as a major cause [71,72].

The second research question came out with the hypothesis that the dominant factors that determine "infiltration friendliness" at the plot scale are tree canopy cover, understory vegetation, litter necromass, and land surface roughness. Our research shows that a number of land cover types had infiltration rates below the required rates at peak rainfall events. Among the four factors tested, tree cover and litter layer necromass could be used to define zone-specific thresholds for infiltration-friendly land use, but understory vegetation and surface roughness could not. Although slopes in the upper watershed are much steeper than in the midstream, the coarser texture and likely higher aggregate stability means that thresholds for canopy cover and litter necromass can be lower. A first "line of defense" of forests is the canopy retention of rainfall, prolonging the time for infiltration, as canopy dripping lasts beyond the rainfall event. Canopy retention of rainfall tends to be relatively high for small (but potentially frequent) rainfall events, and low for high rainfall intensities. Our throughfall results for the two zones corresponded with differences in observed intensity. A five-year study in the Amazon forests of Colombia [73] showed that throughfall ranged from 82 to 87% of gross rainfall in the forests studied (with a canopy cover of 83–91%) and varied with event-level gross rainfall, but also with forest structure, while stemflow contributed, on average, only 1.1% of gross rainfall in all forests. Throughfall is more spatially heterogeneous than rainfall, creating a challenge for its measurement. Roving, rather than fixed, location throughfall gauges led to narrower confidence intervals of throughfall fractions in longer-term studies [74] in lower montane rainforest in Puerto Rico, where throughfall was 75% and stemflow 4.1% of rainfall, with palms responsible for about 3% and other trees 1.1%. Spatial heterogeneity in throughfall can be expected to lead to uneven patterns of deep percolation and groundwater recharge in "patchy" forests [75]. Canopy interception can lead to direct evaporation, throughfall, or stemflow [76]. The ratio between throughfall and stemflow depends on the architecture of leaves (e.g., erect leaves favoring stemflow, pendulous leaves favoring throughfall) and stems. Storage along the stem pathway depends on bark properties [77]. Stemflow accounted for less than 3% of gross rainfall for tropical hardwoods in a study in Panama, while it was high for tall grasses [78]. High stemflow fractions have also been reported for bamboo, bananas, shaded coffee and cocoa, and understory shrubs [79–82]. Canopy interception and direct evaporation tend to be high in coastal areas with frequent light rainfall events, but low where tropical rainstorms are predominant and the canopy storage is rapidly saturated [83,84]. By creating throughfall drops that are larger than those of open-field rainfall, tree canopies may increase sub-canopy erosivity [13,84].

Many authors have emphasized that the key to hydrologic functions is in the soil rather than the aboveground parts of the forest [12]. Still, we found strong and direct relations with canopy cover. Positive effects of canopy cover on infiltration were related to raindrop interception in earlier studies [75]. Interception will (a) reduce the destructive power of rainwater splash on the ground surface (as long as the erosive canopy drips described earlier are avoided), (b) allow more time for infiltration as water reaches the surface more slowly, (c) keep a thin water film on the leaves that will (d) cool the surrounding air when it subsequently evaporates. It reduces the amount of water reaching the soil surface, but by increasing air humidity it also decreases transpiration demand when stomata are open. Coffee gardens close to forest had high macroporosity and infiltration rates relative to more compacted pasture and sugarcane land on volcanic slopes in Costa Rica [85]. Dye infiltration patterns in a comparison of natural forest and rubber plantations in Yunnan (China) showed [86] that the fine roots of understory vegetation promoted subsurface flow and reduced water erosion. The effects of trees on infiltration have been described as a "double-funneling" [87] with stemflow (dependent on the insertion angle of branches on the main stem), bringing water to the soil surface connection point for root-induced preferential flow [88,89].

A comparison of infiltration rates (median K_s values 16–98 mm h^{-1}) in broadleaf, pine-dominated, and mixed community-managed forest in Nepal [90] found the less intensively used pine-dominated site to be more conducive to vertical percolation than the other two forest types. These results were remarkable in relation to the negative local perceptions of the role of pine plantations on declining water resources.

Understory vegetation can theoretically reduce splash impacts on the soil and supports infiltration, as does the litter necromass present. However, the result of this study indicated that the understory shows no significant relationships with the runoff coefficient and soil erosion. This is possibly because surface runoff and erosion are largely controlled by land cover. The growth and development of the understory is determined by canopy cover. Likewise, the tree plantations in each plot are also diverse, so this also affects the diversity of the understory vegetation underneath. The result of this study indicates that the litter layer in the old production forest both upstream and midstream is significantly thicker than that other land uses (Table 3) and there is a significant correlation with the runoff coefficient and soil erosion (Figures 7 and 8, respectively). Litter is the parts of the body of the plant (in the form of leaves, branches, twigs, flowers, and fruit) that die (deciduous or pruned) and lie on the surface of the soil either intact or partially weathered. The role of litter in maintaining infiltration and soil erosion is through: (a) M−maintaining soil looseness by protecting the soil surface from rainwater, so that aggregates and soil macropores are maintained, (b) providing food sources for soil organisms, especially "soil engineers" (e.g., earthworms), so that the organism can live and develop in the soil, thus, the number of macro pores is maintained through the activity of these organisms, and (c) maintaining water quality in the river through the filtering of soil particles carried by surface runoff before entering the river. In a study in North China [91], the presence of the litter of *Quercus variabilis*, representing broadleaf litter, and *Pinus tabulaeformis*, representing needle leaf litter, reduced surface runoff rates by 29.5% and 31.3%, respectively. The overall effect of fast plus slow decomposing surface litter means the protection of the soil surface from splash erosion, surface roughness that reduces sediment entrainment, an energy source for soil biota, and a conducive microclimate [92,93].

Infiltration fractions depend on the scale of measurement and on variations in slope steepness, as overland flow can re-infiltrate on less steep foot-slopes in the case of the upper plots [94] or water infiltrates can re-emerge as surface flow depending on subsoil conductivity [95,96]. Such effects will need to be included if catchment level hydrology is to be predicted from plot-level measurements.

The land surface roughness also contributes to a high infiltration rate, reducing soil erosion. In the upstream, there is no significant different between land uses, but in the midstream, land surface roughness in agroforestry systems with tightly different canopies is significantly higher than rare canopies (Table 3). Without a high canopy cover (Table 3), this roughness was not able to control surface runoff and erosion in the upstream area. This is due to a steep slope in this plot. Both the production forest and agroforestry systems with high canopies maintained a relatively high land surface roughness compared with rare canopies in the midstream area. In the midstream, the land surface roughness was significantly correlated with the runoff coefficient and soil erosion. The role of surface roughness as a sediment filter may depend on frequent regeneration to counter homogenization [97]. Surface roughness in the landscape includes a cavity, the meandering of streams due to the presence of litter, necromass, tree trunks, and rocks, which provide opportunities for water flow to stop for longer periods and experience infiltration. This condition also functions as a sediment filter. This function needs to be managed through land management, so that surface roughness is maintained on the ground.

Shifts in local rainfall patterns between sub-watersheds make it difficult to disentangle the relative importance of land use and climate change through statistical pattern analysis without knowledge of the underlying processes [98,99]. The holy grail of scientific hydrology, connecting overall aggregated flow patterns to local extreme events and possible hysteresis, is still worth searching for even if a general solution might ultimately prove impossible to find [100]. For the deep seepage component of the hillslope and catchment water balance, we can expect that extreme events are less important than gradual changes that influence average flows, but empirical analysis of the uncertainties involved is still a challenge [101].

The third research question is, as an analysis, the answers to the previous two research questions with the hypothesis that it is not always that the upstream watershed area is more sensitive to hydrological disturbance due to changes in land use than the midstream, but the factor of soil properties also determines considerations in watershed hydrological management. From a land use policy perspective, our results suggest that maintaining high (~80%) canopy cover in the mid-slope farmer-controlled landscape under bench terracing, which does not match the slope criteria for designation as watershed protection forest, is important. In Indonesia, protection forest areas have the primary functions of the protection of life support systems to regulate water management, prevent flooding, control soil erosion, and maintain soil fertility [102].

Erosion rates of 9–31 Mg ha^{-1} $year^{-1}$, as estimated here, are a challenge, especially if the 400-year time frame of using up all soil, as used in Equation 7, is replaced by a tolerance equal to the rate of soil formation. A study in a high rainfall area with Inceptisols in Central Java [103] estimated that the rate of chemical weathering was around 0.85 Mg ha^{-1} yr^{-1} and used that as estimate of erosion rates that can be sustained indefinitely without affecting soil depth. Volcanic ash inputs add soil on top of the profile but may also be disproportionately included in what gets removed from the plots. Our measurements in Rejoso suggested that critical thresholds of the degree of canopy cover that is hydrologically desirable depend on soil and climatic conditions, which may vary over a relatively short distance. When the focus is on erosion and net sediment transport, the scale of consideration strongly influenced conclusions in the volcanic Way Besai Watershed in Sumatra as well [104]. With the higher rainfall intensities in midstream Rejoso and more erodible soils upstream, the risks for degradation from a downstream perspective are differentiated by zone. Combining our plot-level results with efforts of hydrologic modeling for the Rejoso catchment as a whole [105,106] can guide further advice to a local watershed forum on the measures and incentives needed to restore and protect the watershed as a whole.

The Indonesian legal requirement of 30% forest cover across all its local government entities [31] is a coarse translation of the hydrologic relations at risk. It clearly matters what the land cover in the "non-forest" parts of the landscape is and how vegetation interacts with soils and geomorphology in shaping rivers and groundwater flows [107,108]. Our findings for the Rejoso Watershed show that, within the agroforestry spectrum, hydrologic the thresholds of infiltration friendliness exist between the systems that are mostly "agro" and those that are mostly "forest", but higher tree cover systems are desirable.

5. Conclusions

Our results demonstrated that vegetation-based thresholds for adequate infiltration, given the existing rainfall intensities, differed between the middle and upper Rejoso Watershed. Despite steep slopes and low tree cover, the upper watershed, with its course soil texture (pseudo-sand/silt), low bulk density due to a high content of amorphic minerals, strong micro-aggregation and individual minerals, sponge-pores typical of Andosols, and land management practices that combine vegetable crops with a tree canopy cover of around 55%, can maintain infiltration and keep erosion at acceptable levels. In the midstream part of the catchment, despite gentle slopes under bench terracing, infiltration-friendly land use on the fine-textured Inceptisols required a canopy cover of 80%. Beyond tree canopy cover, litter layer necromass was found to be a good and easily observed indicator of infiltration rates, while understory vegetation and surface roughness may support infiltration, but are not sufficiently strong indicators.

Author Contributions: D.S., W.W., K.H., and M.v.N. designed the study. N.M. collected data in the midstream, A.L.R. collected data in the upstream, R.M.I. coordinated the data collection in the field, and were academically supervised by D.S., W.W., and K.H. D.S., K.H. and M.v.N. shaped the manuscript, which was approved by all co-authors.

Acknowledgments: Authors thank to the community of the Rejoso Watershed and the "*Rejoso Kita*" Forum. They also thank the Department of Soil Science, Faculty of Agriculture, University of Brawijaya and the Research Group of Tropical Agroforestry for their support of the research. The authors would like to thank the Social Investment Indonesia (SII) organization for connecting the local stakeholders during fieldwork. The manuscript benefitted from the comments of Sampurno Bruijnzeel and an anonymous reviewer.

Appendix A

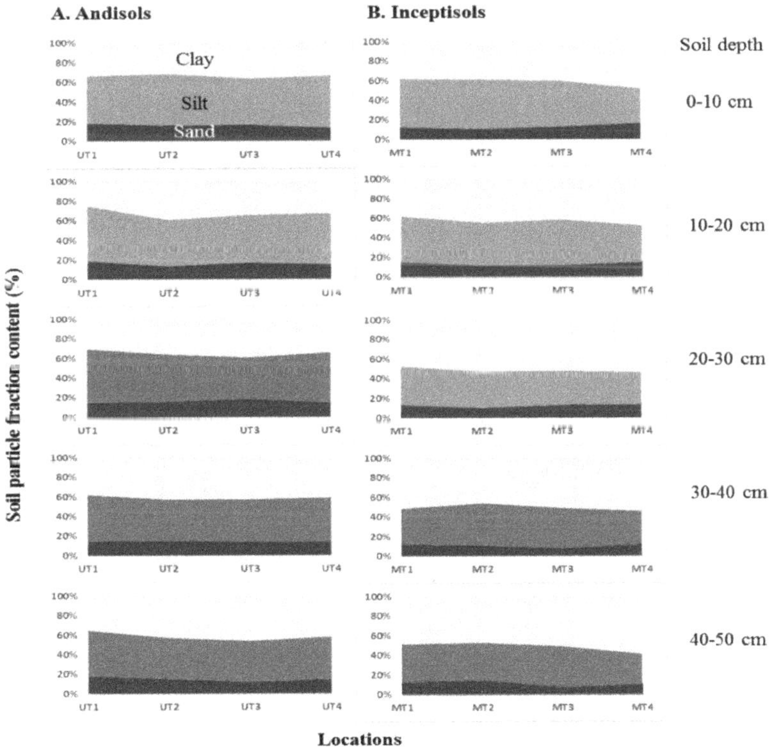

Figure A1. Soil texture in five different layers in runoff plot measurements.

References

1. UN-Water. *2018 UN World Water Development Report, Nature-Based Solutions for Water*; UN-Water: Nairobi, Kenya, 2018.
2. Ostovar, A.L. Investing upstream: Watershed protection in Piura, Peru. *Environ. Sci. Policy* **2019**, *96*, 9–17. [CrossRef]
3. Ali, M.; Hadi, S.; Sulistyantara, B. Study on land cover change of Ciliwung downstream watershed with spatial dynamic approach. CITIES 2015 International Conference, Intelligent Planning towards Smart Cities, CITIES 2015, 3–4 November 2015, Surabaya, Indonesia. *Procedia—Soc. Behav. Sci.* **2016**, *227*, 52–59. [CrossRef]
4. Jacobs, S.R.; Timbe, E.; Weeser, B.; Rufino, M.C.; Butterbach-Bahl, K.; Breuer, L. Assessment of hydrological pathways in East African montane catchments under different land use. *Hydrol. Earth Syst. Sci.* **2018**, *22*, 4981–5000. [CrossRef]
5. Van Noordwijk, M.; Bayala, J.; Hairiah, K.; Lusiana, B.; Muthuri, C.; Khasanah, N.; Mulia, R. Agroforestry solutions for buffering climate variability and adapting to change. In *Climate Change Impact and Adaptation in Agricultural Systems*; Fuhrer, J., Gregory, P.J., Eds.; CAB-International: Wallingford, UK, 2014; pp. 216–232.
6. Van Noordwijk, M.; Bargues-Tobella, A.; Muthuri, C.W.; Gebrekirstos, A.; Maimbo, M.; Leimona, B.; Bayala, J.; Ma, X.; Lasco, R.; Xu, J.; et al. Agroforestry as part of nature-based water management. In *Sustainable Development through Trees on Farms: Agroforestry in Its Fifth Decade*; van Noordwijk, M., Ed.; World Agroforestry (ICRAF): Bogor, Indonesia, 2019; pp. 261–287.
7. Van Noordwijk, M.; Kim, Y.S.; Leimona, B.; Hairiah, K.; Fisher, L.A. Metrics of water security, adaptive capacity, and agroforestry in Indonesia. *Curr. Opin. Environ. Sustain.* **2016**, *21*, 1–8. [CrossRef]
8. Filoso, S.; Bezerra, M.O.; Weiss, K.C.; Palmer, M.A. Impacts of forest restoration on water yield: A systematic review. *PLoS ONE* **2017**, *12*, e0183210. [CrossRef]
9. Bentley, L.; Coomes, D.A. Partial river flow recovery with forest age is rare in the decades following establishment. *Glob. Chang. Biol.* **2020**, *26*, 1458–1473. [CrossRef]
10. Zhang, J. Hydrological Response of a Fire-Climax Grassland and a Reforest Before and After Passage of Typhoon Haiyan (Leyte Island, the Philippines). Ph.D. Thesis, Free University, Amsterdam, The Netherlands, 2018.
11. Bruijnzeel, L.A. Hydrological functions of tropical forests: Not seeing the soil for the trees? *Agric. Ecosyst. Environ.* **2004**, *104*, 185–228. [CrossRef]
12. Peña-Arancibia, J.L.; Bruijnzeel, L.A.; Mulligan, M.; van Dijk, A.I. Forests as 'sponges' and 'pumps': Assessing the impact of deforestation on dry-season flows across the tropics. *J. Hydrol.* **2019**, *574*, 946–963. [CrossRef]
13. Wiersum, K.F. Effects of various vegetation layers of an Acacia auriculiformis forest plantation on surface erosion in Java, Indonesia. In *Soil Erosion and Conservation*; El-Swaify, S.A., Moldenhauer, W.C., Lo, A., Eds.; John Wiley & Sons: Hoboken, NJ, USA, 1985; pp. 79–89.
14. Hairiah, K.; Suprayogo, D.; Widianto, B.; Suhara, E.; Mardiastuning, A.; Widodo, R.H.; Prayogo, C.; Rahayu, S. Forests conversion into agricultural land: Litter thickness, earthworm populations and soil macro-porosity. *Agrivita* **2004**, *26*, 68–80.
15. Suprayogo, D.; Widianto, K.; Hairiah, I.N. *Watershed Management: Hydrological Consequences of Land Use Change during Development*; Universitas Brawijaya Press: Malang, Indonesia, 2017.
16. Creed, I.F.; van Noordwijk, M. *Forest and Water on a Changing Planet: Vulnerability, Adaptation and Governance Opportunities*; A Global Assessment Report. World Series Volume 38; IUFRO: Vienna, Austria, 2018.
17. Jones, J.A.; Wei, X.; Archer, E.; Bishop, K.; Blanco, J.A.; Ellison, D.; Gush, M.; McNulty, S.G.; van Noordwijk, M.; Creed, I.F. Forest-Water Interactions under Global Change. In *Forest-Water Interactions*; Ecological Studies 240; Levia, D.F., Carlyle-Moses, D.E., Michalzik, B., Nanko, K., Tischer, A., Eds.; Springer Nature: Picassopl, Switzerland, 2020; pp. 589–624. [CrossRef]
18. Liu, Y.; Cui, Z.; Huang, Z.; Miao, H.T.; Wu, G.L. The influence of litter crusts on soil properties and hydrological processes in a sandy ecosystem. *Hydrol. Earth Syst. Sci.* **2019**, *23*, 2481–2490. [CrossRef]
19. Zhang, J.; van Meerveld, H.I.; Tripoli, R.; Bruijnzeel, L.A. Runoff response and sediment yield of a landslide-affected fire-climax grassland micro-catchment (Leyte, the Philippines) before and after passage of typhoon Haiyan. *J. Hydrol.* **2018**, *565*, 524–537. [CrossRef]
20. Zhang, J.; Bruijnzeel, L.A.; Quiñones, C.M.; Tripoli, R.; Asio, V.B.; van Meerveld, H.J. Soil physical characteristics of a degraded tropical grassland and a 'reforest': Implications for runoff generation. *Geoderma* **2019**, *333*, 163–177. [CrossRef]

21. Van Noordwijk, M.; Lawson, G.; Hairiah, K.; Wilson, J. Root distribution of trees and crops: Competition and/or complementarity. In *Tree–Crop Interactions: Agroforestry in a Changing Climate*; CABI: Wallingford, UK, 2015; pp. 221–257.
22. Ellison, D.; Morris, C.E.; Locatelli, B.; Sheil, D.; Cohen, J.; Murdiyarso, D.; Gutierrez, V.; van Noordwijk, M.; Creed, I.F.; Pokorny, J. Trees, forests and water: Cool insights for a hot world. *Glob. Environ. Chang.* **2017**, *43*, 51–61. [CrossRef]
23. Ellison, D.; Wang-Erlandsson, L.; van der Ent, R.J.; van Noordwijk, M. Upwind forests: Managing moisture recycling for nature-based resilience. *Unasylva* **2019**, *251*, 13–26.
24. Van Noordwijk, M.; Tanika, L.; Lusiana, B. Flood risk reduction and flow buffering as ecosystem services—Part 1: Theory on flow persistence, flashiness and base flow. *Hydrol. Earth Syst. Sci.* **2017**, *21*, 2321–2340. [CrossRef]
25. Shannon, P.D.; Swanston, C.W.; Janowiak, M.K.; Handler, S.D.; Schmitt, K.M.; Brandt, L.A.; Butler-Leopold, P.R.; Ontl, T. Adaptation strategies and approaches for forested watersheds. *Rev. Artic. Clim. Serv.* **2019**, *13*, 51–64. [CrossRef]
26. Van Noordwijk, M.; Tanika, L.; Lusiana, B. Flood risk reduction and flow buffering as ecosystem services—Part 2: Land use and rainfall intensity effects in Southeast Asia. *Hydrol. Earth Syst. Sci.* **2017**, *21*, 2341–2360. [CrossRef]
27. Coster, C. Wortelstudien in de tropen. V. Gebergtehoutsoorten [Root studies in the Tropics. V. Tree species of the mountain region]. *Tectona* **1935**, *28*, 861–878.
28. Coster, C. De verdamping van verschillende vegetatievormen op Java [The transpiration of different types of vegetation in Java]. *Tectona* **1937**, *30*, 1–124.
29. Gonggrijp, L. De verdamping van gebergtebosch in West Java op 1750–200 m zeehoogte [The transpiration of mountain forest in West Java at an elevation of 1750–2000 m above sea level]. *Tectona* **1941**, *34*, 437–447.
30. Galudra, G.; Sirait, M. A discourse on Dutch colonial forest policy and science in Indonesia at the beginning of the 20th century. *Int. For. Rev.* **2009**, *11*, 524–533. [CrossRef]
31. Government of Indonesia. *Law (Undang-Undang) Number 26 Year 2007 about Spatial Planning*; Government of Indonesia: Pemerintah, Indonesia, 2007.
32. Van Noordwijk, M.; Creed, I.F.; Jones, J.A.; Wei, X.; Gush, M.; Blanco, J.A.; Sullivan, C.A.; Bishop, K.; Murdiyarso, D.; Xu, J.; et al. Climate-forest-water-people relations: Seven system delineations. In *Forest and Water on a Changing Planet: Vulnerability, Adaptation and Governance Opportunities: A Global Assessment Report (No. 38)*; International Union of Forest Research Organizations (IUFRO): Vienna, Austria, 2018; pp. 27–58.
33. Van Noordwijk, M.; Ekadinata, A.; Leimona, B.; Catacutan, D.; Martini, E.; Tata, H.L.; Öborn, I.; Hairiah, K.; Wangpakapattanawong, P.; Mulia, R.; et al. Agroforestry options for degraded landscapes in Southeast Asia. In *Agroforestry for Degraded Landscapes: Recent Advances and Emerging Challenges*; Dagar, J.C., Gupta, S.R., Teketay, D., Eds.; Springer: Singapore, 2020.
34. Sajikumar, N.; Remya, R.S. Impact of land cover and land use change on runoff characteristics. *J. Environ. Manag.* **2015**, *161*, 460–468. [CrossRef] [PubMed]
35. Anache, J.A.; Wendland, E.C.; Oliveira, P.T.; Flanagan, D.C.; Nearing, M.A. Runoff and soil erosion plot-scale studies under natural rainfall: A meta-analysis of the Brazilian experience. *Catena* **2017**, *152*, 29–39. [CrossRef]
36. USGS. Landsat 8 OLI (2019-10-18 (6, 53). U.S. Geological Survey. 2019. Available online: https://earthexplorer.usgs.gov/ (accessed on 9 June 2020).
37. Wischmeier, W.H. The USLE: Some reflections. *J. Soil Water Conserv.* **1984**, *39*, 105–107.
38. Hairsine, P.B.; Rosem, C.W. Modeling water erosion due to overland flow using physical principles: 1. Sheet flow. *Water Resour. Res.* **1992**, *28*, 237–243. [CrossRef]
39. Van Noordwijk, M.; van Roode, M.; McCallie, E.L.; Cadisch, G. Erosion and sedimentation as multiscale, fractal processes: Implications for models, experiments and the real world. In *Soil Erosion at Multiple Scales, Principles and Methods for Assessing Causes and Impacts*; Penning de Vries, F., Agus, F., Kerr, J., Eds.; CAB International: Wallingford, UK, 1998; pp. 223–253.
40. Knox County Tennessee. Stormwater Management Manual, Section on the Rational Method, Volume 2 Technical Guidance. Available online: https://dec.alaska.gov/water/wastewater/stormwater/swppp-dev-runoff-coefficient-values-rm (accessed on 26 April 2020).
41. Widianto, D.S.; Noveras, H.; Widodo, R.H.; Purnomosidhi, P.; van Noordwijk, M. Alih guna lahan hutan menjadi lahan pertanian: Apakah fungsi hidrologis hutan dapat digantikan sistem kopi monokultur? *Agrivita* **2004**, *26*, 47–52.

42. Gee, G.W.; Bauder, J.W. Particle Size Analysis. In *Methods of Soil Analysis Part 1, Physical and Mineralogical Methods*, 2nd ed.; Klute, A., Ed.; American Society of Agronomy: Madison, WI, USA, 1986; Volume 9, pp. 383–411.
43. Blake, G.R.; Hartge, K.H. Bulk density. In *Methods of Soil Analysis. Part I: Physical and Mineralogical Methods*; Klute, A.K., Ed.; American Society of Agronomy—Soil Science Society of America: Madison, WI, USA, 1986; pp. 363–375.
44. Nimmo, J.R. Porosity and pore size distribution. In *Encyclopedia of Soils in the Environment*; Hillel, D., Ed.; Elsevier: London, UK, 2004; pp. 295–303.
45. Walkley, A.; Black, I.A. An examination of the Degtjareff method for determining soil organic matter, and a proposed modification of the chromic acid titration method. *Soil Sci.* **1934**, *37*, 29–38. [CrossRef]
46. Bouwer, H. Intake rate: Cylinder infiltrometer. In *Methods of Soil Analysis*; Klute, A., Ed.; ASA Monograph 9; ASA: Madison, WI, USA, 1986; pp. 825–843.
47. Jennings, S.B.; Brown, N.D.; Sheil, D. Assessing forest canopies and understorey illumination: Canopy closure, canopy cover and other measures. *Forestry* **1999**, *72*, 59–73. [CrossRef]
48. Hairiah, K.; Ekadinata, A.; Sari, R.R.; Rahayu, S. Measurement of Carbon Stock: From land level to landscape. In *Practical Instructions*, 2nd ed.; Bogor, World Agroforestry Centre, ICRAF SEA Regional Office, University of Brawijaya (UB): Malang, Indonesia, 2011; p. 110. Available online: http://apps.worldagroforestry.org/sea/Publications/files/manual/MN0049-11.pdf (accessed on 10 April 2020).
49. Hoechsteetter, S.; Walz, U.; Dang, L.H.; Thinh, N.X. Effect of Topography and Surface Roughness in Analyses of Landscape Structure—A Proposal Modify the Existing Set of Landscape Metrics. *Landsc. Online* **2008**, *3*, 1–14. Available online: https://gfzpublic.gfz-potsdam.de/pubman/item/item_32854 (accessed on 10 April 2020). [CrossRef]
50. Landon, J.R. *Booker Tropical Soil Manual: A Handbook for Soil Survey and Agricultural Land Evaluation in the Tropics and Subtropics*; Routledge: Abingdon, UK, 2014.
51. Gifford, G.F.; Hawkins, R.H. Hydrologic impact of grazing on infiltration: A critical review. *Water Resour. Res.* **1978**, *14*, 305–313. [CrossRef]
52. Widianto, D.S.; Purnomosidhi, P.; Widodo, R.H.; Rusiana, F.; ZAini, Z.; Khasanah, N.; Kusuma, Z. Degradasi sifat fisik tanah sebagai akibat alih guna lahan hutan menjadi sistem kopi monokultur: Kajian perubahan makroporositas tanah. *Agrivita* **2004**, *26*, 60–67.
53. Neris, J.; Jiménez, C.; Fuentes, J.; Morillas, G.; Tejedor, M. Vegetation and land-use effects on soil properties and water infiltration of Andisols in Tenerife (Canary Islands, Spain). *Catena* **2012**, *98*, 55–62. [CrossRef]
54. Ma, Q.; Wang, J.; Zhu, S. Effects of precipitation, soil water content and soil crust on artificial Haloxylon Ammodendron forest. *Acta Ecol. Sin.* **2007**, *27*, 5057–5067. [CrossRef]
55. Sun, D.; Yang, H.; Guan, D.; Yang, M.; Wu, J.; Yuan, F.; Jin, C.; Wang, A.; Zhang, Y. The effects of land use change on soil infiltration capacity in China: A meta-analysis. *Sci. Total Environ.* **2018**, *626*, 1394–1401. [CrossRef] [PubMed]
56. Zimmermann, B.; Elsenbeer, H.; De Moraes, J.M. The influence of land-use changes on soil hydraulic properties: Implications for runoff generation. *For. Ecol. Manag.* **2006**, *222*, 29–38. [CrossRef]
57. Dunkerley, D. A Review of the Effects of Throughfall and Stemflow on Soil Properties and Soil Erosion. In *Precipitation Partitioning by Vegetation*; Springer: Cham, Germany, 2020; pp. 183–214.
58. Colloff, M.J.; Pullen, K.R.; Cunningham, S.A. Restoration of an ecosystem function to revegetation communities: The role of invertebrate macropores in enhancing soil water infiltration. *Restor. Ecol.* **2010**, *18*, 65–72. [CrossRef]
59. Reid, K.D.; Wilcox, B.P.; Breshears, D.D.; MacDonald, L. Runoff and erosion in a pinon–juniper woodland: Influence of vegetation patches. *Soil Sci. Soc. Am. J.* **1999**, *63*, 313–325. [CrossRef]
60. Bhark, E.W.; Small, E.E. Association between plant canopies and the spatial patterns of infiltration in shrubland and grassland of the Chihuahuan Desert, New Mexico. *Ecosystems* **2003**, *6*, 185–196. [CrossRef]
61. Van Noordwijk, M.; Heinen, M.; Hairiah, K. Old tree root channels in acid soils in the humid tropics: Important for crop root penetration, water infiltration and nitrogen management. In *Plant-Soil Interactions at Low pH*; Springer: Dordrecht, The Netherlands, 1991; pp. 423–430.
62. Wu, G.L.; Liu, Y.; Yang, Z.; Cui, Z.; Deng, L.; Chang, X.F.; Shi, Z.H. Root channels to indicate the increase in soil matrix water infiltration capacity of arid reclaimed mine soils. *J. Hydrol.* **2017**, *546*, 133–139. [CrossRef]

63. Zwartendijk, B.W.; Van Meerveld, H.J.; Ghimire, C.P.; Bruijnzeel, L.A.; Ravelona, M.; Jones, J.P.G. Rebuilding soil hydrological functioning after swidden agriculture in eastern Madagascar. *Agric. Ecosyst. Environ.* **2017**, *239*, 101–111. [CrossRef]
64. Van Meerveld, H.J.; Zhang, J.; Tripoli, R.; Bruijnzeel, L.A. Effects of reforestation of a degraded Imperata grassland on dominant flow pathways and streamflow responses in Leyte, the Philippines. *Water Resour. Res.* **2019**, *55*, 4128–4148. [CrossRef]
65. Sanchez, P.A. *Properties and Management of Soils in the Tropics*; Cambridge University Press: Cambridge, UK, 2019.
66. Mwendera, E.J.; Saleem, M.M. Infiltration rates, surface runoff, and soil loss as influenced by grazing pressure in the Ethiopian highlands. *Soil Use Manag.* **1997**, *13*, 29–35. [CrossRef]
67. Martens, D.A.; Frankenberger, W.T. Modification of Infiltration Rates in an Organic-Amended Irrigated. *Agron. J.* **1992**, *84*, 707–717. [CrossRef]
68. Osuji, G.E.; Okon, M.A.; Chukwuma, M.C.; Nwarie, I.I. Infiltration characteristics of soils under selected land use practices in Owerri, Southeastern Nigeria. *World J. Agric. Sci.* **2010**, *6*, 322–326.
69. Bons, C.A. Accelerated erosion due to clearcutting of plantation forest and subsequent Taungya cultivation in upland West Java, Indonesia. *Int. Assoc. Hydrol. Sci. Publ.* **1990**, *192*, 279–288.
70. Rijsdijk, A.; Bruijnzeel, L.S.; Sutoto, C.K. Runoff and sediment yield from rural roads, trails and settlements in the upper Konto catchment, East Java, Indonesia. *Geomorphology* **2007**, *87*, 28–37. [CrossRef]
71. Purwanto, E.; Bruijnzeel, L.A. Soil conservation on rainfed bench terraces in upland West Java, Indonesia: Towards a new paradigm. In *Advances in Geoecology*; Catena: Reiskirchen, Germany, 1998; pp. 1267–1274.
72. Van Dijk, A.I.J.M.; Bruijnzeel, L.A.; Wiegman, S.E. Measurements of rain splash on bench terraces in a humid tropical steepland environment. *Hydrol. Process.* **2003**, *17*, 513–535. [CrossRef]
73. Marin, C.T., Bouten, W.; Sevink, J. Gross rainfall and its partitioning into throughfall, stemflow and evaporation of intercepted water in four forest ecosystems in western Amazonia. *J. Hydrol.* **2000**, *237*, 40–57. [CrossRef]
74. Holwerda, F.; Scatena, F.N.; Bruijnzeel, L.A. Throughfall in a Puerto Rican lower montane rain forest: A comparison of sampling strategies. *J. Hydrol.* **2006**, *327*, 592–602. [CrossRef]
75. Klos, P.Z.; Chain-Guadarrama, A.; Link, T.E.; Finegan, B.; Vierling, L.A.; Chazdon, R. Throughfall heterogeneity in tropical forested landscapes as a focal mechanism for deep percolation. *J. Hydrol.* **2014**, *519*, 2180–2188. [CrossRef]
76. Muzylo, A.; Llorens, P.; Valente, F.; Keizer, J.J.; Domingo, F.; Gash, J.H.C. A review of rainfall interception modelling. *J. Hydrol.* **2009**, *370*, 191–206. [CrossRef]
77. Van Stan, J.T.; Lewis, E.S.; Hildebrandt, A.; Rebmann, C.; Friesen, J. Impact of interacting bark structure and rainfall conditions on stemflow variability in a temperate beech-oak forest, central Germany. *Hydrol. Sci. J.* **2016**, *61*, 2071–2083. [CrossRef]
78. Friesen, P.; Park, A.; Sarmiento-Serrud, A.A. Comparing rainfall interception in plantation trials of six tropical hardwood trees and wild sugar cane *Saccharum spontaneum* L. *Ecohydrology* **2013**, *6*, 765–774.
79. Taniguchi, M.; Tsujimura, M.; Tanaka, T. Significance of stemflow in groundwater recharge. 1: Evaluation of the stemflow contribution to recharge using a mass balance approach. *Hydrol. Process.* **1996**, *10*, 71–80. [CrossRef]
80. Cattan, P.; Bussière, F.; Nouvellon, A. Evidence of large rainfall partitioning patterns by banana and impact on surface runoff generation. *Hydrol. Process.* **2007**, *21*, 2196–2205. [CrossRef]
81. Siles, P.; Vaast, P.; Dreyer, E.; Harmand, J.M. Rainfall partitioning into throughfall, stemflow and interception loss in a coffee (*Coffea arabica* L.) monoculture compared to an agroforestry system with Inga densiflora. *J. Hydrol.* **2010**, *395*, 39–48. [CrossRef]
82. Niether, W.; Armengot, L.; Andres, C.; Schneider, M.; Gerold, G. Shade trees and tree pruning alter throughfall and microclimate in cocoa (*Theobroma cacao* L.) production systems. *Ann. For. Sci.* **2018**, *75*, 38. [CrossRef]
83. Bruijnzeel, L.A.; Wiersum, K.F. Rainfall interception by a young Acacia auriculiformis (A. Cunn) plantation forest in west Java, Indonesia: Application of Gash's analytical model. *Hydrol. Process.* **1987**, *1*, 309–319. [CrossRef]
84. de Almeida, W.S.; Panachuki, E.; de Oliveira, P.T.S.; da Silva Menezes, R.; Sobrinho, T.A.; de Carvalho, D.F. Effect of soil tillage and vegetal cover on soil water infiltration. *Soil Tillage Res.* **2018**, *175*, 130–138. [CrossRef]

85. Toohey, R.C.; Boll, J.; Brooks, E.S.; Jones, J.R. Effects of land use on soil properties and hydrological processes at the point, plot, and catchment scale in volcanic soils near Turrialba, Costa Rica. *Geoderma* **2018**, *315*, 138–148. [CrossRef]
86. Nespoulous, J.; Merino-Martin, L.; Monnier, Y.; Bouchet, D.C.; Ramel, M.; Dombey, R.; Viennois, G.; Mao, Z.; Zhang, J.L.; Cao, K.F.; et al. Tropical forest structure and understorey determine subsurface flow through biopores formed by plant roots. *Catena* **2019**, *181*, 104061. [CrossRef]
87. Johnson, M.S.; Lehmann., J. Double-Funneling of trees: Stemflow and root-induced preferential flow. *Ecoscience* **2006**, *13*, 324–333. [CrossRef]
88. Gonzalez-Ollauri, A.; Stokes, A.; Mickovski, S.B. A novel framework to study the effect of tree architectural traits on stemflow yield and its consequences for soil-water dynamics. *J. Hydrol.* **2020**, *582*, 124448. [CrossRef]
89. Liang, W.L. Effects of Stemflow on Soil Water Dynamics in Forest Stands. In *Forest-Water Interactions*; Springer: Cham, Germany, 2020; pp. 349–370.
90. Badu, M.; Nuberg, I.; Ghimire, C.P.; Bajracharya, R.M.; Meyer, W.S. Negative Trade-Offs between Community Forest Use and Hydrological Benefits in the Forested Catchments of Nepal's Mid-hills. *Mt. Res. Dev.* **2020**, *39*, 822–832. [CrossRef]
91. Li, X.; Niu, J.; Xie, B. The Effect of Leaf Litter Cover on Surface Runoff and Soil Erosion in Northern China. *PLoS ONE* **2014**, *9*, e107789. [CrossRef]
92. Hairiah, K.; Sulistyani, H.; Suprayogo, D.; Purnomosidhi, P.; Widodo, R.H.; Van Noordwijk, M. Litter layer residence time in forest and coffee agroforestry systems in Sumberjaya, West Lampung. *For. Ecol. Manag.* **2006**, *224*, 45–57. [CrossRef]
93. Derpsch, R.; Franzluebbers, A.J.; Duiker, S.W.; Reicosky, D.C.; Koeller, K.; Friedrich, T.; Sturny, W.G.; Sá, J.C.M.; Weiss, K. Why do we need to standardize no-tillage research? *Soil Tillage Res.* **2014**, *137*, 16–22. [CrossRef]
94. Stomph, T.J.; De Ridder, N.; Steenhuis, T.S.; Van de Giesen, N.C. Scale effects of Hortonian overland flow and rainfall-runoff dynamics: Laboratory validation of a process-based model. *Earth Surf. Process. Landf.* **2002**, *27*, 847–855. [CrossRef]
95. Gomi, T.; Sidle, R.C.; Miyata, S.; Kosugi, K.I.; Onda, Y. Dynamic runoff connectivity of overland flow on steep forested hillslopes: Scale effects and runoff transfer. *Water Resour. Res.* **2008**, *44*, 1–16. [CrossRef]
96. Moreno-de las Heras, M.; Nicolau, J.M.; Merino-Martín, L.; Wilcox, B.P. Plot-Scale effects on runoff and erosion along a slope degradation gradient. *Water Resour. Res.* **2010**, *46*, 1–12. [CrossRef]
97. Rodenburg, J.; Stein, A.; van Noordwijk, M.; Ketterings, Q.M. Spatial variability of soil pH and phosphorus in relation to soil run-off following slash-and-burn land clearing in Sumatra, Indonesia. *Soil Tillage Res.* **2003**, *71*, 1–14. [CrossRef]
98. Pradiko, H.; Soewondo, P.; Suryadi, Y. The change of hydrological regime in upper Cikapundung Watershed, West Java Indonesia. *Procedia Eng.* **2015**, *125*, 229–235. [CrossRef]
99. Nurhayati, S.A.; Sabar, A.; Marselina, M. The effect of land use changes to discharge extremities in Cimahi Watershed–Upper Citarum Watershed, West Java. In *E3S Web of Conferences*; EDP Sciences: Ulis, France, 2020; Volume 148, p. 07002.
100. Beven, K. Searching for the Holy Grail of scientific hydrology: Q t=(S, R,? t) a as closure. *Hydrol. Earth Syst. Sci. Discuss.* **2006**, *10*, 609–618. [CrossRef]
101. Graham, C.B.; Van Verseveld, W.; Barnard, H.R.; McDonnell, J.J. Estimating the deep seepage component of the hillslope and catchment water balance within a measurement uncertainty framework. *Hydrol. Process.* **2010**, *24*, 3631–3647. [CrossRef]
102. Government of Indonesia. *Law (Undang-Undang) Number 41 Year 1999 about Forestry*; Government of Indonesia: Pemerintah, Indonesia, 1999.
103. Bruijnzeel, L.A. Evaluation of runoff sources in a forested basin in a wet monsoonal environment: A combined hydrological and hydrochemical approach. *Hydrol. Humid Trop. Reg. IAHS Publ.* **1983**, 140.
104. Verbist, B.; Poesen, J.; van Noordwijk, M.; Suprayogo, D.; Agus, F.; Deckers, J. Factors affecting soil loss at plot scale and sediment yield at catchment scale in a tropical volcanic agroforestry landscape. *Catena* **2010**, *80*, 34–46. [CrossRef]
105. Tanika, L.; Khasanah, N.; Leimona, B. *Simulasi Dampak Perubahan Tutupan Lahan Terhadap Neraca Air Di Das Dan Sub-Das Rejoso Menggunakan Model. Genriver*; Report; World Agroforestry Centre (ICRAF) Southeast Asia Regional Program: Bogor, Indonesia, 2018.

106. Ridwansyah, I.; Pawitan, H.; Sinukaban, N.; Hidayat, Y. Watershed Modeling with ArcSWAT and SUFI2 in Cisadane Catchment Area: Calibration and Validation of River Flow Prediction. *Int. J. Sci. Eng.* **2014**, *6*, 92–101. [CrossRef]
107. Liu, Z.; Ma, D.; Hu, W.; Li, X. Land use dependent variation of soil water infiltration characteristics and their scale-specific controls. *Soil Tillage Res.* **2018**, *178*, 139–149. [CrossRef]
108. Zhao, Y.; Wang, Y.; Wang, L.; Zhang, X.; Yu, Y.; Zhao, J.; Lin, H.; Chen, Y.; Zhou, W.; An, Z. Exploring the role of land restoration in the spatial patterns of deep soil water at watershed scales. *Catena* **2019**, *172*, 387–396. [CrossRef]

Permissions

All chapters in this book were first published in MDPI; hereby published with permission under the Creative Commons Attribution License or equivalent. Every chapter published in this book has been scrutinized by our experts. Their significance has been extensively debated. The topics covered herein carry significant findings which will fuel the growth of the discipline. They may even be implemented as practical applications or may be referred to as a beginning point for another development.

The contributors of this book come from diverse backgrounds, making this book a truly international effort. This book will bring forth new frontiers with its revolutionizing research information and detailed analysis of the nascent developments around the world.

We would like to thank all the contributing authors for lending their expertise to make the book truly unique. They have played a crucial role in the development of this book. Without their invaluable contributions this book wouldn't have been possible. They have made vital efforts to compile up to date information on the varied aspects of this subject to make this book a valuable addition to the collection of many professionals and students.

This book was conceptualized with the vision of imparting up-to-date information and advanced data in this field. To ensure the same, a matchless editorial board was set up. Every individual on the board went through rigorous rounds of assessment to prove their worth. After which they invested a large part of their time researching and compiling the most relevant data for our readers.

The editorial board has been involved in producing this book since its inception. They have spent rigorous hours researching and exploring the diverse topics which have resulted in the successful publishing of this book. They have passed on their knowledge of decades through this book. To expedite this challenging task, the publisher supported the team at every step. A small team of assistant editors was also appointed to further simplify the editing procedure and attain best results for the readers.

Apart from the editorial board, the designing team has also invested a significant amount of their time in understanding the subject and creating the most relevant covers. They scrutinized every image to scout for the most suitable representation of the subject and create an appropriate cover for the book.

The publishing team has been an ardent support to the editorial, designing and production team. Their endless efforts to recruit the best for this project, has resulted in the accomplishment of this book. They are a veteran in the field of academics and their pool of knowledge is as vast as their experience in printing. Their expertise and guidance has proved useful at every step. Their uncompromising quality standards have made this book an exceptional effort. Their encouragement from time to time has been an inspiration for everyone.

The publisher and the editorial board hope that this book will prove to be a valuable piece of knowledge for researchers, students, practitioners and scholars across the globe.

List of Contributors

Jilmar Castañeda-Ccori and Juan Fernández-Manjarrés
Université Paris-Saclay, CNRS, AgroParisTech, Ecologie Systématique et Evolution, F-91405 Orsay, France

Anne-Gaël Bilhaut
Institut Français d'Etudes Andines, Calle Alemania N° 32-188 y Guayanas, Quito 170519, Ecuador

Armelle Mazé
Université Paris-Saclay, INRAE, AgroParisTech, SAD-APT, 75005 Paris, France

Marco Heredia-R
AgSystems, Ceigram, itdUPM, Centro de Innovación en Tecnología para el Desarrollo, Universidad Politécnica de Madrid (UPM), 28040 Madrid, Spain
Universidad Regional Amazónica IKIAM, Km 7 Vía Muyuna, Tena 150150, Ecuador

Carlos G. H. Diaz-Ambrona
AgSystems, Ceigram, itdUPM, Centro de Innovación en Tecnología para el Desarrollo, Universidad Politécnica de Madrid (UPM), 28040 Madrid, Spain

Bolier Torres
Universidad Regional Amazónica IKIAM, Km 7 Vía Muyuna, Tena 150150, Ecuador
Departamento de Ciencias de la Vida, Universidad Estatal Amazónica (UEA), Pastaza 160101, Ecuador

Jhenny Cayambe
School of Agricultural and Environmental Sciences, Pontificia Universidad Católica del Ecuador Sede Ibarra (PUCESI), Imbabura 100112, Ecuador

Nadia Ramos
Andean University Simon Bolivar, Toledo N22-80, Quito 170143, Ecuador

Marcelo Luna
Departamento de Ciencias de la Tierra, Universidad Estatal Amazónica (UEA), Pastaza 160101, Ecuador

Christine Magaju, Leigh Ann Winowiecki, Phosiso Sola, Ibrahim Ochenje, Esther Kiura, Anne Kuria, Jonathan Muriuki and Sammy Carsan
World Agroforestry (ICRAF), UN Avenue, Nairobi 00100, Kenya

Mary Crossland
School of Natural Sciences, Bangor University, Bangor, Gwynedd LL57 2DG, UK

Fergus Sinclair
World Agroforestry (ICRAF), UN Avenue, Nairobi 00100, Kenya
School of Natural Sciences, Bangor University, Bangor, Gwynedd LL57 2DG, UK

Aymen Frija and Hassen Ouerghemmi
International Center for Agricultural Research in the Dry Areas, ICARDA, Rue Hedi Karray, CP 2049 Ariana, Tunisia

Niguse Hagazi and Kiros Hadgu
World Agroforestry (ICRAF), Gurd Shola, Ethiopia

Enrico Bonaiuti
International Center for Agricultural Research in the Dry Areas, ICARDA, 6/106, Osiyo St, Tashkent 100084, Uzbekistan

Mai Phuong Nguyen
World Agroforestry, Vietnam Office, Hanoi 100000, Vietnam

Philippe Vaast
World Agroforestry, Vietnam Office, Hanoi 100000, Vietnam
UMR Eco&Sols, CIRAD, Université Montpellier, 34090 Montpellier, France

Tim Pagella
School of Natural Sciences, Bangor University, Bangor, Gwynedd LL57 2DG, UK

Ylva Nyberg
Department of Crop Production Ecology, Swedish University of Agricultural Sciences (SLU), SE-75007 Uppsala, Sweden

Caroline Musee and Emmanuel Wachiye
Vi Agroforestry, Kitale 30200, Kenya

Mattias Jonsson
Department of Ecology, SLU, SE-75007 Uppsala, Sweden

Johanna Wetterlind
Department of Soil and Environment, SLU, SE-53223 Skara, Sweden

Ingrid Öborn
Department of Crop Production Ecology, Swedish University of Agricultural Sciences (SLU), SE-75007 Uppsala, Sweden
World Agroforestry (ICRAF), UN Avenue, Kenya
Edi Purwanto, Idsert Jelsma and Atiek Widayati
Tropenbos Indonesia, Bogor 16163, Indonesia

Hery Santoso
Java Learning Centre (JAVLEC), Klidon, Yogyakarta 55581, Indonesia

Hunggul Y. S. H. Nugroho
Ministry of Environment and Forestry, Makassar 90243, Indonesia

Abdessamad Fakhech, El Mustapha Outamamat, Soufiane M'Sou and Mohamed Alifriqui
Laboratory of Microbiological Biotechnology, Agricultural Sciences & Environment (BioMAgE), Faculty of Sciences Semlalia, Cadi Ayyad University, Marrakech 40000, Morocco

Didier Genin
IRD, Laboratory Population, Environment, Dévelopment, UMR151 AMU-IRD, 1331 Marseille, France

Mohamed Ait-El-Mokhtar and Abdelilah Meddich
Laboratory of Agri-food, Biotechnologies & Bioresource Valorisation (AGROBIOVAL), Faculté des Sciences Semlalia, Université Cadi Ayyad, Marrakech 40000, Morocco

Mohamed Hafidi
Laboratory of Microbiological Biotechnology, Agricultural Sciences & Environment (BioMAgE), Faculty of Sciences Semlalia, Cadi Ayyad University, Marrakech 40000, Morocco
AgroBioSciences Program, Mohammed IV Polytechnic University (UM6P), Benguerir 43150, Morocco

Edi Dwi Cahyono, Salsabila Fairuzzana, Deltanti Willianto and Eka Pradesti
Faculty of Agriculture, Universitas Brawijaya, Malang 65145, Indonesia

Niall P. McNamara and Rebecca L. Rowe
UK Centre of Ecology & Hydrology, Lancaster LA14AP, UK

Priscilla Wainaina, Peter A. Minang, Eunice Gituku and Lalisa Duguma
World Agroforestry Centre, UN Avenue, Gigiri, Nairobi 00100, Kenya

Ermias Aynekulu, Marta Suber, James M. Roshetko and Todd S. Rosenstock
World Agroforestry (ICRAF), United Nations Avenue, Nairobi 00100, Kenya

Jacobo Arango
International Centre for Tropical Agriculture (CIAT), Km 17, Recta Cali–Palmira CP, Apartado Aéreo 6713, Cali 763537, Colombia

Didik Suprayogo, Kurniatun Hairiah, Nabilla Meilasari, Abdul Lathif Rabbani, Rizki Maulana Ishaq and Widianto Widianto
Soil Science Department, Faculty of Agriculture, Brawijaya University, Jl. Veteran no 1, Malang 65145, Indonesia

Meine van Noordwijk
Soil Science Department, Faculty of Agriculture, Brawijaya University, Jl. Veteran no 1, Malang 65145, Indonesia
World Agroforestry Centre, ICRAF, Indonesia Office, Bogor 16001, Indonesia
Plant Production Systems, Wageningen University and Research, 6708 PB Wageningen, The Netherlands
World Agroforestry (ICRAF), United Nations Avenue, Nairobi 00100, Kenya

Index

www.ingramcontent.com/pod-product-compliance
Lightning Source LLC
Chambersburg PA
CBHW081215180926
51613CB00003B/247